Eco-Friendly and Agile Energy Strategies and Policy Development

Mir Sayed Shah Danish
University of the Ryukyus, Japan

Tomonobu Senjyu
University of the Ryukyus, Japan

A volume in the Practice, Progress, and Proficiency in Sustainability (PPPS) Book Series

IGI Global
PUBLISHER of TIMELY KNOWLEDGE

Published in the United States of America by
IGI Global
Engineering Science Reference (an imprint of IGI Global)
701 E. Chocolate Avenue
Hershey PA, USA 17033
Tel: 717-533-8845
Fax: 717-533-8661
E-mail: cust@igi-global.com
Web site: http://www.igi-global.com

Library of Congress Cataloging-in-Publication Data

Names: Danish, Mir Sayed Shah, 1987- editor. | Senjyu, Tomonobu, 1963- editor.
Title: Eco-friendly and agile energy strategies and policy development / Mir Sayed Shah Danish, and Tomonobu Senjyu, editor.
Description: Hershey, PA : Engineering Science Reference, an imprint of IGI Global, [2022] | Includes bibliographical references and index. | Summary: "This book covers the importance of the state-of-the-art findings of researchers in the field of eco-friendly energy agile strategy and policy development, following processes and their technologies exploitation on a broad theme of sustainable development"-- Provided by publisher.
Identifiers: LCCN 2021047784 (print) | LCCN 2021047785 (ebook) | ISBN 9781799895022 (h/c) | ISBN 9781799895039 (s/c) | ISBN 9781799895046 (ebk)
Subjects: LCSH: Energy policy. | Renewable energy sources. | Sustainable development. | Energy conservation--Government policy.
Classification: LCC HD9502.A2 E216 2022 (print) | LCC HD9502.A2 (ebook) | DDC 333.79--dc23/eng/20211118
LC record available at https://lccn.loc.gov/2021047784
LC ebook record available at https://lccn.loc.gov/2021047785

This book is published in the IGI Global book series Practice, Progress, and Proficiency in Sustainability (PPPS) (ISSN: 2330-3271; eISSN: 2330-328X)

British Cataloguing in Publication Data
A Cataloguing in Publication record for this book is available from the British Library.

Practice, Progress, and Proficiency in Sustainability (PPPS) Book Series

ISSN:2330-3271
EISSN:2330-328X

Editor-in-Chief: Ayman Batisha, International Sustainability Institute, Egypt

MISSION

In a world where traditional business practices are reconsidered and economic activity is performed in a global context, new areas of economic developments are recognized as the key enablers of wealth and income production. This knowledge of information technologies provides infrastructures, systems, and services towards sustainable development.

The **Practices, Progress, and Proficiency in Sustainability (PPPS) Book Series** focuses on the local and global challenges, business opportunities, and societal needs surrounding international collaboration and sustainable development of technology. This series brings together academics, researchers, entrepreneurs, policy makers and government officers aiming to contribute to the progress and proficiency in sustainability.

COVERAGE

- Technological learning
- Socio-Economic
- Sustainable Development
- ICT and knowledge for development
- Intellectual Capital
- E-Development
- Green Technology
- Global Business
- Strategic Management of IT
- Eco-Innovation

IGI Global is currently accepting manuscripts for publication within this series. To submit a proposal for a volume in this series, please contact our Acquisition Editors at Acquisitions@igi-global.com or visit: http://www.igi-global.com/publish/.

The Practice, Progress, and Proficiency in Sustainability (PPPS) Book Series (ISSN 2330-3271) is published by IGI Global, 701 E. Chocolate Avenue, Hershey, PA 17033-1240, USA, www.igi-global.com. This series is composed of titles available for purchase individually; each title is edited to be contextually exclusive from any other title within the series. For pricing and ordering information please visit http://www.igi-global.com/book-series/practice-progress-proficiency-sustainability/73810. Postmaster: Send all address changes to above address.

Titles in this Series

For a list of additional titles in this series, please visit:
http://www.igi-global.com/book-series/practice-progress-proficiency-sustainability/73810

Leadership Approaches to the Science of Water and Sustinability
Kristin Joyce Tardif (University of Arkansas, Fort Smith, USA)
Engineering Science Reference • © 2022 • 300pp • H/C (ISBN: 9781799896913) • US $215.00

Climatic and Environmental Significance of Wetlands Case Studies from Eurasia and North Africa
Abdelkrim Ben Salem (Mohammed V University, Morocco) Laila Rhazi (Mohammed V University, Morocco) and Ahmed Karmaoui (University of Moulay Ismail, Morocco & Southern Center of Culture and Science, Morocco)
Engineering Science Reference • © 2022 • 310pp • H/C (ISBN: 9781799892892) • US $250.00

Handbook of Research on Global Institutional Roles for Inclusive Development
Neeta Baporikar (Namibia University of Science and Technology, Namibia & University of Pune, India)
Information Science Reference • © 2022 • 398pp • H/C (ISBN: 9781668424483) • US $270.00

Analyzing Sustainability in Peripheral, Ultra-Peripheral, and Low-Density Regions
Rui Alexandre Castanho (WSB University, Poland)
Engineering Science Reference • © 2022 • 334pp • H/C (ISBN: 9781668445488) • US $240.00

Frameworks for Sustainable Development Goals to Manage Economic, Social, and Environmental Shocks and Disasters
Cristina Raluca Gh. Popescu (University of Bucharest, Romania & The Bucharest University of Economic Studies, Romania)
Engineering Science Reference • © 2022 • 296pp • H/C (ISBN: 9781668467503) • US $240.00

For an entire list of titles in this series, please visit:
http://www.igi-global.com/book-series/practice-progress-proficiency-sustainability/73810

IGI Global
PUBLISHER of TIMELY KNOWLEDGE

701 East Chocolate Avenue, Hershey, PA 17033, USA
Tel: 717-533-8845 x100 • Fax: 717-533-8661
E-Mail: cust@igi-global.com • www.igi-global.com

Table of Contents

Preface xii

Chapter 1
Energy Strategy and Policy Development Process 1
Mir Sayed Shah Danish, University of the Ryukyus, Japan
Abdul Matin Ibrahimi, University of the Ryukyus, Japan
Mikaeel Ahmadi, University of the Ryukyus, Japan
Shawkatullah Shams, Kabul University, Afghanistan
Habibullah Fedayi, University of the Ryukyus, Japan
Tomonobu Senjyu, University of the Ryukyus, Japan

Chapter 2
Strategies to Green Economics: Analyzing the Renewable Energy Impact in Making the Economy Green 28
Nima Norouzi, Bournemouth University, UK

Chapter 3
Interplay Consequences of COVID-19 on Global Environmental Sustainability 50
Mir Sayed Shah Danish, University of the Ryukyus, Japan
Tomonobu Senjyu, University of the Ryukyus, Japan
Najib Rahman Sabory, University of the Ryukyus, Japan
Alexey Mikhaylov, Financial University Under the Government of the Russian Federation, Russia

Chapter 4
Economic Feasibility of Solar Power Plants in Turkey Based on the PV Module Using Simulation RETScreen Software 70
Emre Akusta, Kirklareli University, Turkey
Raif Cergibozan, Kirklareli University, Turkey

Chapter 5

Lessons From COVID-19 Conferring Environmental Re-Engineering Opportunity 99
Mir Sayed Shah Danish, University of the Ryukyus, Japan
Tomonobu Senjyu, University of the Ryukyus, Japan
Najib Rahman Sabory, University of the Ryukyus, Japan
Alexey Mikhaylov, Financial University Under the Government of the Russian Federation, Russia

Chapter 6
BODIPY Dyes in Solar Energy 119
Seda Cetindere, Gebze Technical University, Turkey

Chapter 7
Do Remittance Inflows Increase Energy Security Risk in the Long Run? Evidence From Selected MENA Countries 143
Alper Karasoy, Afyon Kocatepe University, Turkey

Chapter 8
Renewable Energy in the Framework of a Sustainable Urbanism: Solutions Based on a Spanish Legislation Model 172
María Jesús Garcia García García, University of Valencia, Spain

Chapter 9
Progress and Determinants of Renewable Energy Development in India 190
Sebak Kumar Jana, Vidyasagar University, India
Khushboo Singh, Vidyasagar University, India

Glossary 204

Compilation of References 209

Related References 241

About the Contributors 265

Index 270

Detailed Table of Contents

Preface xii

Chapter 1
Energy Strategy and Policy Development Process 1
Mir Sayed Shah Danish, University of the Ryukyus, Japan
Abdul Matin Ibrahimi, University of the Ryukyus, Japan
Mikaeel Ahmadi, University of the Ryukyus, Japan
Shawkatullah Shams, Kabul University, Afghanistan
Habibullah Fedayi, University of the Ryukyus, Japan
Tomonobu Senjyu, University of the Ryukyus, Japan

This chapter discusses energy policy development at the moment that 169 countries are trying to save the world through ratifying the Paris Climate Change Summit (2016). However, competitive policy planning and design are multidimensional endeavors when most developing nations are faced with the advent of economic and population increase. Some of these countries still do not have energy policies or, for many years, remain unchanged. The proposed framework aims to retrieve nations' socio-economic status onto a full growth path when facing multiple challenges that fit the Sustainable Development Goals (SDGs). This chapter reviews the literature, focusing on a systematic framework that can guide energy policy development. This chapter tries to adopt a new approach for a successful public energy policy development that meets priorities commensurate with the national strategies to meet anticipations and deliver optimum opportunities of technical, technological, political, social, environmental, economic, and institutional benefits.

Chapter 2
Strategies to Green Economics: Analyzing the Renewable Energy Impact in Making the Economy Green 28
Nima Norouzi, Bournemouth University, UK

The crises that threaten countries and human societies are the limited resources of non-renewable (fossil) energy sources and the increasing environmental pollution caused by the excessive consumption of fossil fuels, which are necessary for paying attention to energy resources. The close relationship between economic and environmental issues has led to the emergence of new approaches in international environmental law, one of the most important of which is the green economy. Since one of the most important goals of the green economy is to reduce greenhouse gas emissions, renewable energy sources are a shortcut to the green economy. In this regard, the primary purpose of this chapter is to compare the impact of renewable energy on the green economy in selected middle-income and high-income countries.

Chapter 3
Interplay Consequences of COVID-19 on Global Environmental Sustainability 50
Mir Sayed Shah Danish, University of the Ryukyus, Japan
Tomonobu Senjyu, University of the Ryukyus, Japan
Najib Rahman Sabory, University of the Ryukyus, Japan
Alexey Mikhaylov, Financial University Under the Government of the Russian Federation, Russia

This chapter outlines the essentials of COVID-19 and its relation to environmental mitigation. The COVID-19 pandemic has altered the ranking concern of climate change distress, and it is ranking as the first global priority to be adequately tackled. However, the pandemic demonstrates with economic, social, and cultural constraints. Still, climate change and environmental pollution have been ignored as the utmost precaution while their impact is more severe in the long run. This chapter evaluates available opportunities for environmental sustainability in the pandemic era. At the same time, the most significant aspect of solid waste, especially clinical waste, is critical for limiting pandemics and preventing future consequences of improper waste management. Sustainable production relies on criteria that ensure affordability, accessibility, use efficiency, safety, disparity, and other factors of production, supply, distribution, and consumption that are efficient, cost-effective, and environmentally friendly.

Chapter 4
Economic Feasibility of Solar Power Plants in Turkey Based on the PV Module Using Simulation RETScreen Software 70
Emre Akusta, Kirklareli University, Turkey
Raif Cergibozan, Kirklareli University, Turkey

This chapter aims to analyze the renewable energy production and economic evaluation of five MW solar power plant installations in Turkey using the RETScreen program. For this purpose, all provinces of Turkey are analyzed first in terms of their global radiation values. Provinces with a global radiation value less than 1620 kWh / m2 per year have been eliminated, and the analysis proceeded with 45 provinces. The study includes three different scenarios. The first is based only on the government's guarantee of purchase. The second adds the greenhouse gas reduction incentive to the first scenario. The third scenario adds domestic production incentives to the first and second scenarios. According to the first scenario analysis, the payback period is 5.553 years for Muğla and 6.530 years for Antalya. According to the second scenario, the payback period is 5.230 years for Muğla and 6.184 years for Antalya. Finally, in terms of the third scenario, the payback period is 3.479 years for Muğla and 4.101 years for Antalya.

Chapter 5
Lessons From COVID-19 Conferring Environmental Re-Engineering Opportunity 99
Mir Sayed Shah Danish, University of the Ryukyus, Japan
Tomonobu Senjyu, University of the Ryukyus, Japan
Najib Rahman Sabory, University of the Ryukyus, Japan
Alexey Mikhaylov, Financial University Under the Government of the Russian Federation, Russia

This chapter presents re-engineering environmental sustainability opportunities within constrained pandemic situations that suggest solutions. Climate change disaster and COVID-19 socio-economic impacts are leading nations to dramatic tragedy. Simultaneously, increasing demand for energy are due to population, political competition, and industrial growth globally. At present, the pandemic attracts more attention due to its immediate effect. Ignoring climate change can have a worse consequence not only on the present but for the future generation in the long run. Therefore, re-engineering the current pandemic situation with a futurism outlook for saving the world will enable nations to transform and rethink strategies, policies, procedures, processes, and any actions that can cope with present and possible future pandemics and climate change tragedies.

Chapter 6
BODIPY Dyes in Solar Energy...119
Seda Cetindere, Gebze Technical University, Turkey

Solar energy shows great potential in the future energy market due to its essentially unlimited supply and renewable, cheap, and clean nature. Among the third-generation solar technologies that are able to harvest light and convert it to electricity are dye-sensitized solar cells (DSSCs), organic solar cells (OSCs), and inorganic-organic hybrid solar cells also known as perovskite solar cells (PSCs). Among the various organic dyes, BODIPY (4,4-difluoro-4-bora-3a,4a-diaza-s-indacene) dyes have been recognized as promising candidates for solar cells due to their intrinsic advantages such as sharp and strong absorption near 500 nm, which is leading to efficient light-harvesting capability, diverse modification on the core structure at all positions and with any desired functionality, long excited state lifetimes, excellent photo-stability, and good solubility in organic solvents. This chapter will focuse on studies of BODIPY dyes in solar energy technology.

Chapter 7
Do Remittance Inflows Increase Energy Security Risk in the Long Run? Evidence From Selected MENA Countries 143
Alper Karasoy, Afyon Kocatepe University, Turkey

This chapter examines the long-run effect of remittances on energy security risks of Egypt, Morocco, and Tunisia in a bi-variate setting by utilizing the DOLS, FMOLS, and CCR estimation methods for the 1980-2018 period. The empirical findings of this chapter show that remittances and energy security risks are cointegrated. Additionally, the estimation results show that remittances significantly aggravate energy insecurities in selected MENA countries in the long run. Further, the results of this chapter reveal that remittances' role in increasing energy insecurity should not be overlooked. Therefore, the author proposes that remittance inflows can be utilized as a long-term integral instrument in combating energy insecurity issues and developing new energy policies.

Chapter 8
Renewable Energy in the Framework of a Sustainable Urbanism: Solutions Based on a Spanish Legislation Model 172
María Jesús Garcia García García, University of Valencia, Spain

Urban environments, areas in which multiple activities developed by human beings are brought together, have a strong environmental impact. Therefore, in these areas, measures aimed at rational use of natural resources and efficient consumption of energy must be promoted. To this end, energy efficiency must also be accompanied by a policy of diversifying the sources used in energy production, which opens the door to renewable energy promotion and use. The much discussed energy efficiency cannot consist only of measures aimed at saving and containing demand, but also requires the regulation and management of supply that promotes the introduction of renewable energy sources, which is in turn clean energy given its low level of emissions into the atmosphere.

Chapter 9
Progress and Determinants of Renewable Energy Development in India 190
Sebak Kumar Jana, Vidyasagar University, India
Khushboo Singh, Vidyasagar University, India

For sustainable development of energy, renewable energy can play a pivotal role. In order to meet the huge requirement of electricity in India, massive addition to the installed electricity-generating capacity is required. The major objectives of the present study are (1) to find the progress of renewable energy development in different states of India, (2) to find the determinants of renewable energy development in India, and (3) to assess the scope of renewable energy development in the states of India. The study finds that there is state-wise disparity in the development of renewable energy in India and also reveals that renewable energy potential (MW) and real gross state domestic product of the states are significant factors in state-wise renewable energy development in India.

Glossary 204

Compilation of References 209

Related References 241

About the Contributors 265

Index 270

Preface

A rapid change in technology and lifestyle has led nations to a dramatic increase in energy demand, which has faced the globe with the challenge of ensuring sustainable and clean energy production. Growing energy demand has been the main causer of environmental pollution. Today's cutting-edge renewable technologies must propound sustainability culture in energy systems' planning and design to transform the world with the 2030 Sustainable Development Goals (SDGs).

Efficient exploitation of renewable energy sources and technologies for residential, commercial, industrial, and agricultural sectors offer the opportunity to diminish energy dependence, ensure efficiency and reliability, reduce pollutant emissions, and buoyant economy. This book aims to cover the importance of the state-of-the-art and findings of researchers in the field of eco-friendly energy agile strategy and policy development, following processes and their technologies exploitation on a broad theme of sustainable development. This book establishes interdisciplinary coverage in sustainable energy development by strategic thinking and changing lifestyle by designing agile energy strategies by presenting real-world case studies and examples.

This book collects seven chapters of transboundary research, experiences, and lessons learned that offer integrated conceptual and empirical contributions from different interrelated fields, focusing on energy renewable proper deployment options and scenarios. The book offers viable options for societies' transition in the 21st century by drawing agile strategies and policies to provide affordable access to energy by copping with today's environmental tragedy. Within the scope of this book, strategic policies developed aim to improve well-being life to eradicate poverty, mitigate climate change, promote lifelong learning opportunities, empower sociality, deploy affordable energy, sustain economic growth, offer innovation, reduce inequality, and finally, ensure global sustainability. Therefore, it can be counted as a competitive reference for the researcher, educators, and policymakers in the field.

ORGANIZATION OF THE BOOK

The book is organized into nine chapters. A brief description of each of the chapters follows:

Chapter 1 discusses energy policy development at the moment that 169 countries are trying to save the world through ratifying the Paris Climate Change Summit (2016). However, competitive policy planning and design knew multidimensional endeavors when most developing nations are faced with the advent of economic and population increase day by day. Some of these countries still do not have energy policies or, for many years, remain unchanged. The proposed framework aims to retrieve nations' socio-economic onto a full growth path when facing multiple challenges that fit the sustainable development goals (SDGs). This chapter reviews the literature, focusing on a systematic framework that can guide energy policy development. This chapter tries to adopt a new approach for a successful public energy policy development that meets priorities by commensurate with the national strategies to meet anticipations and deliver optimum opportunities of technical, technological, political, social, environmental, economic, and institutional benefits.

Chapter 2 focuses on the strategic roadmap to ensure socio-economic measures by deploying green economics and renewable energy resources. The crises that threaten countries and human societies are the limited resources of non-renewable (fossil) energy sources and the increasing environmental pollution caused by the excessive consumption of fossil fuels, which are necessary for paying attention to energy resources. The close relationship between economic and environmental issues has led to the emergence of new approaches in international environmental law, one of the most important of which is the green economy. Since one of the most important goals of the green economy is to reduce greenhouse gas emissions, renewable energy sources are a shortcut to the green economy. In this regard, the primary purpose of this chapter is to compare the impact of renewable energy on the green economy in selected middle-income and high-income countries.

Chapter 3 outlines the essentials of COVID-19 and its relation to environmental mitigation. COVID-19 pandemic has altered the ranking concern of climate change distress, and it is ranking as the first global priority to be adequately tackled. However, pandemic demonstrates with economic, social, and cultural constraints. Still, climate change and environmental pollution have been ignored as the utmost precaution while its impact is more severe in the long run. This chapter evaluates available opportunities for environmental sustainability in the pandemic era. At the same time, the most significant aspect of solid waste, especially clinical waste, is critical for limiting pandemics and preventing future consequences of improper waste management. Sustainable production relies on criteria that ensure affordability, accessibility, use efficiency, safety, disparity, and other factors of production, supply, distribution, and consumption in the way of efficient, cost-effective, and environmental-friendly.

Chapter 4 analyzes the renewable energy production and economic evaluation of 5 MW solar power plant installations in Turkey using the RETScreen program. For this purpose, all provinces of Turkey are analyzed first in terms of their global radiation values. Provinces with a global radiation value less than 1620 kWh / m2 per year have been eliminated, and the analysis proceeded with 45 provinces. The study includes three different scenarios. The first is based only on the government's guarantee of purchase. The second adds the greenhouse gas reduction incentive to the first scenario. The third scenario adds domestic production incentives to the first and second scenarios. According to the first scenario analysis, the payback period is 5.553 years for Muğla and 6.530 years for Antalya. According to the second scenario, the payback period is 5.230 years for Muğla and 6.184 years for Antalya. Finally, in terms of the third scenario, the payback period is 3.479 years for Muğla and 4.101 years for Antalya.

Chapter 5 presents re-engineering environmental sustainability opportunities within constrained pandemic situations that suggest solutions. Climate change disaster and COVID-19 socio-economic impacts are leading nations to dramatic tragedy. Simultaneously, increasing demand for energy due to population, political competition, and industrial growth globally. At present, pandemic attracts more attention due to their immediate effect. While climate change is ignored, that can have a worsen consequence not only on the present but for the future generation in the long run. Therefore, re-engineering the current pandemic situation with a futurism outlook for saving the world will enable nations to transform and rethink strategies, policies, procedures, processes, and any actions that can cope with present and possible future pandemics and climate change tragedies.

Chapter 6 presents solar energy technologies focusing on different cells generation efficiencies and application advantages. Solar energy shows great potential in the future energy market due to its essentially unlimited supply and renewable, cheap, and clean nature. Among the third-generation solar technologies that are able to harvest light and convert it to electricity are dye-sensitized solar cells (DSSCs), organic solar cells (OSCs), and inorganic-organic hybrid solar cells also known as perovskite solar cells (PSCs). Among the various organic dyes, BODIPY (4,4-difluoro-4-bora-3a,4a-diaza-s-indacene) dyes have been recognized as promising candidates for solar cells due to their intrinsic advantages such as sharp and strong absorption near 500 nm which is leading to efficient light-harvesting capability, diverse modification on the core structure at all positions and with any desired functionality, long excited state lifetimes, excellent photo-stability and good solubility in organic solvents. This chapter will focus on studies of BODIPY dyes in solar energy technology.

Chapter 7 examines the long-run effect of remittances on energy security risks of Egypt, Morocco, and Tunisia in a bi-variate setting by utilizing the DOLS, FMOLS, and CCR estimation methods for the 1980-2018 period. The empirical findings of this chapter show that remittances and energy security risks are cointegrated. Additionally, the estimation results show that remittances significantly aggravate energy insecurities in selected MENA countries in the long run. Further, the results of this chapter show that remittances' role in increasing energy insecurity should not be overlooked. Therefore, the author proposes that remittance inflows can be utilized as a long-term integral instrument in combating energy insecurity issues and developing new energy policies.

Chapter 8 discussed promoting renewable energy regulation and management in terms of urban electrification using green and efficient energy resources. Urban environments, areas in which multiple activities developed by human beings are brought together, have a strong environmental impact. Therefore, in these areas, measures aimed at rational use of natural resources and efficient consumption of energy must be promoted. To this end, energy efficiency must also be accompanied by a policy of diversifying the sources used in energy production, which opens the door to renewable energy use, promotion, and use. The much discussed energy efficiency cannot consist only of measures aimed at saving and containing demand, but also requires the regulation and management of supply that promotes the introduction of renewable energy sources, which is in turn clean energy given its low level of emissions into the atmosphere.

Chapter 9 deals with exploring renewable energy development in India (a) to find the progress of renewable energy development in different states of India, (b) to find the determinants of renewable energy development in India and (c) to assess the scope of renewable energy development in the states of India. The study finds that there is state-wise disparity in the development of renewable energy in India and also reveals that renewable energy potential (MW) and real gross state domestic product of the states are significant factors in state-wise renewable energy development in India.

Mir Sayed Shah Danish
University of the Ryukyus, Japan

Tomonobu Senjyu
University of the Ryukyus, Japan

Chapter 1
Energy Strategy and Policy Development Process

Mir Sayed Shah Danish
https://orcid.org/0000-0002-5196-2280
University of the Ryukyus, Japan

Abdul Matin Ibrahimi
University of the Ryukyus, Japan

Mikaeel Ahmadi
University of the Ryukyus, Japan

Shawkatullah Shams
Kabul University, Afghanistan

Habibullah Fedayi
University of the Ryukyus, Japan

Tomonobu Senjyu
https://orcid.org/0000-0003-4494-6773
University of the Ryukyus, Japan

ABSTRACT

This chapter discusses energy policy development at the moment that 169 countries are trying to save the world through ratifying the Paris Climate Change Summit (2016). However, competitive policy planning and design are multidimensional endeavors when most developing nations are faced with the advent of economic and population increase. Some of these countries still do not have energy policies or, for many years, remain unchanged. The proposed framework aims to retrieve nations' socio-economic status onto a full growth path when facing multiple challenges that fit the Sustainable Development Goals (SDGs). This chapter reviews the literature, focusing on a systematic framework that can guide energy policy development. This chapter tries to adopt a new approach for a successful public energy policy development that meets priorities commensurate with the national strategies to meet anticipations and deliver optimum opportunities of technical, technological, political, social, environmental, economic, and institutional benefits.

DOI: 10.4018/978-1-7998-9502-2.ch001

INTRODUCTION

Many factors, including population increase, economic growth, and lifestyle changes accelerated energy demand around the globe. From another attitude, the burning of fossil fuels continued for decades affected health and environment. Historically, while energy provided by biomass proved inadequate to supply the growing economies, people turned to hydropower later to coal and then to oil, natural gas, and nuclear power. The twenty-first century is the time for the next transition from fossil fuels towards renewable energy, and climate change is one of the critical driving forces (Matin Ibrahimi et al., 2019). Global average solar radiation per square meter per year can provide the same amount of energy as a barrel of oil, 200 kg of coal, or 140 m3 of natural gas. Recent cost reduction and progress in renewable energy technologies and efficiency as well as implementing policies increased attractions as an alternative to conventional energy sources (Ibrahimi et al., n.d.). According to the report (Simsek et al., 2019), in 2017, nearly 85% of global energy consumption or demand is reported from fossil fuels, including oil, coal, and gas.

Today, introducing various clean energy resources without greenhouse gas (GHG) emissions in the umbrella of smart and robust technologies creates several opportunities for people and their environment. Energy sectors particularly in the electric power system, for instance, the optimal involvement of clean energy resources such as wind and solar, are studied and proved to be encouraging to mitigate the GHG emission crisis and a multi-dimensional contribution on techno-economic growth as well (Ahmadi, Lotfy, Danish, et al., 2019; Ahmadi, Lotfy, Shigenobu, et al., 2019). Based on statistics' findings, the people accessing modern forms of energy in developing countries generally have higher economic and educational opportunities and improved health care services. Therefore, energy is correlated to both economic and social factors such as health care, enhancement of education and knowledge, accessible and clean water, communication services and further. It is not an easy job defining the energy in particular, precise and satisfactory manner to show and indicate its importance to develop societies containing all related activities within them. Energy respect for other elements and aspects in a society has its complexity because of the socioeconomic differentiation. Therefore, making a conceptual framework for energy counts as a critical point and ingredient to respond to the needs of society (Guía Práctica sobre, 2016).

Energy policy is defined as actions in which governments deploy to affect the energy demand and supply (Marcus, 1992). Therefore, these actions pave the way for the governments to cope with supplying energy challenges as well as manipulating energy consumption and economic growth. Establishing energy policy process requires a preliminary consideration in terms of review and assessment. As a fact, energy policy process is the first endeavor that originates a structural theme for energy policy development. The review stage of policy process includes distinguished segments that can be expounded in terms of characterization and weighting value for decision-making (Danish et al., 2019b). These segments depend on policy scale and objectives such as supply and demand, access, sectors, options, renewable energy potential, economy, customers, targets, competitions, and many more characteristics (Danish & Senjyu, 2020; Simsek et al., 2019). Identification and decision-making on importance of these segments are the main points of review process. However, energy policy process review is linked with management principles, while undiscipline analysis can lead to shortcomings in an in-depth analysis (Danish et al., 2018).

Review of a policy process is different from a policy review, which is conducted initially at the initiation and implementation phases, respectively. However, policy review can be assessed at any time for any phase apart from limitations. Authors in (Browne et al., 2018) categorized policy review into main three subjects: traditional approaches, mainstream approaches, and interpretive approaches. Traditional approaches deal with best solution based on objective analysis that focuses on facts and realities. Mainstream approaches mainly rely on the influence of policy actors; and concentrate on values, pressures, politics, trends, etc. And interpretive approaches solve problems with an overall solution that mainly provide descriptive solutions for problems.

POLICY COMPONENTS

In general terms, development of policy entails a series of main components in a broader perspective. Shaping structure of these components to draft policy alike domains can be addressed in terms of sustainable-SMART (shown in Figure 1) mechanism of course of actions fit specific goals (*SMART Goals: How to Make Your Goals Achievable*, 2019).

Figure 1. Policy component based on the SMART concept

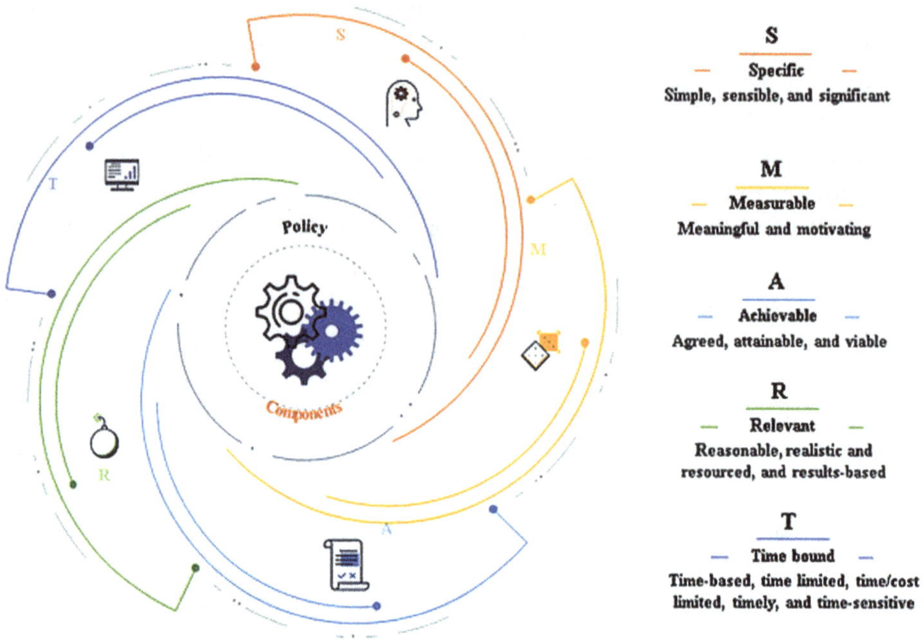

The SMART - Specific, Measurable, Achievable, Realistic, and Timely concept is used as the tool, and policy segments are interpreted/adapted according to the SMART analysis. Therefore, linking the policy to a measurable tool makes it possible to evaluate the performance and impacts of the policy effectively.

The term of specific denotes a defined-triple-constrains (scope, time, cost) scopes. From the project management perspective, ensuring the success of a project mainly relies on the triple constraints, which are scope, time, and cost, to be managed and aligned with the overall objective. Therefore, evaluation of triple constraints with high impacts on the project's outcome requires in-depth attention to be precisely defined.

Therefore, a policy document briefly outlines course of action in a simple statement to a voluminous document. It is worth noting that the proper application of a policy requires various supporting documents backed by management, technical, social, institutional, and related domains. Policies comprise elements or components that are saturated coherently to address specific goals. Standard components are those defined to be necessary for conducting the policy's purposes. These components can be found in all policies, regardless of the topic. There also are some optional components whose relevance varies by policy area and institutional need. However, the number of policy components relate to the nature and objective of a policy. In

(*Writing a Policy: The Structure and Components of a Policy*, 2019) these components are outlined as follows:

- **Statement:** This component demonstrates a big picture of anticipation set forth in a policy in a broader them to be achieved. This component does not consider quantitative and qualitative measures. Policy-makers usually deal with methodological approaches to provide viable procedure mechanisms within regulation, codes, standards, needs, priorities, and so on.
- **Principles and Philosophies:** As an accepted practice of policy development and management, the hierarchy of the components ensures the effectiveness and accurateness of a policy that can differ case by case.
- **Objective:** This component translates scopes into measurable endeavors based on a pre-defined benchmark. This phase breaks down anticipated outcomes into a viable action plan.
- **Strategy:** This component refers to the policy development level. At this point, policy-makers put forward with a specific scheme of policy development. In addition, this component can be used for development phase as well as at the implementation phase. This component addresses general questions such as: How to develop a policy? How to implement it? And so on.
- **Outcomes:** This component aligns actions with objectives to sum up with specific outcomes.
- **Indicators:** This component creates benchmarks to evaluate overall performances of trends with a progressive or regressive tendency (Danish et al., 2019a). For successful policy design and implementation, well-defining of indicators with distinguishing of their hierarchy is indispensable. Some criteria are recounted for indicators such as exhaustiveness, simplicity, validity and adequately, trendability over time, irreversibility, updateability, flexibility, comparability, etc. (Danish et al., 2019b).

Management: The policy management component falls under two separate energy policy development affairs segments and energy-mixed management. Proper management and wisely mixing energy resources are critical for economic prosperity and technical reliability in a policy (Danish et al., 2017a). In (Rahaman & Varis, 2008), energy resource management is defined as the balance of required resources with desired quality (economic, environment, institutional and social) at a proper place (preferred local resources) and adequate time, backed by several supportive principles

POLICY STRUCTURE

A policy is a deliberate system of principles to guide decisions and achieve rational outcomes. Policies could be reactive, which created to respond to a current pressing problem or be proactive, which created a course of actions designed to prevent a problem. Research and consultation are key steps for the policy process. A comprehensive policy is built upon good consultation with those who will be affected (Hardee et al., 2004).

Agenda setting (problem identification) and recognition of certain subject as a problem demanding further attention. Agenda is a list of subjects or problems to which government officials and non-governmental organizations are considering at any given time. It is most often formed by political and policy elites but can also be influenced by non-governmental activist groups and the private sector. Agenda defines target/priority and actions to be implemented by national/local stakeholders. The Agenda has to be firm, transparent and possible to implement, with clarity on achieving the results ("Three Components of Policy Systems," n.d.). Policy formulation is the part of the process by which proposed actions are defined and drafted. Policy formulation includes setting goals and outcomes. The goals and objectives can be general or specific but should express the associated activities and indicators by which they will be obtained and measured.

The goals in a policy describe several expected outcomes or/and achievements where it differs from the objectives, which have measurability characteristics. For example, in a small-scale policy of an electricity utility, energy loss reduction might be one of the goals. Still, maintaining the loss, for instance, less than 20% can be counted as one of their objectives ("Policy Constraints," 2019). Policy analysis is more complicated than problem identification in the process component of the policy circle because policy-makers measure their decisions on several criteria. Policy analysis expands from the technical perspectives of an issue and concentrates on policy reform's political values and benefits. Policymakers tend to make their decisions based on several criteria, including technical advantages; potential effects of the policy on political issues; the impact of the policy change on the government's stability and support; the perceived severity of the problem and whether or not the government is in crisis; and pressure, endorse, or opposition from international aid agencies (Hardee et al., 2004; Wies, 1994).

MODELLING AN ENERGY POLICY

Problem-oriented Approaches

Energy policy development can be pursued with various approaches for different scopes and scales. Problem-oriented policy development is most viable for an average scale application in developing countries for specific outcomes. Problem-oriented policy development deals with many tools and techniques. Herein the analytic hierarchy process methodology is hired for an energy policy development.

An analytic hierarchy process is a methodological approach that applies to make a multi-criteria decision using in term of structural and functional analyses (Heindl & Liefner, 2019). Structural hierarchies approach breakdown a problem based on involved structural characteristics into a realizable piece of information in descending order. Functional hierarchies deal with behavior of break downed information in term of interaction, complexity, and impact of problem segments. In fact, the analytic hierarchy process realizes a problem to descend or ascend order fit with alternatives optimum selection characteristics. This characterization is commonly consisting of three-level interactions as shown in Figure 2.

Figure 2. Problem-oriented approach interactions

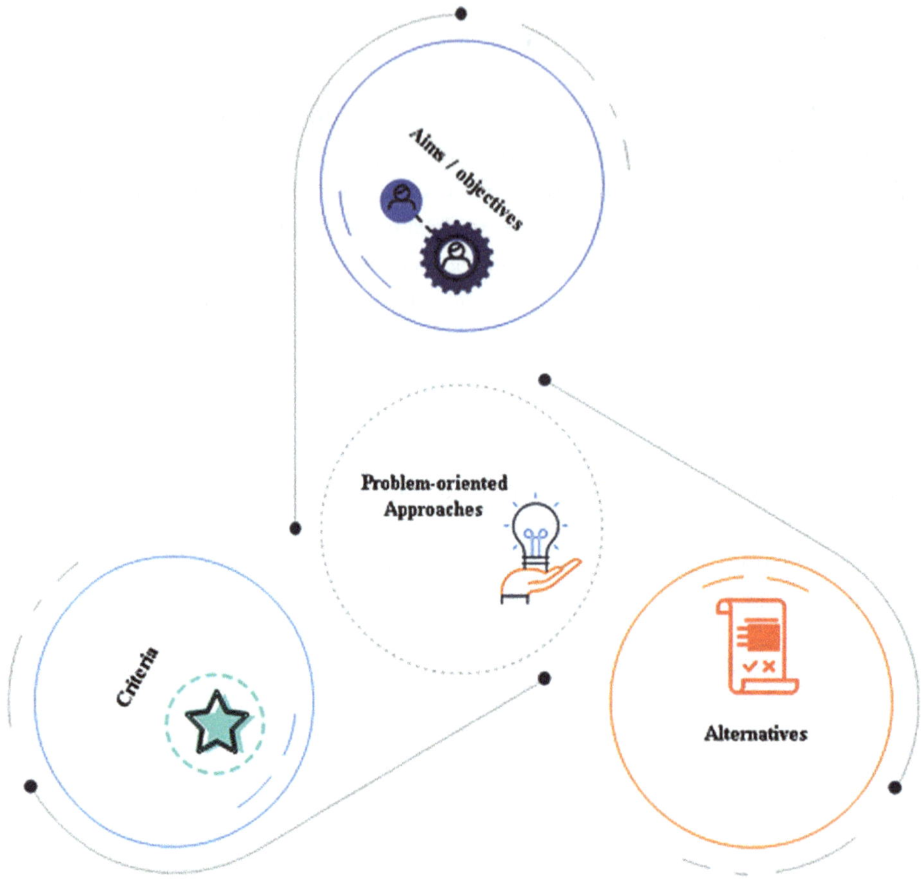

Analytic hierarchy process methodology has various applications due to its simplicity of application in the form of a hierarchical tree structure of goal, criteria, and alternatives. At this point, this methodology can be chiefly hired to assess energy-mixed analysis and determine energy resources share at energy strategy and policy development (Kuzman et al., 2013).

The term "energy-mix" refers to combining various primary energy resources used to match energy demands in a given geographic area. It includes fossil fuels, nuclear energy, non-renewable waste and renewable energies resources. Energy-mix balances countries' share of energy resources considering viable proportions based on national and international competitive benefits. For each region, the composition of the energy mix depends on the availability of usable resources, the type of energy that needs to be met. Finally, policies are determined by historical, economic, social, demographic, environmental, and geopolitical factors (*The Energy Mix and Energy Transition : Items of Content in This Feature Report*, 2015). Optimum implementation of Analytic hierarchy process methodology in the context of energy is applied in 2013 for comparative analysis of passive house construction types that present useful decision-making options through a straightforward analysis. In addition, this methodology comes with a most likely tangible alternative to a decision-making option.

Setting objectives to achieve goals; categorizing criteria and sub-criteria to evaluate impacts, influence, and outcomes; and prioritizing criteria based on objectives, and comparison of alternatives based on specific criteria and importance levels (equal importance, moderate importance, strong importance, very strong importance, extreme importance) (Saaty, 1997). Finally, the analytic hierarchy process of policy development has four axioms as follows (Saaty, 1997):

- **Reciprocity:** reciprocal judgments
- **Homogeneity:** same order, magnitude or hierarchical level
- **Dependency:** hierarchical dependency between the elements of two consecutive levels
- **Consistency:** when pairwise are perfectly consistent.

Analyzing an Energy Policy Model

Energy planners need to abandon their reliance on traditional similarly and instead evaluate conventional and renewable energy sources based on their cost contribution relative to their risk to a mix of generating assets. The extrapolation of past energy consumption trends in energy-mix can be made based on expectations of wealth, population growth and energy conservation measures (Awerbuch, 2006; Weijermars et al., 2012). Energy planning processes need to focus on developing the following generating portfolios (Awerbuch, 2006):

- To minimize expected cost for any given level of risk while minimizing expected risk for every level of the expected cost.
- To reduce generating costs while enhancing energy security by increasing the share of renewables.

Societal significance of future energy choices prompts numerous additional questions like how do we know what a society's energy mix would be in the future? What utility function can be acceptable to all stakeholders and so on. Fortunately, politicians and corporate policy-makers increasingly seek advice on these issues to ensure that they base their resources development and energy vision on thoughtful knowledge (Weijermars et al., 2012).

Conceptualizing a Policy Model

Altogether, policies categorize with respect to defined targets that setting up legitimate procedures to attain objectives. Meanwhile, policies can be classified as fitting the hierarchy of authorization, levels, competitiveness, and so on. Ordinarily, public policies trace a patch toward achieving competitive objectives in adapt with dominancy of national anticipation at the state-level. These policies target socio-economic development under national development policies. In an in-depth prospect, public policy is concerned with state (or a specific territory, province, etc.), national, regional, and international interactions. Public policy is a multidimensional manner that stimulates from needs, initiates from the current situation, and leads to anticipations through procedures, tools, and techniques (Kim & Park, 2018). In other words, a public energy policy put forward to drive targets for achieving objectives by exploring a set of optimum goals within successive changes. Coherently, public energy policies follow a systematic process of analysis and development. A public policy starts with a dynamic means of decision preparation and making, implementation, supervision, and evaluation with a high degree of flexibility and greater temporal consistency (Vedung, n.d.). Usually, public policies are theoretical plans that require supplementary and/or commentary documents to drive on the given sight under certain regulation and condition factors. Some studies on the public policy cycle (Muh et al., 2018; Sokołowski, 2019) outlined some essential features for effective energy policy in terms of public policy, which are listed as follows:

- Political commitment
- Motivating driver
- Institutional arrangements
- Legislative framework
- Applicability measures
- Implementation mechanism
- Financing and incentives
- Public participation
- Managing tools
- Capacity development
- Review and improvement

ENERGY POLICY BASELINING

Baseline study analysis or evaluation is usually developed to identify the opportunities and challenges within the implemented intervention and defined scope. It should be systematically done, taking into account several vital aspects. The baseline assessment has its priority to identifying the objectives as well as defining action plans. The covering area of baseline assessment depends on the field under intervention. The aim of preparing a baseline assessment as a required step in developing energy policy is to enhance, improve, and exploit the strengths and opportunities or discover and correct the exciting weaknesses. When developing energy policy is put forward to be implemented, a comprehensive and all-inclusive analysis of energy sectors is needed. It is to present a detailed status of which and to what extent resources are determined. Due to the extensive area involving energy and several resources as well as different authorities and sub-sectors for energy from pre-production to consumption phase, each one needs to be considered in the detailed analysis. There should be a systematic approach to do and perform each analysis effectively (Guía Práctica sobre, 2016).

In addition, in policy development particularly for a specific region a baseline assessment section contributes significantly to highlight the past and current situations from several perspectives and report the anticipation for the future of the focused issues. In other words, it acts as an examiner to classify several resources in connection to internal and external factors. For example, to understand energy and emission results in a region, a baseline assessment process scrutinizes the related data and results connected to several factors. Economic performance, utilizing efficiency, economic structure, greenhouse gas emitters sector-wise, energy consumption by fuel sources, renewable energy, and energy security are some of the factors contributing to baseline assessment preparation (*Regions for Sustainable Change: Baseline Assessment Report the Regional Climate Confidence Index*, 2010).

More detailed analysis is required to be conducted to go further and pick off one of the factors. For instance, in renewable energy factor, three major points are important to discuss. First, the targets; to what percentage is the target of the specified region or country in energy renewablization comparing to the current situation. Secondly, the potentials; which type of energy resources are available and categorizing the capacity of each type. Different area might ecologically surround different energy resource's availability and capacity in a region or a country. The last could be barriers; what kinds of barriers burdening the available renewable energy resources' penetration. Again, it might encompass financial and market limitations, lack of strategic, political and long-term vision or many other factors (*Energy and Power Baseline Assessment Working Paper*, 2010). In the same way, baseline assessment in connection to each factor generalize and overview the issues and opportunities helping energy policy development and implementation. Baseline assessment tool is considered one of the vital tools in the preparation phase of energy policy development, taking into account national energy-related sectors and global backgrounds and surroundings (Guía Práctica sobre, 2016).

ENERGY POLICY FRAMEWORKING

The current worldwide energy policy framework development is somehow originated from the oil crisis declaration dawn of the seventies and the first oil shock in 1973, which increased the prices to 400%. It was realized that the oil production has its finiteness point and will be over (Pillot et al., 2017). In response to address concerns relating to uncoordinated policy development, a conceptual framework is utilized by local governments in their policy planning process. However, the policy process for various countries is proposed differently; herein, the most viable policy process is shown in Figure 3.

Figure 3. Energy policy developing processes

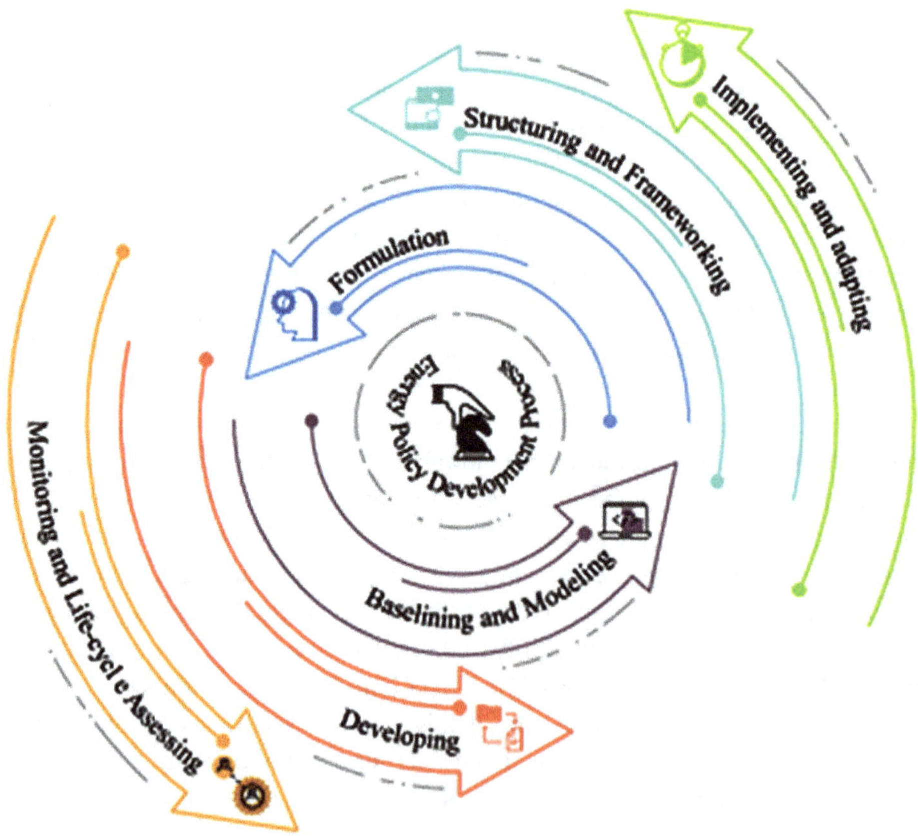

A policy framework outlines the fundamental steps that are significant to the successful development, approval, implementation, and review of policies. A policy should determine the end state and the intermediate states and the activities planned, with objectives, targets, and strategies to reach successive changes on the way toward the desired situation. The legal framework governing the policy planning and policy-making system set the structure for the strategic planning, policy analysis and coordination processes. The methodology of policy analysis and coordination defines the fundamental principles of policy-making. It also elaborates the roles of the players in each of the steps of the decision-making process (Whitford, 2009).

The high cost of energy to serve society is a barrier and a matter of concern that affect the people and environment. Furthermore, exploiting a beneficial form of energy and its supply depends not only on local circumstances but also on international context. When policymakers' experts in the energy field aim to stabilize and justify these gaps by achieving the national goals socially and economically in a secure manner. In other words, efficiency, sustainability, and resiliency will be targeted in the specific policy framework.

A policy development team can be demonstrated with essential skills such as statistical data collection, leading and chairing discussions, writing documents in an appropriate language and manner, and having good communication skills to seek required information from local and international experts.

Implementing a Policy Model

Policy implementation is defined as a process of interaction between stated policy objectives and the actions geared to achieve them. The concept of public energy policy process brings the idea that they are shaped by different types of actors and institutions at all stages. Implementation is inextricably linked to other parts of the policy process, such as governance, policy design and instruments, network studies, outsourcing, public-private partnerships, street-level bureaucracy, management, performance, regulatory enforcement, and compliance (Signé, 2017).

An energy policy implementation process refers to the mechanisms, resources, and relationships that links policies' objectives to action. It supports by an emergence engagement of stakeholders, resources, methodologies, procedures, tools, and techniques that work together to put polices into effect, achieving policy goals. Therefore, understanding the nature of policy implementation is essential. International experience indicates that policies, once adopted, are not always implemented as envisioned and do not necessarily achieve the intended results (A. R. Khan, 2016; Mthethwa, 2012). Designing an energy policy implementation process as part of the entire policy-making ensures the successfulness of a policy. At present, some least developing countries suffer from lack of energy policies and regulations (Danish et al., 2017c, 2017b). Meanwhile, an urgent need for sustainable energy deployment policies follows with open-door immature policies comprising significant economic and technical losses (Danish et al., 2016a). More importantly, energy policies in some developing countries cannot cope with citizen demand and global anticipation in terms of sustainable development (Chen et al., 2014; Cox et al., 2019; Erdiwansyah et al., 2019; Gungah et al., 2019; Savvidis et al., 2019).

The progress of an adopted policy depends on how successfully it is implemented. Different factors such as the content of a policy, nature of the process, actors involved in the process, and context influence energy policy development and implementation. To recall the relation of strategic management with energy strategy and policies development, energy process implementation predominantly depends on projecting and programming methodologies in term of implementation. Considering the scale and scope domain of a policy, it can be managed as a small project or a big program involving different endeavors. Apart from the type of management methodology, the process of execution management deals with all aspects of policy implantation. However, establishing links and adaption of the management education process into the policy implementation phases requires in-depth analysis. Since a policy implementation does not follow a specific roadmap due to verity in type, scope, authority, resources, etc., to shape the policy implementation process in the form of a tangible and manageable task, a management standard methodology definitely comes in action.

Various influential factors can be involved in development of energy policy. Among all factors, energy chain as lifecycle of the eco-energy-system is crucial to be well-addressed. In a general consensus from energy resources standpoint, energy chain consisting primary energy resources, transformation of these resources in the form of secondary source of energy, and final energy consumption. Each element of the energy resources chain contains different processes, inputs, and outputs that can be named energy or energy resources import and export, supply (in primary and secondary form), losses (technical, economic, transformation, and so on). To align the management and policy development process at implementation phase, influential of top, middle and low levels process coordination and adaptions are needed.

One of the problems on the way of successful implementation of a policy is lack of directions and guidelines on how to put in action. Another bottleneck is to convince involved parties and gain their awareness and acceptance. It is essential that policy application also comprises the sensibility of citizens, politicians, journalists and experts before, during and after implementing measures. Understanding success or failure by elaborating on factors that affect the implementation process will help policymakers and implementers to evaluate the processes that influence and establish the outcome of a public policy (R. A. Khan & Khandaker, 2016).

THE PROPOSED IMPLEMENTATION MODEL

The top-down implementation models use the decision of authorities as a starting point, identifying the compressibility of the challenges and the ability to structure implementation and the non-statutory variables affecting implementation. Most of the top-down models advise the clear and consistent goals, limit the size of change necessary, and place the responsibility for implementation. These models neglect prior and political aspects, as the implementation were only a matter of government, depending only on the availability of resources. Moreover, these models tend to concentrate on the entity crafting the policy rather than those affected. It also ignores the role of policy opponents, who make demands in the policy process. The top-down implementation models have been subject to extensive critique, declared that the days of solving significant problems through an 'engineering' approach have ended. Modern communities are pluralistic rather than uniform, and not amenable to top-down general solutions. Social groups increasingly present differences in aspirations, values, and perspectives that confound the possibility of such models. (Cerna, 2013; Signé, 2017).

In bottom-up implementation models, policy inspects from the perspective of the target population and the service deliverers. In this theory, the discretion of those who are relied upon to implement policy in the field is a key factor in successful implementation. Bottom-up theorists tend to believe that top-down decision-making is poorly adapted to local conditions. in this model, those responsible for implementation (e.g., the front-line service deliverers) are more important to success than top-down administration. Bottom-up implementation models are usually criticized as overemphasizing local autonomy and favoring administrative accountability over the ability of policy leaders to structure local behaviors. The difference between the top-down and bottom-up methods is the metrics that are chosen for evaluation. Top-down models typically see legislative objectives as the metric that should determine success or failure. Bottom-up approaches also use legislative objectives to assess policy success or failure but explain that the gap between legislative objectives and resources in top-down models changes the nature of implementation (Maynard-Moody et al., 1990; Signé, 2017).

For any model, government commitment, the capacity of implementing agency and political support is essential for successful policy implementation. Without proper consensus and support (from government and opposition parties), it is unable to implement the policy successfully. Social awareness is another important determinant for successful policy implementation. Media attention to the problem, public support or their voice, attitudes, and resources of constituency groups are identified as important factors for the policy implementation (Rajapakshe, 2017).

Developing a practical and implementable energy policy is not an easy task as it is a significant framework to build or improve energy-related infrastructures. Towards successful expected output to be covered in terms of short-term targets, medium-term objectives, and long-term goals as policy components this framework of policy particularly, energy policy requires great background history and anticipation for future need and status. Therefore, a multi-aspectual knowledge base assessment before addressing the objectives and their prioritization is the first process in conduction energy policy preparation or development. After setting the objectives and classifying their importance based on their time scale manner, introducing action plans should be put forward to implement the determined policies and strategies.

An action plan provides the decisions moving from the current situation to the desired situation within the strategic line. However, without specific all-inclusive tools making strategic path is not achievable. Therefore, tools help the strategic path to be operational. Lastly, to complete an action plan, proper activities are selected for the designated tools or tools. Finally, monitoring and controlling tools' performance during energy policy implementation is counted to ensure nothing is going wrong or preventing major failures.

It is worth mentioning that these major processes cannot be as simple as described above. The details of each process to maintain positive outcomes need to be analyzed based on the actual status of energy and related sectors and incorporate background experiences in other regions and review several implemented energy policies.

THE PROPOSED FRAMEWORK

First and foremost, the proposed framework in Figure 4. covers all sustainability pillars, which are essential for national energy policy. The concept of sustainability in consensus defines using energy resources in an optimum way that can be sufficient for now and does not compromise generations' ability to meet their needs (Danish et al., 2016b). Sustainability demonstrated with some essential contributes such as accessibility, affordability, disparity, safety, use efðciency, supply and production efðciency, cost-effectiveness, and environmental impacts on air, water, and soil quality (O'Neill-Carrillo et al., 2008) that consisting five pillars: Technical sustainability, economic sustainability, institutional sustainability, environmental sustainability, social sustainability (Danish et al., 2017a). For a successful policy development, financial sustainability (transparency, consistency, accountability, leadership, and time-bound and budgeted), environmental and social sustainability (climate change, data management, etc.) along with indicators and baselines are known exigence.

Figure 4. The proposed framework

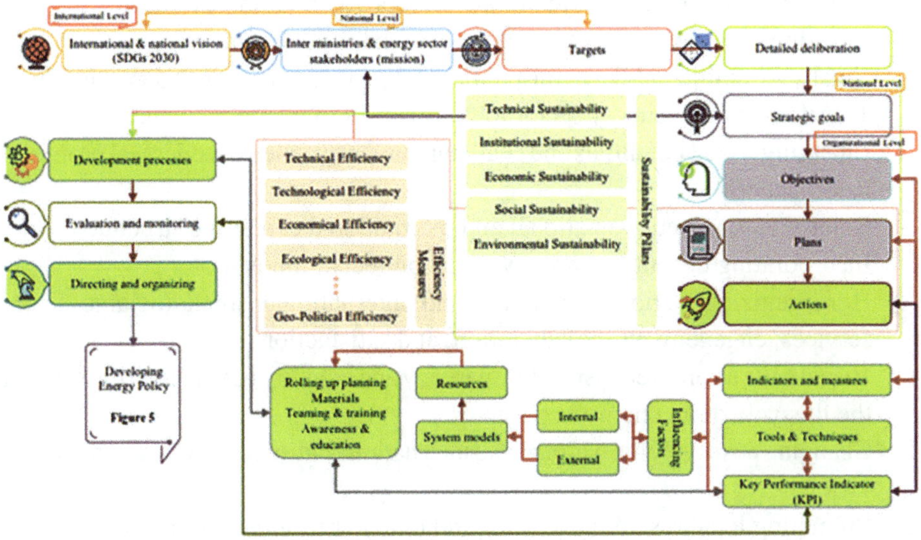

All Strategies refer to a collection of related process, methodologies, actions, and overall endeavors that stakeholders improve their service, product, or any deliverables in term of performances. Competitive-prosperous planning necessitates indispensable tools and techniques in conformity of scientific and practitioner principles.

In archaic literature, strategies are classified into three main themes (rationality: comprehensive and bounded, vision and involvement), and eleven typologies (Liu et al., 2013). A sustained competitive strategy enables an energy utility or an organization to maintain above-average profitability in view of long-term services (Magnusson & Berggren, 2018; Xie, 2016). Strategy deals with goals to present plans in action with the flexibility that is usually developed by the top and middle levels of management as an extroverted approach. While the policy is an action principle (the basis for guiding the actions) with limited flexibility of decision in terms of introverted approach (Surbhi, 2015).

One of the main drawbacks of policy development has been ambiguous strategy hierarchy in the context of policy involvement—also, lack of national strategy as a baseline for energy policy development. At any hierarchy of national strategies, energy policy depends and is affected by these national (particularly, energy involved strategies and sectors). The proposed framework supposed national policies as a prerequisite in the context of the baseline roadmap.

Some internal and external factors are the key points to evaluate the effectiveness of energy policy in terms of socio-economic prosperity as following:

1. Enhancing industrial and agriculture productivity
2. Adjusting governmental and private sectors effectual involvement
3. Ensuring dynamic corporate and private sector governance
4. Creating flexible public, semi-public, and private investment opportunities
5. Transforming the energy sector to win-win socio-economic zones
6. Emerging social potentials and challenges as an excellent intellectual property-based benefit
7. Reforming competitive national and international economic structures
8. Incorporating decent technology based on recent innovations
9. Homogenizing regional, national, and international small to large scales services, engagements, collaboration, and satisfaction
10. Visualizing a comprehensive transparency and equality mechanism throughout the lifecycle of an energy services
11. Changing positively the national and international attitudes toward a cooperative vision
12. Promoting local resources (energy and human resources) utilization
13. Realizing life-cycling in terms of adaptability, viability, and updatability.

Some barriers are also highlighted in the literature that reported from South Korea, Taiwan, Japan, Malaysia, Indonesia, Brunei, Singapore, Canada, Thailand, Myanmar, Philippines, Cambodia, Vietnam, and Laos countries such as lack of supporting policies, legitimation and time-consuming process of permissions, systems topologies, imbalance trade, and inequity issue, and social, economic, political pressures (Chen et al., 2014; Erdiwansyah et al., 2019; Krupa, 2012; Shum, 2017). A critical barrier for successful policy development and implementation is a lack of national strategy as a baseline. This framework tries to address these lessons learned and sum up with a comprehensive solution, which are briefed in Table 1.

Table 1. The main process details, tools, techniques and methods

Process	Description	Tool, Technique and Method
International & national vision (SDGs 2030)	• Identify issues and benefits available, aligned with the UN 17 Sustainable Development Goals (SDGs) and 169 targets at international and national levels.	SWOT analysis, root and cause analysis, etc.
Inter ministries & energy sector stakeholders (mission)	• Define scope and clear image of purposes • Implementation strategy • Involved sectors integration	
Targets	• Predict and qualify goals and objectives • Align goals and objectives with the plans and with the Key Performance Indicator (KPI) • Justify the main reason to pursue the target • Prioritize objectives and goals	
Detailed deliberation	• In-depth analysis and interpretation of targets to strategic goals • Archive and analysis of information and decisions • Identify and analysis of internal and external factors	Macroappraisal (Cook, 2005; Sassoon, 2007)
Strategic goals	• Explore requirement and type of required structures/documents (strategy, policies, mythologies, procedures, acts, etc.)	
Objectives		
Plan	• Background and historical data and information analysis • Rolling up planning approaches • Ensure public acceptance	Maturity model (Hart, 1992; Jansen, 2020)
Action	• Limit bureaucracy and autocracy	Hilbert–Samuel (Taylor, 2018)
Development Process		
Evaluation and monitoring	• Leverage resources • Record and analysis of lessons learned • Adapts codes and standards • Assess and evaluate overall process and progress	
Directing and organizing	• Hire best management practices	PMBOK (Rose, 2013)

CONCLUSION

Viable energy policy can be measured by its broadness in terms of national and international that how it adapts priorities and responding opportunities to give it political and social visibility. The process of transforming the concept to a measurable and trackable lifecycle that enables rethinking of optimum challenges handling to opportunity is the main target of a successful policy. Rethinking process can be achieved by setting aims, objectives, goals, and targets aligned with the utmost anticipation of the energy policy. The proposed framework in this study deals with sustainability requirements, which are essential for national energy policy. These pillars demonstrate some essential contributes such as accessibility, affordability, disparity, safety, use efðciency, supply and production efðciency, cost-effectiveness, and environmental impacts on air, water, and soil quality that basically discussed into five categories technical sustainability, economic sustainability, institutional sustainability, environmental sustainability, social sustainability. Also, for successful policy development, financial sustainability (transparency, consistency, accountability, leadership, and time-bound and budgeted), environmental and social sustainability (climate change, data management, etc.) along with indicators and baselines are discussed in this study. This paper adopts a new approach for a successful public energy policy development that meets priorities by commensurate with the national strategies to meet anticipations and deliver optimum opportunities of technical, technological, political, social, environmental, economic, and institutional benefits. In reference to a broad range of international organizations and policy-makers, a strategic policy development, ensuring clear milestones for reaching decisions is shown in Figure 5. The process components can be undertaken based on need and priority.

Figure 5. Policy development processes

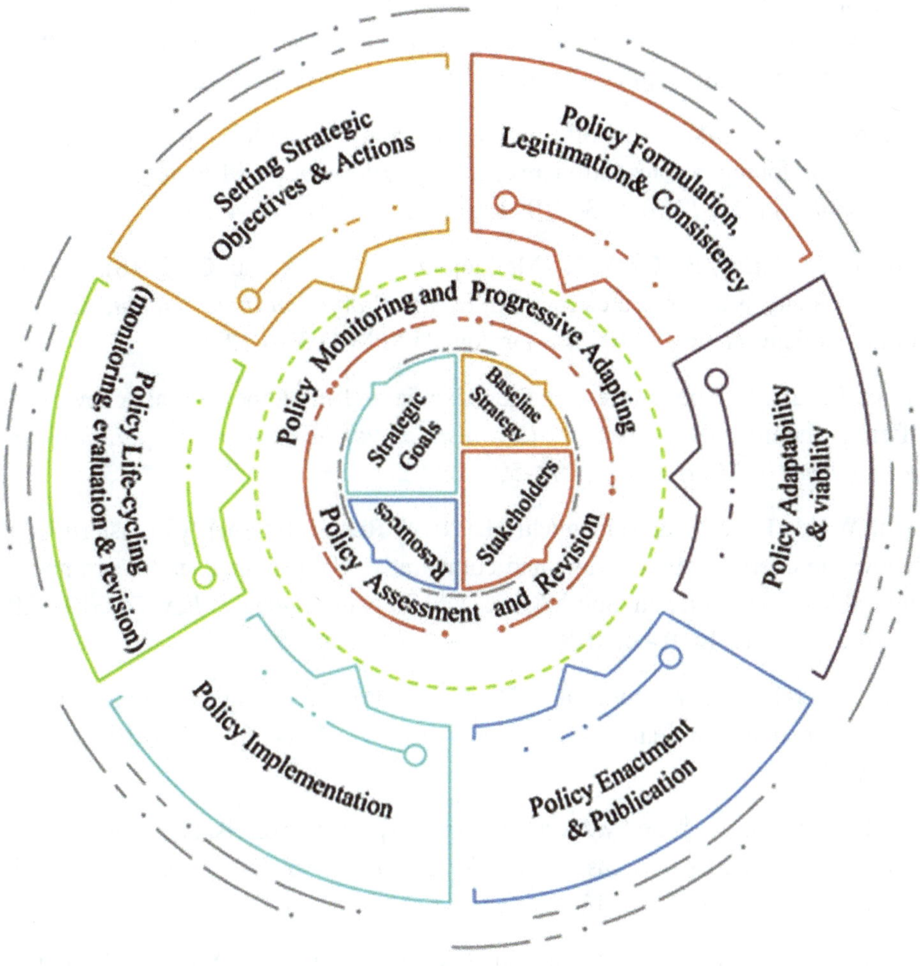

REFERENCES

Ahmadi, M., Lotfy, M. E., Danish, M. S. S., Ryuto, S., Yona, A., & Senjyu, T. (2019). Optimal multi-configuration and allocation of SVR, capacitor, centralised wind farm, and energy storage system: A multi-objective approach in a real distribution network. *IET Renewable Power Generation*, *13*(5), 762–773. Advance online publication. doi:10.1049/iet-rpg.2018.5057

Ahmadi, M., Lotfy, M. E., Shigenobu, R., Howlader, A. M., & Senjyu, T. (2019). Optimal Sizing of Multiple Renewable Energy Resources and PV Inverter Reactive Power Control Encompassing Environmental, Technical, and Economic Issues. *IEEE Systems Journal, 13*(3), 3026–3037. doi:10.1109/JSYST.2019.2918185

Awerbuch, S. (2006). Portfolio-based electricity generation planning: Policy implications for renewables and energy security. *Mitigation and Adaptation Strategies for Global Change, 11*(3), 693–710. doi:10.100711027-006-4754-4

Browne, J., Coffey, B., Cook, K., Meiklejohn, S., & Palermo, C. (2018). A guide to policy analysis as a research method. *Health Promotion International.* Advance online publication. doi:10.1093/heapro/day052 PMID:30101276

Cerna, L. (2013). The nature of policy change and implementation: A review of different theoretical approaches. *Organisation for Economic Cooperation and Development (OECD) Report*, 492–502.

Chen, W.-M., Kim, H., & Yamaguchi, H. (2014). Renewable energy in eastern Asia: Renewable energy policy review and comparative SWOT analysis for promoting renewable energy in Japan, South Korea, and Taiwan. *Energy Policy, 74*, 319–329. doi:10.1016/j.enpol.2014.08.019

Cook, T. (2005). Macroappraisal in Theory and Practice: Origins, Characteristics, and Implementation in Canada. *Archival Science, 5*(2), 101–161. doi:10.100710502-005-9010-2

Cox, E., Royston, S., & Selby, J. (2019). From exports to exercise: How non-energy policies affect energy systems. *Energy Research & Social Science, 55*, 179–188. doi:10.1016/j.erss.2019.05.016

Danish, M. S. S., Sabory, N. R., Danish, S. M. S., Ludin, G. A., Yona, A., & Senjyu, T. (2016a). An Open-door Immature Policy for Rural Electrification: A Case Study of Afghanistan. *International Journal of Sustainable and Green Energy, 6*(3), 8–13.

Danish, M. S. S., Sabory, N. R., Danish, S. M. S., Senjyu, T., Ludin, G. A., Noorzad, A. S., & Yona, A. (2017c). Electricity Sector Development Trends in an After-war Country: Afghanistan Aspiration for an Independent Energy Country. *Journal of Energy and Power Engineering, 11*, 553–557.

Danish, M. S. S., Sabory, N. R., Ershad, A. M., Danish, S. M. S., Ohta, R., Sediqi, M., Ahmadi, M., & Senjyu, T. (2016b). The Least Developed Countries Needs for Changing the Passive Trend of Renewable Energy Exploitation to a Proactive Trend. *International Journal of Energy and Power Engineering, 5*(6), 215–221. doi:10.11648/j.ijepe.20160506.17

Danish, M. S. S., & Senjyu, T. (2020). Green Building Efficiency and Sustainability Indicators. In *Green Building Management and Smart Automation* (1st ed., pp. 128–145). IGI Global. doi:10.4018/978-1-5225-9754-4.ch006

Danish, M. S. S., Senjyu, T., Funabashi, T., Ahmadi, M., Ibrahimi, A. M., Ohta, R., Howlader, H. O. R., Zaheb, H., Sabory, N. R., & Sediqi, M. M. (2018). A sustainable microgrid: A sustainability and management-oriented approach. *Applied Energy Symposium and Forum, Renewable Energy Integration with Mini/Microgrids, 159*, 160–167. 10.1016/j.egypro.2018.12.045

Danish, M. S. S., Senjyu, T., Sabory, N. R., Danish, S. M. S., Ludin, G. A., Noorzad, A. S., & Yona, A. (2017a). Afghanistan's aspirations for energy independence: Water resources and hydropower energy. *Renewable Energy, 113*, 1276–1287. doi:10.1016/j.renene.2017.06.090

Danish, M. S. S., Senjyu, T., Sabory, N. R., Danish, S. M. S., Ludin, G. A., Noorzad, A. S., & Yona, A. (2017b). Afghanistan's aspirations for energy independence: Water resources and hydropower energy. *Renewable Energy, 113*, 1276–1287. doi:10.1016/j.renene.2017.06.090

Danish, M. S. S., Senjyu, T., Zaheb, H., Sabory, N. R., Ibrahimi, A. M., & Matayoshi, H. (2019a). A Novel Transdisciplinary Paradigm for Solid Waste to Sustainable Energy. *Journal of Cleaner Production*. Advance online publication. doi:10.1016/j.jclepro.2019.05.402

Danish, M. S. S., Zaheb, H., Sabory, N. R., Karimy, H., Faiq, A. B., Fedayi, H., & Senjyu, T. (2019b, June). The road ahead for municipal solid waste management in the 21st century: A novel-standardized simulated paradigm. *IOP Conference Series. Earth and Environmental Science, 291*(1), 012009. doi:10.1088/1755-1315/291/1/012009

Energy and power baseline assessment working paper. (2010). *International Centre for Environmental Management (ICEM)*. https://www.mrcmekong.org/assets/Publications/Consultations/SEA-Hydropower/2-Energy-and-Power-baseline-assessment.pdf

Erdiwansyah, M., Mamat, R., Sani, M. S. M., & Sudhakar, K. (2019). Renewable energy in Southeast Asia: Policies and recommendations. *The Science of the Total Environment, 670*, 1095–1102. doi:10.1016/j.scitotenv.2019.03.273 PMID:31018425

Guía Práctica Sobre, P. E. (2016). *Energy policy: A practical Guidebook* (1st ed.). Latin American Energy Organization (OLADE).

Gungah, A., Emodi, N. V., & Dioha, M. O. (2019). Improving Nigeria's renewable energy policy design: A case study approach. *Energy Policy*, *130*, 89–100. doi:10.1016/j.enpol.2019.03.059

Hardee, K., Feranil, I., Boezwinkle, J., & Clark, B. (2004). *The policy circle: A framework for analyzing the components of family planning reproductive health maternal health and HIV/AIDS policies*. Academic Press.

Hart, S. L. (1992). An Integrative Framework for Strategy-Making Processes. *Academy of Management Review*, *17*(2), 327–351. doi:10.5465/amr.1992.4279547

Heindl, A.-B., & Liefner, I. (2019). The Analytic Hierarchy Process as a methodological contribution to improve regional innovation system research: Explored through comparative research in China. *Technology in Society*, *59*, 101197. doi:10.1016/j.techsoc.2019.101197

Ibrahimi, A. M., Howlader, H. O. R., Danish, M. S. S., Shigenobu, R., Sediqi, M. M., & Senjyu, T. (n.d.). Optimal Unit Commitment with Concentrated Solar Power and Thermal Energy Storage in Afghanistan Electrical System. *International Journal of Emerging Electric Power Systems*.

Jansen, S. (2020). A focus area maturity model for software ecosystem governance. *Information and Software Technology*, *118*, 106219. doi:10.1016/j.infsof.2019.106219 PMID:32490064

Khan, A. R. (2016). Policy implementation: Some aspects and issues. *Journal of Community Positive Practices*, *13*(3), 3–12.

Khan, R. A., & Khandaker, S. (2016). A critical insight into policy implementation and implementation performance. *Public Policy and Administration*, *15*(4), 538–548. doi:10.13165/VPA-16-15-4-02

Kim, J., & Park, S. (2018). A contingent approach to energy mix policy. *Energy Policy*, *123*, 749–758. doi:10.1016/j.enpol.2018.08.028

Krupa, J. (2012). Identifying barriers to aboriginal renewable energy deployment in Canada. *Energy Policy*, *42*, 710–714. doi:10.1016/j.enpol.2011.12.051

Kuzman, M. K., Grošelj, P., Ayrilmis, N., & Zbašnik-Senegačnik, M. (2013). Comparison of passive house construction types using analytic hierarchy process. *Energy and Building*, *64*, 258–263. doi:10.1016/j.enbuild.2013.05.020

Liu, W., Wang, C., & Mol, A. P. J. (2013). Rural public acceptance of renewable energy deployment: The case of Shandong in China. *Applied Energy*, *102*, 1187–1196. doi:10.1016/j.apenergy.2012.06.057

Magnusson, T., & Berggren, C. (2018). Competing innovation systems and the need for redeployment in sustainability transitions. *Technological Forecasting and Social Change*, *126*, 217–230. doi:10.1016/j.techfore.2017.08.014

Marcus, A. A. (1992). *Controversial issues in energy policy* (1st ed., Vol. 2). SAGE Publications. doi:10.4135/9781483325729

Matin Ibrahimi, A., Masih Sediqi, M., Or Rashid Howlader, H., Sayed Shah Danish, M., Chakraborty, S., & Senjyu, T. (2019). Generation expansion planning considering renewable energy integration and optimal unit commitment: A case study of Afghanistan. AIMS Energy, 7(4), 441–464. doi:10.3934/energy.2019.4.441

Maynard-Moody, S., Musheno, M., & Palumbo, D. (1990). Street-wise social policy: Resolving the dilemma of street-level influence and successful implementation. *The Western Political Quarterly*, *43*(4), 833–848. doi:10.1177/106591299004300409

Mthethwa, R. M. (2012). Critical dimensions for policy implementation. *African Journal of Public Affairs*, 1–12.

Muh, E., Amara, S., & Tabet, F. (2018). Sustainable energy policies in Cameroon: A holistic overview. *Renewable & Sustainable Energy Reviews*, *82*, 3420–3429. doi:10.1016/j.rser.2017.10.049

O'Neill-Carrillo, E., Irizarry-Rivera, A. A., Colucci-Rios, J. A., Perez-Lugo, M., & Ortiz-Garcia, C. (2008). Sustainable Energy: Balancing the Economic, Environmental and Social Dimensions of Energy. *2008 IEEE Energy 2030 Conference*, 1–7. doi:10.1109/ENERGY.2008.4781010

Pillot, B., Muselli, M., Poggi, P., & Dias, J. B. (2017). *On the impact of the global energy policy framework on the development and sustainability of renewable power systems in Sub-Saharan Africa: The case of solar PV*. Academic Press.

Policy Constraints. (2019). *Online Learning for Sports Management*. http://www.leoisaac.com/policy/top124.htm

Rahaman, M. M., & Varis, O. (2008). *Central Asian waters: Social, economic, environmental and governance puzzle* (1st ed.). Helsinki University of Technology.

Rajapakshe, W. (2017). Factors Affecting Human Resources Management Policy Implementation in Small and Medium Enterprises (SMEs) in Sri Lanka. *International Journal of Academic Research in Business & Social Sciences*, *7*(12), 1129–1144.

Regions for Sustainable Change: Baseline assessment report the regional climate confidence index. (2010). *INTERREG IVC*. http://www.rscproject.org/docs/RSC_Baseline_Assessment_Report.pdf

Rose, K. H. (2013). A Guide to the Project Management Body of Knowledge (PMBOK® Guide)—Fifth Edition. Project Management Journal, 44(3). doi:10.1002/pmj.21345

Saaty, T. L. (1997). That is not the analytic hierarchy process: What the AHP is and what it is not. *Journal of Multi-Criteria Decision Analysis*, *6*(6), 324–335. doi:10.1002/(SICI)1099-1360(199711)6:6<324::AID-MCDA167>3.0.CO;2-Q

Sassoon, J. (2007). Sharing our story. An archaeology of archival thought. *ECU University Archives and Manuscripts*, *2*(35), 1–40.

Savvidis, G., Siala, K., Weissbart, C., Schmidt, L., Borggrefe, F., Kumar, S., Pittel, K., Madlener, R., & Hufendiek, K. (2019). The gap between energy policy challenges and model capabilities. *Energy Policy*, *125*, 503–520. doi:10.1016/j.enpol.2018.10.033

Shum, K. L. (2017). Renewable energy deployment policy: A transition management perspective. *Renewable & Sustainable Energy Reviews*, *73*, 1380–1388. doi:10.1016/j.rser.2017.01.005

Signé, L. (2017). Policy implementation: A synthesis of the study of policy implementation and the causes of policy failure. OCP Policy Center.

Simsek, Y., Lorca, Á., Urmee, T., Bahri, P. A., & Escobar, R. (2019). Review and assessment of energy policy developments in Chile. *Energy Policy*, *127*, 87–101. doi:10.1016/j.enpol.2018.11.058

SMART Goals: How to make your goals achievable. (2019). *Mind Tools*. https://www.mindtools.com/pages/article/smart-goals.htm

Sokołowski, M. M. (2019). When black meets green: A review of the four pillars of India's energy policy. *Energy Policy*, *130*, 60–68. doi:10.1016/j.enpol.2019.03.051

Surbhi, S. (2015, June 22). Difference between strategy and policy (with comparison chart). *Key Differences*. https://keydifferences.com/difference-between-strategy-and-policy.html

Taylor, W. D. (2018). Interpolating between Hilbert–Samuel and Hilbert–Kunz multiplicity. *Journal of Algebra*, *509*, 212–239. doi:10.1016/j.jalgebra.2018.05.015

The energy mix and energy transition: Items of content in this feature report. (2015). *Planete Energies*. https://www.planete-energies.com/en/medias/close/about-energy-mix

Three components of policy systems. (n.d.). *AGRIlink2019*. Retrieved November 20, 2019, from https://www.agrilinks.org/post/three-components-policy-systems

Vedung, E. (n.d.). *Public Policy and Program Evaluation* (1st ed.). Tylor & Francis Group. doi:10.4324/9781315127767

Weijermars, R., Taylor, P., Bahn, O., Das, S. R., & Wei, Y.-M. (2012). Review of models and actors in energy mix optimization–can leader visions and decisions align with optimum model strategies for our future energy systems? *Energy Strategy Reviews*, *1*(1), 5–18. doi:10.1016/j.esr.2011.10.001

Whitford, M. (2009). A framework for the development of event public policy: Facilitating regional development. *Tourism Management*, *30*(5), 674–682. doi:10.1016/j.tourman.2008.10.018

Wies, R. (1994). Policy Definition and Classification: Aspects, Criteria and Examples. *Proceedings of the IFIP/IEEE International Workshop on Distributed Systems: Operation and Management*, 10–12.

Writing a policy: The structure and components of a policy. (2019). *Online Learning for Sports Management*. http://www.leoisaac.com/policy/top125.htm

Xie, G. (2016). Cooperative strategies for sustainability in a decentralized supply chain with competing suppliers. *Journal of Cleaner Production*, *113*, 807–821. doi:10.1016/j.jclepro.2015.11.013

Chapter 2
Strategies to Green Economics:
Analyzing the Renewable Energy Impact in Making the Economy Green

Nima Norouzi
https://orcid.org/0000-0002-2546-4288
Bournemouth University, UK

ABSTRACT

The crises that threaten countries and human societies are the limited resources of non-renewable (fossil) energy sources and the increasing environmental pollution caused by the excessive consumption of fossil fuels, which are necessary for paying attention to energy resources. The close relationship between economic and environmental issues has led to the emergence of new approaches in international environmental law, one of the most important of which is the green economy. Since one of the most important goals of the green economy is to reduce greenhouse gas emissions, renewable energy sources are a shortcut to the green economy. In this regard, the primary purpose of this chapter is to compare the impact of renewable energy on the green economy in selected middle-income and high-income countries.

INTRODUCTION

The close connection between economic and environmental issues has led to the emergence of new approaches in international environmental law, one of the most prominent of which is the green economy. It is possible to move beyond the traditional economics approach and achieve a green economy by observing the principle of fairness and environmental integration. In other words, the traditional economy is

DOI: 10.4018/978-1-7998-9502-2.ch002

based more on the excessive use of natural resources and disregard for the rights of present and future generations. The effects of such an economy can be irreversible in practice. Meeting the environmental challenges of the world requires moving towards a green economy. Therefore, turning to a green and ecological economy should be to reduce greenhouse gas emissions (Aldieri & Vinci, 2018), protection of natural resources, the realization of social and individual justice to fight inequalities that reducing greenhouse gas emissions is one of the most important goals of the green economy (Bovenberg & van der Ploeg, 1994).

Increasing the supply of energy from renewable sources, in addition to the benefits of reducing greenhouse gas emissions, reduces the risks of rising fossil fuel prices. The energy sector is responsible for two-thirds of greenhouse gas emissions. Estimates show that by 2030, the cost of climate change in terms of climate adaptation will increase from $ 50 billion to $ 170 billion, only half of which can be borne by developing countries. As importers of crude oil, many countries have also been challenged by rising fossil fuel prices. For example, oil accounts for 10 to 15% of total African imports and more than 30% of the average, attract export revenue. Some African countries, such as Kenya and Senegal, spend more than half of their export earnings on energy imports. Investment in renewable resources, which are also locally available, can in many cases significantly increase energy security along with development, economic and financial security (Bailey & Caprotti, 2014).

In recent years, various developed and developing countries have paid much attention to renewable energy, and rising fossil fuel prices, environmental considerations, energy security, petrochemical use, technological advances, and economic justification are largely decisive. The future has been renewable energy. Renewable energy is essentially environmentally friendly and keeps the environment healthy, and as a result, can reduce the major greenhouse gas emissions that impose high costs on society (Barbier, 2011). Therefore, we should look for alternatives to fossil fuels, such as renewable energy. Renewable energies (new energies) such as wind, solar, hydropower, geothermal, biogas and biomass are compatible with nature and do not pollute the environment, and from the major greenhouse gas emissions impose many costs on society. In summary, the top three characteristics of renewable energy are (Bordeianu, 1995):

- Renewable energy sources have a long life and natural cycles and, unlike non-renewable energy sources, such as fossil fuels, are not finite, which ensures the continuity of energy consumption for future generations;
- Renewable energy sources, especially wind and solar energy, have significant potential in energy production due to their abundance and convenient geographical facilities, and their use can save fossil fuel consumption;

- The unique use of fossil fuel power plants will create a focus on energy production areas, while renewable energy sources can easily be used to generate energy in any location with suitable geographical conditions. This leads to decentralized energy production in sparsely populated areas such as villages (Cuomo et al., 2016).

With the increase in world population and limited energy resources, all countries face energy consumption. The crises that threaten countries and human societies are the limited resources of non-renewable or fossil energy sources such as oil, gas, and coal, and the other is the increase in environmental pollution caused by excessive consumption of fossil fuels. Destructive and irreversible effects of the greenhouse effect, acid rain, have increased carbon dioxide emissions, which have endangered human life and living organisms (Claessens & Yurtoglu, 2013). Preventing excessive degradation of the environment and subsequent ozone depletion, climate change and changing the rhythm of the seasons, global warming, reduction of agricultural areas and reduction of non-renewable natural resources, protection of natural resources for future generations is essential and shows the importance of paying attention to the use of renewable energy sources.

Therefore, the main purpose of this study is to compare the impact of renewable energy consumption on the green economy in selected middle-income and high-income countries in the period 2005-2016 using data panel models. In this regard, these hypotheses are raised that renewable energy consumption has a negative and significant effect on carbon dioxide emissions as an indicator of a green economy in the group of selected countries. Also, the impact of renewable energy on the green economy in the group of selected middle-income countries is higher than the group of selected high-income countries. Then, after reviewing the theoretical foundations and research history, the structure of the model used is introduced and estimated, and finally, conclusions and suggestions are presented.

BACKGROUND

Environmental protection is one of the major concerns of the international community. To protect non-renewable natural resources and respect the rights of current and future generations, the issue of sustainable development has been considered by governments and international organizations (Djankov et al., 2004). The goal of sustainable development is to strike a balance between economic, social, and environmental dimensions. The traditional approach of the brown economy and the maximum use of natural resources, regardless of the rights of current and future generations, leads to the destruction and pollution of the environment, in contrast to the green economy option that some governments and international institutions consider, not only It is a response to the challenges facing the international community, but it has also helped to realize the concept of sustainable development. On the other hand, the greenness of the economy and reasonable and fair productivity of resources will help their sustainability and renewability. The third dimension of the concept of sustainable development, which is environmental protection, will be strengthened (Fani & Norouzi, 2020). The green economy is a type of economy in which economic growth and development are based on the ecological balance of the environment, assuming that the two goals of economic development and environmental protection can be achieved simultaneously.

Since one of the most important goals of the green economy is to reduce greenhouse gas emissions, the use of renewable energy sources is a shortcut to the green economy. The green economy is recognized as the key to sustainable development. In other words, achieving sustainable development is impossible without a green economy. Therefore, the transition to a green economy is one of the requirements for sustainable development (Hamdouch & Depret, 2010).

The United Nations Environment Program (UNEP) has recently expanded the concept of clean production, including resource efficiency, which is a key element in the transition to a green economy (Guo et al., 2020). The development and expansion of renewable energy help achieve economic, social, and environmental development of countries, which is one of the key factors in achieving sustainable development. The use of renewable energy can reduce dependence on fossil fuels, reduce emissions from energy production and consumption, and reduce greenhouse gas emissions, which significantly impact global warming. Renewable, clean (clean), abundant and reliable energy, and, if properly developed, can play an important role as a sustainable energy source in achieving the goals of sustainable development of countries (Hu & Wang, 2020). The use of renewable energy and low carbon technologies is the most important sub-sector of the green economy. Because on the one hand, most of the emissions occur during energy production, and on the other hand, resources are needed to invest in this sector. It is a big financial thing that is difficult or impossible without the support of the public sector. The cost of producing energy from renewable sources is much higher compared to fossil fuels. As a result, recognizing the existing capacities, evaluating the optimal option, and appropriate policy in the energy sector can help accelerate the movement towards low carbon products and processes (Hayek, 1945).

The approach of the green economy is to pay attention to environmental constraints and its vulnerabilities while addressing the traditional and main goals of economics, the use of renewable resources and non-excessive use of non-renewable resources and consequently the irreparable damage to the environment, etc. have been gradually considered with the pervasiveness of the green economy. Since a green economy is an economic development based on sustainable development and is based on environmental compatibility, its goal is to reduce environmental risks. Enables the peaceful coexistence of human beings with nature, which requires increasing the optimal use of renewable energy. The use of renewable energy reduces dependence on fossil resources and reduces the greenhouse effect, which is one of the goals of the green economy to reduce greenhouse gas emissions (Jalilian et al., 2007). In this regard, studies in the field of renewable energy and green economy can be classified into three categories:

The first group of studies that have addressed the importance of renewable energy and the green economy. Such as Khoshnava et al. (2019), Kiviaho et al. (2004), Kaufmann et al. (2014), Kaufmann et al. (2005), Li & Lin (2016), La Porta et al. (1997). La Porta et al. (2002) and Milani (2000) explained that the green economy framework has helped harmonize environmental and socio-economic goals. The environmental dimension covers various environmental issues (e.g., climate change, renewables, energy, natural capital), while the economic dimension includes various economic aspects such as development, growth, cost, or competition. The social dimension is less visible. Emphasis on these three aspects of sustainability shows the strong link between a green economy and sustainability. The green economy can often be used to reduce resource pressures, climate change, and emissions while guaranteeing economic growth and employment(Norouzi et al., 2020a). In this regard, the role of governments in implementing the framework of long-term green economy regulations It is very important that the development of renewable resources can create short-term socio-economic benefits that indicate green growth (Norouzi et al., 2020b).

Norouzi (2021a), Norouzi & Kalantari (2020), and Norouzi et al. (2021) explained in studies that renewable and clean energy could be the first option due to limited fossil fuel resources and environmental pollution. Be converted to produce energy. The use of renewable energy plays an important role in preserving the natural and human environment, reducing global warming, and achieving sustainable development goals (Norouzi & Fani, 2020). Renewable energy is more compatible with nature and the environment, and its production leads to little environmental pollution; due to their renewability, the resources of this type of energy are not limited and depleted. In general, it is recommended that communities seeking sustainable development use renewable energy sources.

Norouzi & Fani (2021a) and Norouzi et al. (2021b) described in studies that after the Rio Conference in 1992; Sustainable development has become one of the main goals of the international community. Green growth requires programs to achieve economic growth and prosperity with less consumption of resources and waste for food production, transportation, construction, housing, and energy. Green growth is a precondition for building a green economy. Sustainable environmental development in Iran can be provided by introducing the green economy versus the brown economy and the green productivity approach.

The second category of studies examines the causal relationship between renewable energy and green economy using the Granger causality test. Such as the studies of Norouzi & Ataei (2021), and Norouzi & Fani (2020b). Porter & Van der Linde (1995) examined the causal relationship between renewable energy and carbon dioxide emissions in Pakistan in the period 1970-1990, found that there is a two-way causal relationship between renewable energy consumption and CO_2 emissions. Stiglitz (1998) examined this relationship in Denmark during the period 1972-1995. The results showed a one-way causal relationship between renewable energy consumption and greenhouse gas emissions. Szyja et al. (2016) also showed that in 19 countries in 1984-2007, there is a two-way causal relationship between renewable energy consumption and carbon dioxide emissions.

The third group of studies examined the impact of renewable energy on the green economy and other economic variables using time series methods. In this regard, we can refer to the studies of Vuola et al. (2020). Vukovic et al. (2019) explained that renewable energy consumption had harmed pollution in Central and Eastern Europe, Western Europe, East Asia, the Pacific, South Asia, and the United States. At the same time, renewable energy consumption has no significant effect on pollution in the Middle East, North Africa, and sub-Saharan Africa. Zhao et al.'s (2020) study in China in 1975-2005 showed a direct relationship between economic growth and carbon dioxide emissions. Pahle et al.(2016), in the period 2003-2004, explained that the increase in per capita income has a positive and significant effect on per capita renewable energy consumption and the long-term price elasticity of per capita renewable energy consumption is -0.70 (23). Loiseau et al. (2016), in a study in Germany in the period 1971-1999, described Germany as the largest economy in Europe and the leader in the consumption of renewable energy throughout the continent. Also, the consumption of renewable energy has had a significant impact on the economic growth of Germany, so that a 1% increase in the consumption of renewable energy has increased the economic growth of Germany by 0.2194% (Zhang et al., 2017).

METHODS AND MATERIALS

In this paper, using theoretical foundations and empirical study of Apergis et al. (2010) to investigate the impact of renewable energy on the green economy in the group of selected countries with modifications, model (1) has been used.

$$\left(LCO_2\right)_{it} = \beta_0 + \beta_1 LRENEWABLE_{it} + \beta_2 LGDP_{it} + \beta_3 LHU_{CA_{it}} \\ +\beta_4 LICT_{it} + \beta_5 LOPENNESS_{it} + \beta_6 RU\_LAW_{it} + U_{it} \tag{1}$$

In the regression equation (1), LCO2 is the logarithm of carbon dioxide emissions (the main greenhouse gas) as an indicator of green economy, LRENEWABLE logarithm of renewable energy consumption, LGDP is the logarithm of GDP changes Fixed per capita price in 2010 as an indicator of economic growth, LHU_CA stands for the logarithm of gross enrollment rate in higher education as an indicator of human capital, LICT is the logarithm of Internet penetration coefficient as an indicator Indicates information and communication technology (ICT), LOPENNESS is the logarithm of the ratio of total exports and imports of goods and services to GDP as an indicator of the degree of openness of the economy, the RU_LAW is the rule of law, including equation error and, i and t represent the country and time(Al-Mulali et al., 2016).

In terms of identifying and collecting information and statistical sources, the method used in this article is written documentation, library, receipt, and electronic information. This article collects data from the statistical information available in the World Bank at www.worldbank.org, governance site at www.govindicators.org, BP energy data site at www.bp.com, and the International Renewable Energy Agency (IRENA) at www.irena.org. The statistical population is selected according to the criteria of the World Bank. The World Bank classifies countries by geographical area or by income level. The countries selected in this study are selected based on income level. The World Bank classifies countries into low-income, middle-income, and high-income countries based on per capita income. In the World Bank classification, Iran is a middle-income country(Akin, 2014).

Therefore, in this article, middle-income countries (including Iran) have been selected. This group includes Argentina, South Africa, Ukraine, Ecuador, Indonesia, Iran, Brazil, Peru, Thailand, Turkey, China, Russia, Romania, Philippines, Kazakhstan, Colombia, Malaysia, Egypt, Mexico, Venezuela, Vietnam, India. Also, to compare the impact of renewable energy on the green economy in two different structures, another group has been selected in this article. This group includes countries with high incomes. Includes Germany, Austria, Australia, Spain, United Kingdom, USA, Italy, Belgium, Portugal, Czech Republic, Denmark, Japan, Sweden, Switzerland, Chile, France, Finland, Canada, South Korea, Poland, Norway, New Zealand, The Netherlands, Costa Rica, and Greece. In selecting the selected countries, middle-income and high-income countries have been selected that produce and consume renewable energy. Statistical data of the variables used in this paper were also available in the period under review (Ollo-López & Aramendía-Muneta, 2012).

RESULTS AND DISCUSSION

To estimate the model, it is first necessary to determine the type of estimation method for the specific type of panel data. Therefore, the F-Limer statistic was first used to determine the presence (absence) of width from a separate origin for each country. According to the amount of F-Limer statistics calculated in Table 1, the null hypothesis of the test based on the use of the ordinary least squares method is rejected. As a result, constrained regression (ordinary least squares) is not valid, and the width of different sources (fixed or random effects method) should be considered in the model. Then, the Hausman test was used to test the model using the fixed or random effects method. According to the number of statistics obtained in Table 1, the fixed effects method was used to estimate the model.

Table 1. Results of estimating the effect of renewable energy consumption on the green economy in the group of selected countries (dependent variables: logarithm of carbon dioxide emissions)

Mid-income Countries	High-income Countries	Variable
Coefficient [t-stat](prob.)	Coefficient [t-stat](prob.)	
136.224 [4.881] (0.000)	93.653 [3.812] (0.000)	C
-2.303[-5.536](0.000)	-1.413[-14.523](0.000)	LRENEWABLE
4.192E-9[14.443](0.000)	2.861E-10[13.892](0.000)	LGDP
-2.278[-10.096](0.000)	1.304E-5[-1.754](0.077)	LHU_CA
-0.963[-2.732](0.006)	-0.778[-4.474](0.000)	LICT
0.576[2.773](0.006)	36.962[1.775](0.077)	LOPENNESS
-110.122[-3.956](0.000)	-34.654[-2.489](0.013)	RU_LAW
0.997	0.996	R^2
0.663	1.134	D-W
508.642(0.000)	243.291(0.000)	F-statistic(Prob.)
5.286(0.000)	422.331(0.000)	F-limer stat
38.864(0.000)	36.245(0.000)	Hausman stat

Estimating the regression model using the fixed effects method in the group of selected middle-income and high-income countries in 2000-2020 is shown in Table 1.

The logarithm of renewable energy consumption has a negative and significant effect on carbon dioxide emissions as an indicator of the green economy in selected countries. Therefore, the hypothesis about the negative and significant impact of renewable energy consumption on the green economy in the group of selected countries can not be rejected.

The impact of renewable energy consumption on the green economy in middle-income countries is higher than that of high-income countries. Therefore, the hypothesis about the greater impact of renewable energy consumption on the green economy in middle-income countries than in the group of high-income countries can not be rejected.

Economic growth has a positive and significant effect on carbon dioxide emissions as an indicator of the green economy in selected countries. Increasing economic growth requires more natural resources and energy, especially fossil fuels, followed by releasing large amounts of pollutants that cause environmental degradation and serious damage to the environment. The results of this study are consistent with the studies of Cole (2004) and Cuomo et al. (2016).

The logarithm of the gross rate of enrollment in higher education as an indicator of human capital has a negative and significant effect on carbon dioxide emissions as an indicator of the green economy in the group of selected countries. Human resources are literate and aware of the rules and regulations, with the correct use of energy resources and the protection of the environment around it, reduces environmental pollution, and maintains its quality.

The logarithm of Internet penetration coefficient as an indicator of information and communication technology (ICT) has a negative and significant effect on carbon dioxide emissions as an indicator of the green economy in the group of selected countries. Proper use of ICT services will reduce traffic and consequently reduce the consumption of transportation fuels and greenhouse gas emissions, as well as the use of ICT services, useful information, and access to various issues raised in communities; it increases people's awareness and then implements effective strategies to improve the quality of the environment and reduce pollutants. Alopez and Aramandia Monta (2012) showed that information and communication technology reduces greenhouse gas emissions (Cristea et al., 2013).

The logarithm of the ratio of total exports and imports of goods and services to GDP as an indicator of the degree of openness of the economy has a positive and significant effect on carbon dioxide emissions as an indicator of the green economy in the selected countries. Environmental pollution has received a lot of attention in recent years. On the other hand, trade liberalization can affect the emission of pollution by increasing productivity by increasing the productivity of products, according to the principle of comparative advantage of countries in the production of goods in which they have a comparative advantage(Barbier, 2011). Increasing trade affects pollution emissions in two ways: 1) by increasing carbon dioxide emissions from international transportation, and 2) by transferring carbon dioxide emissions from the importing country to the exporting country, while Increasing the production of polluting goods in exporting countries may increase pollution in these countries (Hamdouch & Depret, 2010). The rule of law has a significant negative impact on carbon dioxide emissions as an indicator of the green economy in the group of selected countries. The rule of law makes the transparency of actions and

The activities of governments and the improvement of the decision-making process of governments are done through popular participation. The transparency of the laws related to the activities of companies and economic institutions and the proper functioning of the government increase investment and economic growth (Guo et al., 2020). The value of the coefficient of determination in the model shows that more than 90% of the changes in carbon dioxide emissions as an indicator of the green economy in the group of selected countries are explained by the explanatory variables of the model.

Climate change due to increased greenhouse gas emissions is one of the important factors in climate change and events, environmental hazards, natural disasters, and serious damage to the economy, which expands renewable energy as a source of clean energy. It can play an important role in reducing environmental pollution and the major emissions of polluting gases such as carbon dioxide and other greenhouse gases and prevent the imposition of high costs on society. In other words, renewable energy, clean (clean), abundant and reliable, and if properly developed, can play an important role as a sustainable energy source in achieving sustainable development goals. The main reason for emphasizing renewable energy and helping to solve environmental problems and prevent the loss of fossil fuels is the protection of natural resources for future generations, which undoubtedly is renewable energy due to the simplicity of its technology. They are important in the green economy and sustainable economic development. The results of this study are consistent with the studies of Khoshnava et al. (2019), Kiviaho et al. (2014), and Loiseau et al. (2016). However, it is not consistent with the results of the study of Vuola et al. (2020) in the Middle East, North Africa, and sub-Saharan Africa. Ollo-López et al. (2012) in Pakistan in 1970-1990 showed that renewable energy had played an important role in reducing carbon dioxide emissions. Pahle et al. (2016) In 1996-2014 showed that the intensity of carbon dioxide emissions had decreased significantly due to the use of more efficient and cleaner technologies and the tendency to use cleaner fuels. Barbier (2011) in Denmark 1972-2007 showed a one-way causal relationship between the consumption of renewable energy and greenhouse gas emissions. The study by Cristea et al. (2013) Also showed that renewable energy consumption had not had a significant impact on reducing pollution in the Middle East, North Africa, and sub-Saharan Africa. Therefore, based on empirical evidence, there is no consensus on how renewable energy affects carbon dioxide emissions as an indicator of the green economy in different countries. This study also showed that the impact of renewable energy consumption on the green economy in selected middle-income countries is higher than the group of selected high-income countries. This indicates the lack of adequate and efficient use of renewable energy in the group of selected middle-income countries. The expansion of its use compared with the group of selected high-income countries has been able to have more effects on reducing carbon dioxide emissions as an indicator of a green economy.

SUCCESSFUL EXPERIENCES IN THE GREE ECONOMICS

Green economics or environmental economics has come to the attention of many governments in recent decades. The interactions between economics and the environment have become an undeniable reality; Just as economic policies affect the environment, the economy is also affected by environmental change. Environmental pollution is one of the most important challenges of societies. In developed countries, by investing in renewable energy, new technologies in controlling environmental pollution, and improving energy efficiency, steps have been taken to improve it, but it has not been considered in developing countries. The GGEI is the first green economy index introduced in 2010 and is widely used today by policymakers, international organizations, civil society, and the private sector. A leading report compiled by the Deputy Minister of Economic Affairs of the Ministry of Economic examines the experience of five countries - Sweden, Norway, Costa Rica, Germany, and Denmark - in the green economy and, in particular, the development of renewable energy(Nima, 2021).

Green Economics in Sweden

According to the Global Green Economic Index in 2014, Sweden ranked first to have the best performance among countries in the world. Sweden currently has the highest rate of renewable energy use in the European Union; More than 45% of Sweden's energy supply is renewable, relying on hydropower and biofuels. More than 12 percent of electricity generation comes from cogeneration plants and 2 percent from wind farms. Since 2013, Sweden has helped to encourage the use of second-generation biofuels with a tax exemption on hydrogenated vegetable and animal fats and oils up to 15% by volume of diesel fuel. Diesel provided sustainable conditions for biofuel producers and distributors and helped promote the use of renewable energy. Other strengths of Sweden include developing environmentally friendly technologies in recycling, renewable energy (wind, solar, hydropower, biofuels), information technology, green transport, electric motors, green chemistry, lighting, and many energy-efficient industries.

The government invested in environmental technologies from 2011 to 2014, thus supporting its commitment to working with China, India, and Russia on environmental technologies. In general, Sweden has been a leader in organic farming, renewable energy use, per capita investment in green technologies, and sustainable development research. Given that Sweden was the first country in the world to impose a heavy tax on fossil fuels in 1991 to develop green energy resources and to increase the efficiency of this type of tax, it reduced its exemption in 2013 and 2015, the government of this country To further protect the environment, it intends to eliminate fossil fuels from all cars by 2030 and to eliminate carbon by 2050. The country was able to take effective steps to increase energy efficiency by enacting a new planning and construction law in 2011 and working on smart grids(Guo et al., 2020).

Given the scarcity of raw materials and the need to increase resource efficiency, Sweden implements a minerals and resource efficiency strategy. Sweden has also made significant changes in waste recycling in recent decades; More than 99% of household waste is recycled. In 1975, only 38% of household waste was recycled. Because fuel waste is relatively cheap, the Swedes use it efficiently and profitably. Overall, Sweden is a leader in reducing environmental impact compared to neighboring countries and is unique in converting waste into electricity using high-power incinerators, which has led to the import of waste from other countries. The Swedish Environmental Protection Agency has also developed a practical plan to prevent waste, including encouraging manufacturers to make products with longer lifespans and is considering offering tax breaks to repair some goods.

Green Economics in Norway

Norway has set ambitious environmental policy goals for sustainable development. These goals are supported by a strong analytical framework on environmental, social, and economic issues that focus on managing human, natural, productive, and financial capital. Measures such as simplification of regulations, decentralization of environmental responsibilities, and intelligent use of economic tools have contributed to the successful implementation of many Norwegian environmental policies. Also, the requirement of all projects to conduct environmental impact assessments and better inform the affected people about these projects, special attention to issues such as air pollution, water and sewage infrastructure, and river management are among the most important environmental measures in Norway. Norway is also one of the world leaders in financing clean climate projects. It helps developing countries reduce deforestation, expand renewable energy and adapt to climate change(Hu & Wang, 2020).

Green Economics in Costa Rica

Costa Rica is one of the countries in Central America and is ranked third in terms of best performance in the global green economy index in 2014, committed to becoming a carbon-free country by 2021. To increase the use of renewable energy, the country is offering new incentives for the construction of renewable energy plants of seven megawatts or larger, including the import of materials without customs duties, exemption from operating taxes for some time, and the possibility of issuing carbon offsets. Pointed to be of considerable value for increasing investment attractiveness. Doing so has generated more than 90 percent of Costa Rica's electricity from renewable energy sources such as hydropower, geothermal energy, and wind power. Costa Rica also provides tax incentives to the biofuel industry. With the high supply of palm oil, biodiesel production is expected to become the leading biofuel in the country(Khoshnava et al., 2019).

Green Economics in Germany

According to the Global Green Economy Performance Index in 2014, Germany ranks fourth. It is one of the pioneers in promoting renewable energy protection policies. The Renewable Energy Law, first implemented in Germany in 2000, is one of the factors in the country's success in the environment. Within the framework of this law, the policy of encouraging cogeneration power plants, limited emission exchange system, energy tax reform, etc., are included. The enactment of the Renewable Energy Law and the development of incentive tariffs for wind, solar, hydro, geothermal, and biomass energy in this law have been effective in forming a 29% share of net electricity consumption from renewable energies. Nevertheless, Germany is reviewing the Renewable Energy Act to motivate access to and expand electricity networks, marine wind energy, and technologies for peak consumption management and storage. Energy pricing through taxes and other financial instruments plays an important role in the composition of German energy policy.

The German parliament passed the Environmental Tax Reform Act in 1999, which gradually increased oil and gas tax rates and introduced a new electricity tax. Germany also has a good position in exporting green technologies. According to the German Solar Energy Association, photovoltaic exports increased from 14% in 2004 to 55% in 2011 and 65% in 2013, with 80% targeting 2020. Also, according to the German Wind Energy Association, the share of current exports of the wind industry is between 65 and 70%. Germany plays a key role in the renewable energy market and in the market for products that increase energy efficiency; In 2004, Germany accounted for 17% of the global efficiency market and even larger shares than the United States, Japan, and Italy. One of Germany's goals is to increase renewable energy consumption to 18% of final energy consumption by 2020 and 80% of electricity consumption by 2050(Kiviaho et al., 2014).

Green Economics in Denmark

Denmark is one of the pioneers in the implementation of appropriate policies in the field of renewable energy, energy efficiency, and climate change in the member countries of the Organization for Economic Cooperation and Development (OECD); According to the Global Green Economy Index (GGEI) in 2014, the country ranked fifth among countries in the world. Enacting appropriate tax laws in the field of environment is one of the factors influencing the proper position of this country in the green economy; In 1992, Denmark was the second country after Sweden to impose a carbon tax on some types of energy consumption by households and industries, and in 2012 it collected the highest energy tax among EU countries. Carbon tax rates in Denmark vary for different purposes; In such a way that the home sector pays the highest taxes and the energy industries pay the lowest carbon tax in addition to tax rebates to facilitate competition. To develop environmental innovations, the country has allocated carbon tax revenue to subsidize the area.

In addition, incentive tariffs on the wind, biomass, geothermal, hydroelectric, and solar energy are another action taken by the Danish government to reduce carbon emissions. Denmark is one of the pioneers in the development of wind energy in Europe. On average, the country's wind farms supply more than a quarter of Denmark's electricity needs. Electricity generation from renewable sources is supported at a price above market price. Total market price and surplus price guarantee a stable income for the producer. All subsidy costs are transferred to the consumer as equal public service requirements. Under Denmark's energy program, the Danish government, as the first European country, provided many subsidies to the fledgling wind industry, which was also successfully implemented in Germany. The rapid decline in lead emissions in Denmark in the early 1990s due to laws banning the sale of lead fuels for transportation resulted from regulations for catalytic converters in automobile exhaust systems and nitrate-free units in thermal power plants (Loiseau et al., 2016).

In 2001, Denmark introduced a limited emission exchange system for electricity generation, according to which free licenses were granted to firms in proportion to the number of pollutants emitted by firms in the past. Denmark's limited edition exchange system was developed in 2003 and replaced in 2005 by the EU Emissions Exchange Scheme. Denmark has also transformed its car power system to eliminate fossil fuels by 2050 and supply all renewable energy sources. In general, the most important measures taken in Denmark to improve performance in the field of the green economy can be such as environmental impact assessment and attention in policy decisions in all sectors, funding for the use of new technologies, providing information to consumers on how to influence Their choices focused on the environment and the regulation of resource use and pollution mitigation through price and tax tools.

CONCLUSION

According to the results obtained in line with this study, it can be concluded that measures to provide access to renewable and clean energy in remote and rural areas create job opportunities, reduce poverty and establish social justice. Creating appropriate educational and advertising fields in expanding the use of renewable energy Allocating sufficient funds to implement renewable energy technologies for electricity generation of residential, commercial buildings, factories, transportation industries, and water treatment that reduce environmental pollution Followed by the achievement of a green economy. Government policies and investments to use environmentally friendly technologies, renewable energy, implement and finance industrial research projects, train efficient staff and comply with existing laws and regulations, and implement treaties and agreements International will create a green economy in the communities. As a strategic policy, environmental taxes can require producers to use environmental laws and standards and prevent environmental degradation. Creating suitable grounds for using software and hardware in line with environmental goals and creating the necessary culture can significantly impact environmental degradation.

REFERENCES

Akin, C. S. (2014). The impact of foreign trade, energy consumption and income on CO2 emissions. *International Journal of Energy Economics and Policy*, *4*(3), 465.

Al-Mulali, U., Ozturk, I., & Solarin, S. A. (2016). Investigating the environmental Kuznets curve hypothesis in seven regions: The role of renewable energy. *Ecological Indicators*, *67*, 267–282. doi:10.1016/j.ecolind.2016.02.059

Aldieri, L., & Vinci, C. P. (2018). Green economy and sustainable development: The economic impact of innovation on employment. *Sustainability*, *10*(10), 3541. doi:10.3390u10103541

Apergis, N., Payne, J. E., Menyah, K., & Wolde-Rufael, Y. (2010). On the causal dynamics between emissions, nuclear energy, renewable energy, and economic growth. *Ecological Economics*, *69*(11), 2255–2260. doi:10.1016/j.ecolecon.2010.06.014

Bailey, I., & Caprotti, F. (2014). The green economy: Functional domains and theoretical directions of enquiry. *Environment & Planning A*, *46*(8), 1797–1813. doi:10.1068/a130102p

Barbier, E. (2011). The policy challenges for green economy and sustainable economic development. *Natural Resources Forum*, *35*(3), 233–245. doi:10.1111/j.1477-8947.2011.01397.x

Bordeianu, S. (1995). Political risk services and political risk yearbook: Edited William D. Coplin and Michael K. O'Leary. Syracuse, NY: Political Risk Services, 1994. 7 vols. $5,570.00 commercial subscribers (to the services), $750.00 (to the Yearbook). Journal of Government Information, 22(2), 186-188.

Bovenberg, A. L., & van der Ploeg, F. (1994). Environmental policy, public finance and the labour market in a second-best world. *Journal of Public Economics*, *55*(3), 349–390. doi:10.1016/0047-2727(93)01398-T

Claessens, S., & Yurtoglu, B. B. (2013). Corporate governance in emerging markets: A survey. *Emerging Markets Review*, *15*, 1–33. doi:10.1016/j.ememar.2012.03.002

Cole, M. A. (2004). Trade, the pollution haven hypothesis and the environmental Kuznets curve: Examining the linkages. *Ecological Economics*, *48*(1), 71–81. doi:10.1016/j.ecolecon.2003.09.007

Cristea, A., Hummels, D., Puzzello, L., & Avetisyan, M. (2013). Trade and the greenhouse gas emissions from international freight transport. *Journal of Environmental Economics and Management*, *65*(1), 153–173. doi:10.1016/j.jeem.2012.06.002

Cuomo, F., Mallin, C., & Zattoni, A. (2016). Corporate governance codes: A review and research agenda. *Corporate Governance*, *24*(3), 222–241. doi:10.1111/corg.12148

Djankov, S., McLiesh, C., & Klein, M. U. (Eds.). (2004). *Doing business in 2004: understanding regulation* (Vol. 1). World Bank Publications.

Fani, M., & Norouzi, N. (2020). Using Social and Economic Indicators for Modeling, Sensitivity Analysis and Forecasting the Gasoline Demand in the Transportation Sector: An ANN Approach in case study for Tehran metropolis. *Iranian Journal of Energy*, *23*(2), 71–91.

Guo, M., Nowakowska-Grunt, J., Gorbanyov, V., & Egorova, M. (2020). Green technology and sustainable development: Assessment and green growth frameworks. *Sustainability*, *12*(16), 6571. doi:10.3390u12166571

Hamdouch, A., & Depret, M. H. (2010). Policy integration strategy and the development of the 'green economy': Foundations and implementation patterns. *Journal of Environmental Planning and Management*, *53*(4), 473–490. doi:10.1080/09640561003703889

Hayek, F. A. (1945). The use of knowledge in society. *The American Economic Review*, *35*(4), 519–530.

Hu, W., & Wang, D. (2020). How does environmental regulation influence China's carbon productivity? An empirical analysis based on the spatial spillover effect. *Journal of Cleaner Production*, *257*, 120484. doi:10.1016/j.jclepro.2020.120484

Jalilian, H., Kirkpatrick, C., & Parker, D. (2007). The impact of regulation on economic growth in developing countries: A cross-country analysis. *World Development*, *35*(1), 87–103. doi:10.1016/j.worlddev.2006.09.005

Kaufmann, D., Kraay, A., & Mastruzzi, M. (2004). Governance matters III: Governance indicators for 1996, 1998, 2000, and 2002. *The World Bank Economic Review*, *18*(2), 253–287. doi:10.1093/wber/lhh041

Kaufmann, D., Kraay, A., & Mastruzzi, M. (2005). Governance matters IV: governance indicators for 1996-2004. *World bank policy research working paper series*, (3630).

Khoshnava, S. M., Rostami, R., Zin, R. M., Štreimikienė, D., Yousefpour, A., Strielkowski, W., & Mardani, A. (2019). Aligning the criteria of green economy (GE) and sustainable development goals (SDGs) to implement sustainable development. *Sustainability*, *11*(17), 4615. doi:10.3390u11174615

Kiviaho, J., Nikkinen, J., Piljak, V., & Rothovius, T. (2014). The co-movement dynamics of European frontier stock markets. *European Financial Management*, *20*(3), 574–595. doi:10.1111/j.1468-036X.2012.00646.x

La Porta, R., Lopez-de-Silanes, F., Shleifer, A., & Vishny, R. (2002). Investor protection and corporate valuation. *The Journal of Finance*, *57*(3), 1147–1170. doi:10.1111/1540-6261.00457

La Porta, R., Lopez-de-Silanes, F., Shleifer, A., & Vishny, R. W. (1997). Legal determinants of external finance. *The Journal of Finance*, *52*(3), 1131–1150. doi:10.1111/j.1540-6261.1997.tb02727.x

Li, J., & Lin, B. (2016). Green economy performance and green productivity growth in China's cities: Measures and policy implication. *Sustainability*, *8*(9), 947. doi:10.3390u8090947

Loiseau, E., Saikku, L., Antikainen, R., Droste, N., Hansjürgens, B., Pitkänen, K., Leskinen, P., Kuikman, P., & Thomsen, M. (2016). Green economy and related concepts: An overview. *Journal of Cleaner Production*, *139*, 361–371. doi:10.1016/j.jclepro.2016.08.024

Milani, B. (2000). *Designing the green economy: The postindustrial alternative to corporate globalization*. Rowman & Littlefield.

Norouzi, N. (2021a). The Pahlev Reliability Index: A measurement for the resilience of power generation technologies versus climate change. *Nuclear Engineering and Technology*, *53*(5), 1658–1663. doi:10.1016/j.net.2020.10.013

Norouzi, N. (2021b). Post-COVID-19 and globalization of oil and natural gas trade: Challenges, opportunities, lessons, regulations, and strategies. *International Journal of Energy Research*, *45*(10), 14338–14356. doi:10.1002/er.6762 PMID:34219899

Norouzi, N., & Ataei, E. (2021). Covid-19 Crisis and Environmental law: Opportunities and challenges. *Hasanuddin Law Review*, *7*(1), 46–60. doi:10.20956/halrev.v7i1.2772

Norouzi, N., de Rubens, G. Z., Choupanpiesheh, S., & Enevoldsen, P. (2020a). When pandemics impact economies and climate change: Exploring the impacts of COVID-19 on oil and electricity demand in China. *Energy Research & Social Science*, *68*, 101654. doi:10.1016/j.erss.2020.101654 PMID:32839693

Norouzi, N., & Fani, M. (2020a). The impacts of the novel corona virus on the oil and electricity demand in Iran and China. *Journal of Energy Management and Technology*, *4*(4), 36–48.

Norouzi, N., & Fani, M. (2020b). Black gold falls, black plague arise-An Opec crude oil price forecast using a gray prediction model. *Upstream Oil and Gas Technology*, *5*, 100015. doi:10.1016/j.upstre.2020.100015

Norouzi, N., & Fani, M. (2021). The prioritization and feasibility study over renewable technologies using fuzzy logic: A case study for Takestan plains. *Journal of Energy Management and Technology*, *5*(2), 12–22.

Norouzi, N., Fani, M., & Ziarani, Z. K. (2020b). The fall of oil Age: A scenario planning approach over the last peak oil of human history by 2040. *Journal of Petroleum Science Engineering*, *188*, 106827. doi:10.1016/j.petrol.2019.106827

Norouzi, N., & Kalantari, G. (2020). The sun food-water-energy nexus governance model a case study for Iran. *Water-Energy Nexus*, *3*, 72–80. doi:10.1016/j.wen.2020.05.005

Norouzi, N., Khanmohammadi, H. U., & Ataei, E. (2021b). The Law in the Face of the COVID-19 Pandemic: Early Lessons from Uruguay. *Hasanuddin Law Review, 7*(2), 75–88. doi:10.20956/halrev.v7i2.2827

Norouzi, N., Zarazua de Rubens, G. Z., Enevoldsen, P., & Behzadi Forough, A. (2021a). The impact of COVID-19 on the electricity sector in Spain: An econometric approach based on prices. *International Journal of Energy Research, 45*(4), 6320–6332. doi:10.1002/er.6259

Ollo-López, A., & Aramendía-Muneta, M. E. (2012). ICT impact on competitiveness, innovation and environment. *Telematics and Informatics, 29*(2), 204–210. doi:10.1016/j.tele.2011.08.002

Pahle, M., Pachauri, S., & Steinbacher, K. (2016). Can the Green Economy deliver it all? Experiences of renewable energy policies with socio-economic objectives. *Applied Energy, 179*, 1331–1341. doi:10.1016/j.apenergy.2016.06.073

Porter, M. E., & Van der Linde, C. (1995). Toward a new conception of the environment-competitiveness relationship. *The Journal of Economic Perspectives, 9*(4), 97–118. doi:10.1257/jep.9.4.97

Stiglitz, J. (1998). The private uses of public interests: Incentives and institutions. *The Journal of Economic Perspectives, 12*(2), 3–22. doi:10.1257/jep.12.2.3

Szyja, P. (2016). The role of the state in creating green economy. *Oeconomia Copernicana, 7*(2), 207–222. doi:10.12775/OeC.2016.013

Vukovic, N., Pobedinsky, V., Mityagin, S., Drozhzhin, A., & Mingaleva, Z. (2019). A study on green economy indicators and modeling: Russian context. *Sustainability, 11*(17), 4629. doi:10.3390u11174629

Vuola, M., Korkeakoski, M., Vähäkari, N., Dwyer, M. B., Hogarth, N. J., Kaivo-oja, J., Luukkanen, J., Chea, E., Thuon, T., & Phonhalath, K. (2020). What is a green economy? Review of national-level green economy policies in Cambodia and Lao PDR. *Sustainability, 12*(16), 6664. doi:10.3390u12166664

Zhang, B., Wang, B., & Wang, Z. (2017). Role of renewable energy and non-renewable energy consumption on EKC: Evidence from Pakistan. *Journal of Cleaner Production, 156*, 855–864. doi:10.1016/j.jclepro.2017.03.203

Zhao, M., Liu, F., Song, Y., & Geng, J. (2020). Impact of air pollution regulation and technological investment on sustainable development of green economy in Eastern China: Empirical analysis with panel data approach. *Sustainability, 12*(8), 3073. doi:10.3390u12083073

ADDITIONAL READING

Anderson, V. (2006). Turning economics inside out. *International Journal of Green Economics*, *1*(1-2), 11–22. doi:10.1504/IJGE.2006.009334

Cato, M. S. (2012). *Green economics: an introduction to theory, policy and practice*. Routledge. doi:10.4324/9781849771528

Eichholtz, P., Kok, N., & Quigley, J. M. (2013). The economics of green building. *The Review of Economics and Statistics*, *95*(1), 50–63. doi:10.1162/REST_a_00291

Green, D. (2003). *Silent revolution: The rise and crisis of market economics in Latin America*. NYU Press.

Hahnel, R. (2014). *Green economics: Confronting the ecological crisis*. Routledge. doi:10.4324/9781315703947

Hydes, K. R., & Creech, L. (2000). Reducing mechanical equipment cost: The economics of green design. *Building Research and Information*, *28*(5-6), 403–407. doi:10.1080/096132100418555

Pearce, D. (1992). Green economics. *Environmental Values*, *1*(1), 3–13. doi:10.3197/096327192776680179

Smith, R., & World Economics Association. (2016). *Green capitalism: The god that failed*. London: College Publications.

KEY TERMS AND DEFINITIONS

Circularity: A circular economy (also referred to as “circularity”) is an economic system that tackles global challenges like climate change, biodiversity loss, waste, and pollution. Most linear economy businesses take a natural resource and turn it into a product that is ultimately destined to become waste because it has been designed and made. This process is often summarised by “take, make, waste.” By contrast, a circular economy uses reuse, sharing, repair, refurbishment, remanufacturing, and recycling to create a closed-loop system, minimize resource inputs, and create waste, pollution, and carbon emissions. The circular economy aims to keep products, materials, equipment, and infrastructure in use for longer, thus improving the productivity of these resources. Waste materials and energy should become input for other processes through waste valorization: either as a component or recovered resource for another industrial process or as regenerative resources for nature (e.g., compost). This regenerative approach contrasts with the traditional linear economy, which has a "take, make, dispose of" production model.

Eco Commerce: Eco commerce is a business, investment, and technology-development model that employs market-based solutions to balancing the world's energy needs and environmental integrity. Through green trading and green finance, eco-commerce promotes the further development of "clean technologies" such as wind power, solar power, biomass, and hydropower.

Eco-Tariffs: An eco-tariff, also known as an environmental tariff, is a trade barrier erected to reduce pollution and improve the environment. These trade barriers may take the form of import or export taxes on products with a large carbon footprint or imported from countries with lax environmental regulations.

Emissions Trading: Emissions trading (also known as cap and trade, emissions trading scheme, or ETS) is a market-based approach to controlling pollution by providing economic incentives for reducing the emissions of pollutants.

Environmental Enterprise: An environmental enterprise is an environmentally friendly/compatible business. Specifically, an environmental enterprise is a business that produces value in the same manner which an ecosystem does, neither producing waste nor consuming unsustainable resources. In addition, an environmental enterprise rather finds alternative ways to produce one's products instead of taking advantage of animals for the sake of human profits. To be closer to being an environmentally friendly company, some environmental enterprises invest their money to develop or improve their technologies which are also environmentally friendly. In addition, environmental enterprises usually try to reduce global warming, so some companies use environmentally friendly materials to build their stores. They also set in environmentally friendly place regulations. All these efforts of the environmental enterprises can bring positive effects both for nature and people. The concept is rooted in the well-enumerated theories of natural capital, the eco-economy, and cradle-to-cradle design. Examples of environmental enterprises would be Seventh Generation, Inc., and Whole Foods.

Green Economy: A green economy is an economy that aims at reducing environmental risks and ecological scarcities and that aims for sustainable development without degrading the environment. It is closely related to ecological economics but has a more politically applied focus. The 2011 UNEP Green Economy Report argues "that to be green, and an economy must be not only efficient but also fair. Fairness implies recognizing global and country-level equity dimensions, particularly in assuring a Just Transition to an economy that is low-carbon, resource-efficient, and socially inclusive."

Green Politics: Green politics, or ecopolitics, is a political ideology that aims to foster an ecologically sustainable society often, but not always, rooted in environmentalism, nonviolence, social justice, and grassroots democracy. It began taking shape in the western world in the 1970s; since then, Green parties have developed and established themselves in many countries around the globe and have achieved some electoral success.

Low-Carbon Economy: A low-carbon economy (LCE) or decarbonized economy is based on low-carbon power sources with minimal greenhouse gas (GHG) emissions into the atmosphere, specifically carbon dioxide. GHG emissions due to anthropogenic (human) activity are the dominant cause of observed climate change since the mid-20th century. Continued emission of greenhouse gases may cause long-lasting changes worldwide, increasing the likelihood of severe, pervasive, and irreversible effects for people and ecosystems.

Natural Resource Economics: Natural resource economics deals with the supply, demand, and allocation of the Earth's natural resources. One main objective of natural resource economics is to understand better the role of natural resources in the economy to develop more sustainable methods of managing those resources to ensure their future generations. Resource economists study interactions between economic and natural systems intending to develop a sustainable and efficient economy.

Sustainable Development: Sustainable development is an organizing principle for meeting human development goals while simultaneously sustaining the ability of natural systems to provide the natural resources and ecosystem services on which the economy and society depend. The desired result is a state of society where living conditions and resources are used to continue to meet human needs without undermining the integrity and stability of the natural system. Sustainable development can be defined as development that meets the needs of the present without compromising the ability of future generations to meet their own needs. Sustainability goals, such as the current UN-level Sustainable Development Goals, address the global challenges, including poverty, inequality, climate change, environmental degradation, peace, and justice.

Chapter 3
Interplay Consequences of COVID-19 on Global Environmental Sustainability

Mir Sayed Shah Danish
https://orcid.org/0000-0002-5196-2280
University of the Ryukyus, Japan

Tomonobu Senjyu
https://orcid.org/0000-0003-4494-6773
University of the Ryukyus, Japan

Najib Rahman Sabory
University of the Ryukyus, Japan

Alexey Mikhaylov
https://orcid.org/0000-0003-2478-0307
Financial University Under the Government of the Russian Federation, Russia

ABSTRACT

This chapter outlines the essentials of COVID-19 and its relation to environmental mitigation. The COVID-19 pandemic has altered the ranking concern of climate change distress, and it is ranking as the first global priority to be adequately tackled. However, the pandemic demonstrates with economic, social, and cultural constraints. Still, climate change and environmental pollution have been ignored as the utmost precaution while their impact is more severe in the long run. This chapter evaluates available opportunities for environmental sustainability in the pandemic era. At the same time, the most significant aspect of solid waste, especially clinical waste, is critical for limiting pandemics and preventing future consequences of improper waste management. Sustainable production relies on criteria that ensure affordability, accessibility, use efficiency, safety, disparity, and other factors of production, supply, distribution, and consumption that are efficient, cost-effective, and environmentally friendly.

DOI: 10.4018/978-1-7998-9502-2.ch003

INTRODUCTION

In January 2020, World Health Organization (WHO) was announced Covid-19 (SARS-CoV-2 virus), a pandemic affecting the respiratory tract, which was first reported in Wuhan city in China (Zhu et al., 2020). Currently, COVID-19 poses a threat to the human race, similar to how environmental changes threaten the extinction of all living organisms. Because of continued environmental changes driven by human activities and natural ecological processes that bring about risk in COVID-19 infection, evidence-based research should be cultivated to look at the global relationship between the two. COVID-19 is caused by a novel coronavirus previously in animals whose occurrence and progression worldwide have been driven by human activities.

To fight the spreading coronavirus, governments worldwide imposed lockdown measures and curfews on their citizens. This influenced a short-term positive shift in sustaining clean environment (El Keshky et al., 2020). For example, states have adopted a lockdown policy to prevent the transmission of COVID-19, causing a reduction in economic activities and a fall in transport industrial emissions, improving air quality (Geneva Environment Network, 2022). Prior research have indicated that persons constantly exposed to increased air pollution are severely affected by infections of the respiratory tract in comparison to those breathing clean air. Such persons usually require greater medical attention and support when infected with coronavirus since their health is already compromised. Regions worldwide experiencing warmer weather are more likely to curtail COVID-19 transmission than regions experiencing cold weather.

The lack of proper waste management has created ideal conditions for the spread of zoonotic infections like coronavirus. During pandemic health emergencies, weak healthcare waste management systems will cost lives, putting healthcare workers and patients at significant risk of infection invasion. Clinical and other wastes generated from medical care are preferably disposed of by burying in a sanitary landfill or waste-burning to produce usable energy (Singh et al., 2020). Evaluating the impact of COVID-19 leads to a contradictory outlook mitigating atmospheric pollutions and increasing organic and nonorganic waste generation that consequences in social, technical, technological, political, and environmental concerns in communities. This phenomenon has caused a significant change in economic circulation balance and climate change mitigation with ignorable impact. Its physiological and cultural impacts are the turning point that could change habits and influence social relations with unknown timelines to overcome this pandemic. Infectious diseases are prone to happen through human interactions and from animals to humans. Managing COVID-19 waste from hospitals, healthcare facilities and individuals require maximizing the use of available waste solution circumventing each possible prolonged influence against the earth. Responding to COVID-19 emergency has had countries producing more waste volume than usual, such as personal protective equipment, masks, gloves, testing kits, empty pharmaceutical products and other wastes that could be infected with the virus. Countries will require more diverse systems and current waste solution technologies to deal with massive surge waste for waste collection and management. Global participation in segregation, timely collection, and effective coronavirus waste disposal are essential elements of appropriate handling of healthcare waste. Environmental decline due to human activity is one of the factors blamed for the emergence of zoonoses. Continued deforestation has certainly escalated global pandemics (Hub, 2021).

BACKGROUND

Over the years, infectious diseases have emerged that enter humans from infected wild animals or domesticated animals (Abdelzaher et al., 2020). Since people like to travel globally and live in urban centers where the population is higher, infections like coronavirus are likely to be transmitted in seconds through touching, sneezing, and coughing. Since all viruses naturally mutate over time, Coronavirus variants and mutations are being reported in some countries worldwide, including the latest one called 'Mu' variant reported in the UK with increasing transmissibility and mortality rate. Once the coronavirus vaccine was discovered, global guidelines, regulations, and manufacturers requirements were established with regard to how vaccines must be handled and transported. Controlled temperatures are one of the crucial requirements to guarantee and maintain the quality of vaccines being transported for distribution in countries, to their medical facilities and eventually to the people. Natural disasters can cause power outages that affect vaccine storage systems and hinder vaccine transportation when transport systems are destroyed. The persistent effects of the pandemic are evident not only in our socio-economic and healthcare domains but also environmental in the form of spiked solid and hazardous wastes from COVID-19 medical devices and personal protective equipment. Preventing and mitigating the spread of COVID-19 not only constitutes vaccination, isolation, and quarantining, wearing masks, social distancing and washing hands, it also involves how we take care of solid and hazardous waste from our households to our healthcare facilities. Training medical professionals while facing such challenges is important to have guidelines on how to manage vaccines exposed to extreme temperatures. And globally, distribution, accessibility, injection rate, etc. are remained challenging.

Historically, scientific breakthrough in understanding the transmission mode of disease-bearing pathogens from animals to humans has facilitated the establishment of preventive and control measures for infections. Public health measures implemented to curb the spread of COVID-19 include isolation of infected persons from healthy individuals, quarantine of individuals thought to have interacted with confirmed infected persons, essential handwashing to maintain hygiene, lockdown of areas with high infection rates, and border control to limit physical interaction between countries. The measures were strictly observed at a time when medical research was still in progress with no recommended pharmaceutical intervention. Controlling COVID-19 Pandemic still requires such containment methods together with its newly discovered vaccine. It is necessary to enhance the quick detection of zoonotic spillovers that threaten the global well-being of the public sector. Vaccine development breakthrough that could be effective against infectious diseases with pandemic potential has been developed by the University of North Carolina Gillings School of Global Public Health. The vaccine is effective against COVID-19 and its mutative variants to build immunity from future coronavirus family pandemics (Kirkendoll, 2021). While other viral outbreaks are bound to occur, the coronavirus remains a long-term threat responsible for SARS in 2003 and the COVID-19 pandemic in 2019. Identifying gaps in practices and knowledge related to vaccine development involves intersectoral collaboration, engagement, and mobilization of global funding bodies, dedicated donors, and biological researchers.

Addressing human-led factors is critical to contain the contraction and spread of zoonotic diseases. Lessons from COVID-19 must be part of the solutions to end future pandemics. The lessons include:

- **Confronting inequalities** – Socio-economic inequalities borne out of structural inequities are a global challenge from social services to infrastructure which is deployed unevenly. This affected health services' inclusivity during the pandemic.
- **Global Solidarity** – Global accountability and responsibility towards strengthening and enforcing the existing international conventions supporting Sustainable Development Goals should be adhered to.
- **Promote Universal Health** – Equitable distribution of resilient health systems among all communities is crucial in enduring shocks and continued care to affected communities (Kiogora, 2020). The high infection rate has overstretched the health sector reducing its capability to provide quality health care, especially to remote areas and isolated environments.
- **Involving Communities** – This is about social mobilization and collaborative effort to engage communities, the general public, and individuals in fostering cooperation towards common goals (Hertel & Keil, 2021).
- **Promote Human Rights Approaches** – Focusing on nature as part of the COVID-19 recovery plan to lower the probability of future pandemics by

creating extra sustainable and unbiassed societies for all. Recovery to current and future pandemics should be based on human rights principles.

- **Promoting Scientific Innovations** – Investing in scientific research to improve early detection of potentially contagious diseases, develop reliable control measures, and create a viral genetic library that can be referenced to sufficiently identify new pathogens to inhibit their spread. Their spread can be controlled through effective wildlife traded monitoring and supporting technologies. Education and mass awareness on modes of disease transmission, safe handling of animals, improve sanitation standards, promotion of sustainable wildlife conservation and management, and supportive programs to indigenous people reliant on wildlife for food (Dobson et al., 2020).

There has been an increase in the global frequency of zoonotic diseases linked mainly to the high rate of disruption of natural habitats through human activities. The pressure we put on nature is a catalyst of zoonotic spillover. Throughout history, the emergence and transmission of zoonotic infectious diseases capable of infecting the civilian population to a pandemic level have been recorded, with HIV/AIDS being one of the most dreaded diseases. Other recognized significant pandemics and epidemics include swine flu (H1N1) in 2009, bird flu (H5N1) in 2004, cholera, tuberculosis, malaria, plague, viral fevers like Ebola in 2014, Middle East Respiratory Syndrome Coronavirus (MERS-CoV), and Severe Acute Respiratory Syndrome Coronavirus (SARS-CoV) have already afflicted humanity (Piret & Boivin, 2021). Coronavirus disease has been ravaging the world since 2019 (COVID-19) with the possibility of sustained future pandemic risk.

COVID-19 crises led to heightened demand for requisite medical supplies as the few available from low scale suppliers were reserved well off countries for managing COVID-19 patients and controlling transmissions among populaces. The equipment includes opioids, ventilators, oxygen tanks, and anesthetics. A crucial strategic objective is to empower local industries' production capacity to produce basic medical equipment and drugs necessary to address future pandemics (Alami et al., 2021). Medical equipment like masks, gloves, screening kits, PPEs, and mechanical ventilators needs to be manufactured independently by individual countries to meet demand against the frequency of use by public health workers to prevent a shortage.

SUSTAINABLE DEVELOPMENT

From many definitions in the literature, the sustainability concept defines optimally resources utilization and reservation for the asking to preserve these resources in a way that does not compromise the ability of future generations to meet their needs (Danish et al., 2017a). Sustainable services, products, or any advantages to humankind demonstrated with essential criteria such as accessibility, affordability, disparity, safety, use efficiency, supply and production efficiency, cost-effectiveness, and environmental impacts on air, water, and soil quality to attain Sustainable Development Goals (Danish et al., 2021f). Sustainable Development Goals aim to lower vulnerabilities and enhance the capacity of needy groups, vulnerable societies, and communities to prevent and control losses to life, physical property in affected areas, and the surrounding environment. SDG also targets alleviating human suffering and accelerating development. It builds on existing infrastructural capacities that can effectively aid in reducing vulnerabilities. The ultimate goal is realizing resilient communities empowered to face imminent disasters. In a general term, the concept of energy and environment sustainability presents within the below listed pillars that require to be in harmony with each other without missing a single one (Danish et al., 2019b):

- Environmental sustainability
- Economic sustainability
- Technical sustainability
- Institutional sustainability
- Social sustainability

The Agenda for Sustainable Development Goals (SDGs) was enacted in 2015 by the United Nations General Assembly to encompass humanity and our planet challenges by 2030 by adopting sustainability 17 goals and 169 sub-targets (Sadiq et al., 2020). Active country's involvement in SDG should help a country gain their confidence back and guarantee that their needs and expectations can be met. The scope of SDG includes interventions, infrastructural measures, activities, progressive development projects, and reliable programs to lower disaster risks based on disaster response needs. Timely data with recommendations are urgently needed to monitor and share SDG progress around the world (Sachs et al., 2021). The pandemic has challenged the value of timely and disaggregated data to inform targeted actions by practitioners and institutions to save lives. In the future, the focus is on decision-makers who will formulate SDGs policies to help align with the vision 2030 development goals, management of uncertainty in governance, social protection of civilians, and continued conservation of a sustainable green economy to combat the COVID-19 pandemic. The pandemic presents a significant hurdle that impedes efforts to realize the 2030 economic Agenda alongside achieving desired SDGs (United Nations Development Programme (UNDP), 2020). The data will give communities autonomy to make informed development decisions that affect their future.

There are some common environmental and sustainability measures and indicators to be observed by policymakers, environmental activists, researchers and scholars as listed below, which are described in the Term Section at the end of chapter (Danish et al., 2021c):

- Environmental Sustainability Index (ESI)
- Dashboard of Sustainability (DS)
- Ecological Footprint (EF)
- Living Planet Index (LPI)
- Gross Domestic Product (GDP)
- Human Development Index (HDI)
- Salience, credibility, and legitimacy of GDP and HDI
- Well-being Index (WI)
- Direct Material Consumption (DMC)

NATURAL RESOURCES AND ENVIRONMENTAL SUSTAINABILITY

The survival of the human race and animals depends on how we care for the environment and natural resources utilization. Among environmental pollutants, energy production is listed at the top, which can make renewable energy projects cost-effective by optimum and multi-purpose use of natural resources (Danish et al., 2017b). For example, recirculating aquaponic systems (RASs) can produce fish and plants with minimum water loss using a recirculating water system. These systems are flexible and affordable for different environments for various purposes, ensuring food security and ecosystem stability. More importantly, by proper design and implementation of aquaponics systems, five Sustainable Development Goals (SDGs) can be achieved together as follows (detail for SDG 2,7,8,12, and 14 are given at the end of chapter) (Danish et al., 2021b, 2021h):

- SDG2: Zero hunger
- SDG7: Energy for everyone and clean
- SDG8: Both job satisfaction and economic growth
- SDG12: Responsibility to create responsibility to use
- SDG14: Protect the richness of the sea

Their survival is now at risk with the current climate change characterized by the production of new oil fossils that release harmful carbon emissions worsening global warming (World Health Organization (WHO), 2022). Raising climate change concerns is a goal for various international organizations led by environmental activists such as the human extinction rebellion group. This group broke out in October 2019, intending to urge the government and the public to address climate changes. There is nothing normal about natural disasters like wildfire also occurring simultaneously with the pandemic. While changing to clean and green energy seems huge capital investment and a strong economic foundation, most developing countries will not meet the target except the deployment of micro-scale clean energy projects (Danish et al., 2014). Economic losses in the world rising from a natural disaster are instigated by a lack of preparedness in investments to reduce the effects of the crisis. With the covid-19 pandemic, three aspects of Sustainable Development Goals: economic, social, and environmental have been impacted. Every government's highest priority is to suppress pandemics to create room for economic recovery post-COVID-19. The critical proposal of SDGs is to reduce global challenges, including pandemics, financial inequalities, climate change, and the biodiversity crisis by 2030.

Tackling climate change has corporations at the forefront of contributing towards a net-zero world where major oil fossils businesses like the Shell Global Company limit carbon emissions in the atmosphere in support of the Paris Agreement: reduce the rise in mean temperature of the world (*The Energy Transformation Scenarios*, 2020). This commitment is a pure example of corporate responsibility to the environment and its inhabitants. However, efforts like the Paris Summit 2015 on climate change "to strengthen the global response to the threat of climate change by keeping global temperature rise this century well below 2 degrees Celsius above pre-industrial levels and to pursue efforts to limit the temperature increase even further to 1.5 degrees Celsius." (Danish et al., 2021h) remains limited with no significant impact on nature prosperity. For example, day by day forest and agricultural landscape reduction has exerted pressure on biodiversity, and several displaced species are potentially facing a risk of extinction. Food production accounts for one-quarter of the global greenhouse gas emissions (Danish et al., 2021d).

Strengthening disasters risk governance for an effective and efficient disasters management should be a distributed governance of disaster risks at various levels such as national level for each country, regional level for countries geographically neighboring each other, and at the global level. Distributed governance is necessary to foster collaboration and partnership to reduce disaster risks and foster sustainable development. Investment in proper disaster risk management, enhancing resilience at cost-effective measures is vital for saving lives, averting and reducing resulting consequential losses, and ensuring proper recovery, restoration, and rehabilitation through private and public investments in disaster risk and prevention. Post-disaster assessment is carried out to assess effects and impacts to the then determined recovery needs. The governments mostly conduct them to demonstrate leadership, timely action, and integrity in publicly addressing the needs of citizens after a disaster.

The COVID-19 pandemic disrupted livelihoods attributed to the effects of measures taken to control the outbreak. The measures include the closure of public places, restrictions on the operation of businesses, limitation on movement, and reduced sitting capacity on travelers (Deshpande et al., 2020).

WATER RESOURCES SUSTAINABILITY

Hygienic practices decrease the transmission of COVID-19, such as hand hygiene significantly reduces the spread of coronavirus and is crucial to people's safety and health (*Water, Sanitation, Hygiene, and Waste Management for the COVID-19 Virus*, 2020). Today, many world dwellers have no access to safe and clean water, proper sanitation coupled with inadequate funding. Communities living near shared water bodies are experiencing water shortages because of increased population resulting from social migration during pandemic restrictions. Developing nations are concerned about access to clean water resources that can increase pandemic intensity. Lifted COVID-19 restrictions to allow sea opening and its reinvestment will promote economic sustainability and lower risk of infections.

Water and sanitation are key for pandemic outbreak prevention. Limited access to sufficient water, improper sanitation standards, and inadequate hygiene lead to the outbreak of zoonotic infections like coronavirus, increasing the incidence of future pandemics. Poverty-stricken communities suffer disproportionate effects of the pandemic due to communal sharing of basic social amenities like toilets and watering points: collection, processing, storage, and adequate water supply. In addition, wastewater treatment, reuse, and conservation are needed to ensure reliable water availability. Creating awareness on water bodies to help protect the blue life and human health is essential (Cooper, 2020, p. 19).

Unsuitable clinical waste discarding in underdeveloped nations characterizes environmental issues that are evidently observed (Ejaz et al., 2010). In developing countries, COVID-19 prevalence of COVID-19 is exposing fragility in waste management strategies, especially with unregulated urban growth without assemblage, carriage, treatment and disposal guidelines that pollute the atmosphere and water resources. The limited support by governments has made it difficult to achieve proper solid and hazardous waste disposal because of failure to implement strict waste collection and disposal guidelines. Low investment in modern technology systems and infrastructure to treat solid and hazardous waste has also contributed to this predicament in low-income countries. Studies suggest addressing future water problems by wastewater treatment and reuse besides reserving water resources. It can be combined for electricity production and heat generation utilizing wastewater and controlling water pollution using digester gas ignition with high heat energy potential

(Danish et al., 2021c). Some studies conducted research on wastewater purification by removal of organic contaminants from wastewater using different treatment technology: photocatalyst to degrade water pollutants, metal oxides application, etc. for environmental remediation as a way to restore water that has been used and or contaminated back to a desirable quality (Danish et al., 2020, 2021a, 2021e).

WASTE MANAGEMENT

According to (Danish et al., 2019c) among solid waste potentials, municipal waste ranked the second most available feedstock potential that can be used by 2030 with 39% greenhouse gas emission. "A more stringent scenario arbitrary set to a 40.0% reduction by 2030 compared to 1990 levels". Improper disposal and mismanagement of solid and hazardous waste lead to public health and environmental issues like air pollution from the foul smell produced when rotting garbage is dumped while being left for decomposition (*Source and Types of Waste*, 2022). Burning of hazardous waste releases poisonous gas into the air. Pathogens and bacteria are likely to cause infection to find a conducive breeding ground for different types of infections and water pollution when chemical and toxic wastes from industries are released into the environment, polluting soil, blocking drainage systems, and polluting underground water resources when they drain. This issue is mostly practiced in developing countries with a high population intensity (Pollution Index - PI) and weak waste management system. The top ten pollutant countries with low air quality are reported as follows (Danish et al., 2019d):

- Accra, Ghana, PI 97.88
- Tetovo, Macedonia, PI 96.85
- Faridabad, India, PI 96.24
- Ghaziabad, India, PI 96.10
- Kathmandu, Nepal, PI 95.54
- Ulaanbaatar, Mongolia, PI 94.93
- Kabul, Afghanistan, PI 94.84
- Cairo, Egypt, PI 93.79
- Dhaka, Bangladesh, PI 92.96
- Ho Chi Minh City, Vietnam, PI 92.79

Also, indirect disease transmission due to waste exposures to infectious pathogens poses serious hazards among health workers, patients, waste workers, and the community. This impact is escalated by improper treatment and disposal of healthcare waste. Without adequate pollution control, some waste disposal methods like incineration or open burning cause the emission of toxic contaminants in air risking the health of waste workers and the surrounding community. This treat has been worsened with pandemic. Vaccination has allowed for the employment of medical personnel and waste workers. Even though these groups come into contact with solid and hazardous waste, the medical repercussions will be scaled back with the support of high functioning health systems relying on accessible and pertinent use of prudent, effectual and excellent waste management products.

Concerning the pandemic era, opportune dependable data on the shift in the public health position, health care capital, utility, and human medical resources are expository policymakers' ability to appraise health structure performance and model suitable initiatives related to protecting healthcare workers during bio-medical waste management. In hospitals, policy guidelines for waste management should be determined by types and quantities of waste generated, medical waste management personnel, and current medical waste disposal methods.

Among vulnerable places against pandemic, prisons-COVID-19 in prisons are exceptionally dangerous due to their highly transmittable nature. The spread and management of such an infectious disease create general distress in confined spaces where a high concentration of people favor rapid spread of diseases. Adequate hygiene is one of the directions issued to law enforcement institutions to control the current global pandemic.

FUTURE OUTLOOK

COVID-19 pandemic was a disaster, and now it's an emergency. It is still a disaster that 72 billion people in the world are managing. One measure to tackle COVID-19 and to lower its economic impact is prioritizing health projects by the governments of the affected nations to flatten the COVID-19 infection trend curve, ensure adequate food supplements, sufficient water to maintain basic hygiene, and empowerment of the vulnerable informal sector, especially those who rely on daily wages. This has thrown the implementation of SDG into jeopardy in developing countries. Leaders of several international organizations have presented insight on key elements affecting sustainability following the COVID-19 outbreak. The insight further describes the possible impact to the 2030s Agenda for Sustainable Development alongside the concerns for the environmental goals as documented in the agreement on climate change in Paris. The international organizations include United Nations Environment Programme (UNEP), World Bank, Organization for Economic Co-operation and Development (OECD), and Global Green Growth Institute (GGGI) (GGKP News, 2020).

Understanding disaster risks policies are required to be grounded on appreciating disaster in all aspects of diversity, understanding the vulnerability damage to assets, geographical location of risks, affected population, and economic losses. This information is important for prevention, mitigation, and implementing policies, appropriate preparedness, and effective responses to disasters. We can start by addressing the COVID-19 pandemic challenges individual countries face by restoring the balance between people and the planet. The significance of development accommodates equitable economic growth and a focus on the conservative utilization of the existing resources, which has perhaps never been more in focus than at this time of the COVID-19 Pandemic (Asian Development Bank, 2021).

Designing goals that promote health security among countries can potentially reduce morbidity rate and mortality rates. In turn, the goals aid in improving the countrywide well-being of all the citizens (Blinken & Becerra, 2021). Adopting policies that strengthen health systems as practical and accessible through financial investment can reinforce lasting resilience of the health system.

The major role of Scientific Technology Innovation in managing COVID-19 and the resultant economic crisis could provide fresh impetus for preparing for future pandemics and identify strategies to enhance preparedness to future shocks and challenges, including climate change. This could lead to significant increases in public investments in STI for research (OECD Policy Responses to Coronavirus (COVID-19), 2021). Telemedicine provides the capability to apply for advances within networks spanning beyond geographical barriers and regional confines (Bashshur et al., 2020). Continued growth in smartphone usage and application will foresee the creation of applications for monitoring the spread and potentially aid in curbing the virus without threats to privacy and abuse.

CONCLUSION

COVID-19 pandemic has had unforeseen impacts for which many existing disaster preparedness plans need new formulation tailored for climate disasters and biological crises. In the meantime, we are learning the right lessons from the pandemic that will build our resilience for the next global crisis. Countries have an opportunity to strengthen Sustainable Development Goals Post-COVID-19 pandemic (Pricewaterhouse Coopers, n.d.). By enhancing international cooperation between rich countries, and low-income countries through international, regional, and bilateral initiatives, the ability of low-income countries to reduce the overall impact of the pandemic, including harm to civil infrastructure and interruption of essential public services such as health care facilities and educational institutions, will be strengthened. The ongoing ravaging global COVID-19 virus attack presents valuable lessons to be observed for preparedness for future health disasters. It has offered an opportunity to reevaluate existing plans, development strategies, and the proper organization of cities and other populated human settlements (Feleke et al., 2021). We have seen governments allocating funds in the healthcare and educational sectors to buy personal protective equipment and ensure that all sectors have or can access the virus prevention guidelines to develop resilience.

Combating global poverty, one of the goals of sustainable development, has been incredibly threatened by the COVID-19 pandemic destroying years of its progress. Economic growth is important to reduce pandemic-related financial collapse because these zoonotic diseases threaten global health and destroy economies. International efforts limiting exposure to wildlife meats and markets are vital to physically and economically safeguard people. Thus, people remain employed and in good health. Focusing policy choices that aim at re-investing in civil social services will neutralize the devastating impact of the virus among the vulnerable populaces (Prakash Narain et al., 2021).

The provision of foreign aid to low-income countries has developed a culture of dependency where agreements have been reached between well-off developed countries and low-income countries. The arrangements are either bilateral or multilateral but are faced with challenges evident in the current response to the pandemic. The aid system is mainly considered to have been overstretched, with low-income countries being largely left out in the campaign to curb COVID-19. This results from leaders lacking opportunity policies that spur democracy, creating jobs as a priority, and ensuring economic engagements at the development level. Most of the developmental approach and the ensuing humanitarian response to the pandemic deliberately focused on incremental enhancements to reevaluate and repurpose the current aid approaches.

REFERENCES

Abdelzaher, H., Saleh, B. M., Ismail, H. A., Hafiz, M., Gabal, M. A., Mahmoud, M., Hashish, S., Gawad, R. M. A., Gharieb, R. Y., & Abdelnaser, A. (2020). COVID-19 genetic and environmental risk factors: A look at the evidence. *Frontiers in Pharmacology, 11*, 579415. doi:10.3389/fphar.2020.579415 PMID:33117174

Alami, H., Lehoux, P., Fleet, R., Fortin, J.-P., Liu, J., Attieh, R., Cadeddu, S. B. M., Abdoulaye Samri, M., Savoldelli, M., & Ag Ahmed, M. A. (2021). How can health systems better prepare for the next pandemic? Lessons learned from the management of COVID-19 in Quebec (Canada). *Frontiers in Public Health, 9*, 1–15. doi:10.3389/fpubh.2021.671833 PMID:34222176

Asian Development Bank. (2021). *Accelerating sustainable development after COVID-19: The role of SDG bonds*. Asian Development Bank. https://www.adb.org/publications/sustainable-development-after-covid-19-sdg-bonds

Bashshur, R., Doarn, C. R., Frenk, J. M., Kvedar, J. C., & Woolliscroft, J. O. (2020). Telemedicine and the COVID-19 pandemic, lessons for the future. *Telemedicine Journal and e-Health*, *26*(5), 571–573. doi:10.1089/tmj.2020.29040.rb PMID:32275485

Blinken, A. J., & Becerra, X. (2021). Strengthening global health security and reforming the international health regulations: Making the world safer from future pandemics. *Journal of the American Medical Association*, *326*(13), 1255–1256. doi:10.1001/jama.2021.15611 PMID:34464446

Cooper, R. (2020). *Water security beyond Covid-19*. Institute of Development Studies.

Danish, M. S. S., Bhattacharya, A., Stepanova, D., Mikhaylov, A., Grilli, M. L., Khosravy, M., & Senjyu, T. (2020). A systematic review of metal oxide applications for energy and environmental sustainability. *Metals*, *10*(12), 1604. doi:10.3390/met10121604

Danish, M. S. S., Estrella, L. L., Alemaida, I. M. A., Lisin, A., Moiseev, N., Ahmadi, M., Nazari, M., Wali, M., Zaheb, H., & Senjyu, T. (2021a). Photocatalytic applications of metal oxides for sustainable environmental remediation. *Metals*, *11*(1), 80. doi:10.3390/met11010080

Danish, M. S. S., Ibrahimi, A. M., Yaqobi, M. A., Udagawa, S., Mikhaylov, A., Faisal, N., & Senjyu, T. (2021b). Afghanistan aquaculture and fishery sectors—A foresight outlooks. *Journal of Ecoscience and Plant Revolution*, *2*(1), 17–37. doi:10.37357/1068/jepr.2.1.03

Danish, M. S. S., Matayoshi, H., Howlader, H. O. R., Chakraborty, S., Mandal, P., & Senjyu, T. (2019a). Microgrid Planning and Design: Resilience to Sustainability. *2019 IEEE PES GTD Grand International Conference and Exposition Asia (GTD Asia)*, 253–258. 10.1109/GTDAsia.2019.8716010

Danish, M. S. S., Sabory, N. R., Ershad, A. M., Danish, S. M. S., Ohta, R., Sediqi, M. M., Ahmadi, M., & Senjyu, T. (2017a). The least developed countries need for changing the passive trend of renewable energy exploitation to a proactive trend. *International Journal of Energy and Power Engineering*, *5*(6), 215–221. doi:10.11648/j.ijepe.20160506.17

Danish, M. S. S., Sabory, N. R., Ershad, A. M., Danish, S. M. S., Yona, A., & Senjyu, T. (2016). Sustainable Architecture and Urban Planning trough Exploitation of Renewable Energy. *International Journal of Sustainable and Green Energy*, *6*(3), 1. doi:10.11648/j.ijrse.s.2017060301.11

Danish, M. S. S., Senjyu, T., Ahmadi, M., Ludin, G. A., Ahadi, M. H., Karimy, H., & Khosravy, M. (2021c). A review on energy efficiency for pathetic environmental trends mitigation. *Journal of Sustainability Outreach*, *2*(1), 1–8. doi:10.37357/1068/jso.2.1.01

Danish, M. S. S., Senjyu, T., Faisal, N., Stannikzai, M. Z., Nazari, A. M., & Vargas-Hernández, J. G. (2021d). A review on environmental-friendly energy multidisciplinary exposition from goals to action. *Journal of Environmental Science Revolution*, *2*(1), 1–9. doi:10.37357/1068/jesr.2.1.01

Danish, M. S. S., Senjyu, T., Ibrahimi, A. M., Ahmadi, M., & Howlader, A. M. (2019b). A managed framework for energy-efficient building. *Journal of Building Engineering*, *21*, 120–128. doi:10.1016/j.jobe.2018.10.013

Danish, M. S. S., Senjyu, T., Ibrahimi, A. M., Bhattacharya, A., Nazari, Z., Danish, S. M. S., & Ahmadi, M. (2021e). Sustaining energy systems using metal oxide composites as photocatalysts. *Journal of Sustainable Energy Revolution*, *2*(1), 6–15. doi:10.37357/1068/jser.2.1.02

Danish, M. S. S., Senjyu, T., Nazari, M., Zaheb, H., Nassor, T. S., Danish, S. M. S., & Karimy, H. (2021f). Smart and sustainable building appraisal. *Journal of Sustainable Energy Revolution*, *2*(1), 1–5. doi:10.37357/1068/jser.2.1.01

Danish, M. S. S., Senjyu, T., Sabory, N. R., Danish, S. M. S., Ludin, G. A., Noorzad, A. S., & Yona, A. (2017b). Afghanistan's aspirations for energy independence: Water resources and hydropower energy. *Renewable Energy*, *113*, 1276–1287. doi:10.1016/j.renene.2017.06.090

Danish, M. S. S., Senjyu, T., Sabory, N. R., Khosravy, M., Grilli, M. L., Mikhaylov, A., & Majidi, H. (2021g). A forefront framework for sustainable aquaponics modeling and design. *Sustainability*, *13*(16), 9313. doi:10.3390u13169313

Danish, M. S. S., Senjyu, T., Urasaki, N., Rahmany, N. A., Ershad, A. M., Sabory, N. R., Zarabi, K., Anwarzai, M. A., Karimy, H., & Zaheb, H. (2021h). Developing nations as a foremost allotment for achieving 2030 SDGs—A case study. *Journal of Engineering and Technology Revolution*, *2*(1), 1–10. doi:10.37357/1068/jetr.2.1.01

Danish, M. S. S., Senjyu, T., Zaheb, H., Sabory, N. R., Ibrahimi, A. M., & Matayoshi, H. (2019c). A novel transdisciplinary paradigm for municipal solid waste to energy. *Journal of Cleaner Production, 233*, 880–892. doi:10.1016/j.jclepro.2019.05.402

Danish, M. S. S., Yona, A., & Senjyu, T. (2014). Pre-design and life cycle cost analysis of a hybrid power system for rural and remote communities in Afghanistan. *The Journal of Engineering-IET, 2014*(8), 438–444. doi:10.1049/joe.2014.0172

Danish, M. S. S., Zaheb, H., Sabory, N. R., Karimy, H., Faiq, A. B., Fedayi, H., & Senjyu, T. (2019d). The Road Ahead for Municipal Solid Waste Management in the 21st Century: A Novel-standardized Simulated Paradigm. *IOP Conference Series. Earth and Environmental Science, 291*(1), 1–5. doi:10.1088/1755-1315/291/1/012009

Deshpande, R., Kibe, J., & Kaaria, L. (2020, April 15). *COVID-19 exposes risks and opportunities in Kenya's gig economy*. CGAP. https://www.cgap.org/blog/covid-19-exposes-risks-and-opportunities-kenyas-gig-economy

Dobson, A. P., Pimm, S. L., Hannah, L., Kaufman, L., Ahumada, J. A., Ando, A. W., Bernstein, A., Busch, J., Daszak, P., Engelmann, J., Kinnaird, M. F., Li, B. V., Loch-Temzelides, T., Lovejoy, T., Nowak, K., Roehrdanz, P. R., & Vale, M. M. (2020). Ecology and economics for pandemic prevention. *Science, 369*(6502), 379–381. Advance online publication. doi:10.1126cience.abc3189 PMID:32703868

Ejaz, N., Akhtar, H., Nisar, U., & Ali, N. (2010). *Environmental impacts of improper solid waste management in developing countries: A case study of Rawalpindi city*. doi:10.2495/SW100351

El Keshky, M. E. S., Basyouni, S. S., & Al Sabban, A. M. (2020). Getting through covid-19: The pandemic's impact on the psychology of sustainability, quality of life, and the global economy – A systematic review. *Frontiers in Psychology, 11*, 1–12. doi:10.3389/fpsyg.2020.585897 PMID:33281683

Feleke, A., Intrator, K., Palmer, C., & Giles, M. (2021, June 10). *Five lessons for participatory planning during a pandemic*. ReliefWeb. https://reliefweb.int/report/world/five-lessons-participatory-planning-during-pandemic

Geneva Environment Network. (2022, January 20). *COVID-19 and the environment*. https://www.genevaenvironmentnetwork.org/resources/updates/updates-on-covid-19-and-the-environment/

GGKP News. (2020, April 29). *What does COVID-19 mean for sustainable development?* Green Growth Knowledge Platform. https://www.greengrowthknowledge.org/news/what-does-covid-19-mean-sustainable-development

Hertel, S., & Keil, R. (2021, January 1). *After isolation: Urban planning and the Covid-19 pandemic*. Ontario Professional Planners Institute. https://ontarioplanners.ca/blog/planning-exchange/january-2021/after-isolation-urban-planning-and-the-covid-19-pandemic

Hub, I. S. K. (2021, July 6). *Guest article: Preventing future pandemics starts with protecting our forests*. SDG Knowledge Hub. https://sdg.iisd.org:443/commentary/guest-articles/preventing-future-pandemics-starts-with-protecting-our-forests/

Kiogora, M. (2020, December 9). *Why universal health care is crucial in managing future pandemics*. United Nations Kenya. https://kenya.un.org/en/109033-why-universal-health-care-crucial-managing-future-pandemics

Kirkendoll, S. M. (2021, July 17). *New universal coronavirus vaccine could prevent future pandemics*. The University of North Carolina at Chapel Hill. https://www.unc.edu/posts/2021/07/07/new-universal-coronavirus-vaccine-could-prevent-future-pandemics/

OECD Policy Responses to Coronavirus (COVID-19). (2021). How will COVID-19 reshape science, technology and innovation? *OECD Science, Technology and Industry Policy Papers, 107*, 1–10. doi:10.1787/23074957

Piret, J., & Boivin, G. (2021). Pandemics throughout history. *Frontiers in Microbiology, 11*, 1–16. doi:10.3389/fmicb.2020.631736 PMID:33584597

Prakash Narain, J., Sodani, P. R., & Kant, L. (2021). COVID-19 pandemic: Lessons for the health systems. *Journal of Health Management, 23*(1), 74–84. doi:10.1177/0972063421994982

Pricewaterhouse Coopers. (n.d.). *Post-Covid business recovery: Realising new opportunities*. PwC. Retrieved January 21, 2022, from https://www.pwc.com/gx/en/issues/crisis-solutions/covid-19.html

RobinsonJ. (2020, April 2). *Pandemics*. WebMD. https://www.webmd.com/cold-and-flu/what-are-epidemics-pandemics-outbreaks

Sachs, J., Schmidt-Traub, G., Kroll, C., Lafortune, G., & Gray. (2021). *The sustainable development goals and Covid-19: Includes the SDG index and dashboards*. Cambridge University Press.

Sadiq, M. A. F., Sabory, N. R., Danish, M. S. S., & Senjyu, T. (2020). Role of micro-hydropower plants in socio-economic development of rural Afghanistan. *Journal of Sustainable Energy Revolution, 1*(1), 1–7. doi:10.37357/1068/jser.1.1.01

Singh, N., Tang, Y., Zhang, Z., & Zheng, C. (2020). COVID-19 waste management: Effective and successful measures in Wuhan, China. *Resources, Conservation and Recycling*, *163*, 105071. doi:10.1016/j.resconrec.2020.105071 PMID:32834489

Source and types of waste. (2022, January 20). *BYJUS*. https://byjus.com/chemistry/waste/

The energy transformation scenarios. (2020). *Shell International B.V.* https://www.shell.com/promos/energy-and-innovation/download-full-report/_jcr_content.stream/1627553067906/fba2959d9759c5ae806a03acfb187f1c33409a91/energy-transformation-scenarios.pdf

United Nations Development Programme (UNDP). (2020). *COVID-19 and the SDGs*. https://feature.undp.org/covid-19-and-the-sdgs/

Water, sanitation, hygiene, and waste management for the COVID-19 virus. (2020). *World Health Organization and the United Nations Children's Fund (UNICEF)*. https://apps.who.int/iris/bitstream/handle/10665/331846/WHO-2019-nCoV-%20IPC_WASH-2020.3-eng.pdf?sequence=1&isAllowed=y

World Health Organization (WHO). (2022, January 20). *Climate change and health*. https://www.who.int/news-room/fact-sheets/detail/climate-change-and-health

Zhu, H., Wei, L., & Niu, P. (2020). The novel coronavirus outbreak in Wuhan, China. *Global Health Research and Policy*, *5*(1), 6. doi:10.118641256-020-00135-6 PMID:32226823

KEY TERMS AND DEFINITIONS

Dashboard of Sustainability (DS): Is software developed by the European Union's Joint Research Centre at Ispra and represents the complicated relationship between environmental, economic, and social issues. Moreover, it provides information in a way palatable to decision-makers and the general public.

Direct Material Consumption (DMC): Measures the total sum of the domestic extraction flows, including imported but excludes exported.

Disaster: Is an unexpected event whose occurrence results in extensive damage, including properties, the environment, animals, and humans inhabiting the affected region. Since disasters occur unexpectedly, there is a need to understand and determine strategies to minimize the effects caused by such disasters.

Ecological Footprint (EF): Is the area of productive land needed to maintain its current consumption ratio while using the prevailing technology to absorb wastes and calculated for a specific population.

Environmental Sustainability Index (ESI): Measures overall progress towards environmental sustainability for 142 countries measured by the World Economic Forum. It provides a more analytical approach to environmental decision-making and allows comparison of the progress among the nations.

Gross Domestic Product (GDP): Gives the total value of goods and services produced in a specific time frame, usually a year. It is a popular indicator in the economic sphere but does not apply in the social sphere.

Hazard: Is the potential for a sudden event to occur and whose occurrence can lead to unforeseen consequences and emergencies.

Human Development Index (HDI): Measures the average achievement in the three primary aspects of human development; knowledge, longevity, and improved living standards. Life expectancy defines longevity, enrollment in schools, and adult literacy provides awareness, and the GDP per capita defines the living standard. The UN development program published it.

Living Planet Index (LPI): Assesses the overall global state of the ecosystem using national and international data on the impact of human activities on the environment.

Mitigation: Is measures taken before a disaster happens to neutralize or decrease its impact on society or the environment.

Pandemic: Is a disease outbreak that spreads across countries or continents that affects more people and takes more lives than an epidemic (Robinson, 2020).

Risk: Is the probability, depending on how high or low a hazard will cause harm. Risk is determined by vulnerable conditions associated with physical surroundings, social setup, environmental factors, or economic factors.

Salience, Credibility, and Legitimacy of GDP and HDI: Salience means that the indicators are useful, applicable, and attractive to the user. Credibility means that the pointers are valid and make scientific sense. Finally, legitimacy touches pointers' perception from the perspectives of users, stakeholders, businesses, trade unions, and environmental non-governmental organizations.

Sustainability in Terms of Energy and Environment: Defines developing affordable and sustainable protocols for various applications of clean and green energy aligned with environmental requirements with minimum greenhouse gas emission over time with contemplating some indicators such as deployment diversity, policy developments, technology costs, and investment in renewable energy (Danish et al., 2016).

Sustainability Pillars: Have counted the parameters to be analyzed to balance the proposed system or solution in accordance with resiliency and sustainability pillars within these criteria: technical sustainability, economic sustainability, institutional sustainability, environmental sustainability, social sustainability, etc. (Danish et al., 2019a).

Well-Being Index (WI): Is called stress index, combines two types of indicators then combines them. The first type has thirty-six pointers for health, population, wealth, freedom, peace, crime, equity, communication, and education. The other type has fifty-one land, water, air, and energy.

Chapter 4
Economic Feasibility of Solar Power Plants in Turkey Based on the PV Module Using Simulation RETScreen Software

Emre Akusta
https://orcid.org/0000-0002-6147-5443
Kirklareli University, Turkey

Raif Cergibozan
https://orcid.org/0000-0001-7557-5309
Kirklareli University, Turkey

ABSTRACT

This chapter aims to analyze the renewable energy production and economic evaluation of five MW solar power plant installations in Turkey using the RETScreen program. For this purpose, all provinces of Turkey are analyzed first in terms of their global radiation values. Provinces with a global radiation value less than 1620 kWh / m2 per year have been eliminated, and the analysis proceeded with 45 provinces. The study includes three different scenarios. The first is based only on the government's guarantee of purchase. The second adds the greenhouse gas reduction incentive to the first scenario. The third scenario adds domestic production incentives to the first and second scenarios. According to the first scenario analysis, the payback period is 5.553 years for Muğla and 6.530 years for Antalya. According to the second scenario, the payback period is 5.230 years for Muğla and 6.184 years for Antalya. Finally, in terms of the third scenario, the payback period is 3.479 years for Muğla and 4.101 years for Antalya.

DOI: 10.4018/978-1-7998-9502-2.ch004

INTRODUCTION

Energy is one of the most important factors determining economic and political developments. In particular, a rise in the need for energy with production has increased the value of energy resources even more. Although energy consumption is so important for countries, it also has some negative aspects because energy production is mainly based on fossil fuels, which cause environmental problems such as carbon emission and global warming. Energy use increased significantly after the Industrial Revolution, and the environmental problems caused by fossil fuels were ignored during this period. In subsequent periods, environmental problems and climate change have caused obligatory measures to be taken for fossil fuels. For this purpose, the Kyoto Protocol, established in December 1997 and enacted in 2005, requires countries to reduce their greenhouse gas emissions. For this reason, countries have begun to move towards environmentally friendly renewable energy sources that reduce the environmental problems from fossil fuels. In addition, despite the uneven distribution of fossil fuels in the world, the more even and fair distribution of renewable energy sources is another reason to turn again to renewable energy sources. For these reasons, renewable energy sources have gained importance, and their usage has become widespread over the last twenty years.

With the increase of environmental awareness around the world, countries' attention has been directed towards environmentally friendly alternative energy sources that reduce greenhouse gas emissions instead of the traditional energy methods obtained from fossil fuels, and this interest increases day by day. Air pollution is one of Turkey's most important environmental problems as most countries in the world. The total greenhouse gas emission in 2018 is calculated as 520.9 million tons (Mt) CO2 equivalent. In 2018, the largest share in CO2 equivalent emissions was energy-related emissions with 71.6%, followed by industrial processes and product use with 12.5%, agricultural activities with 12.5% and waste with 3.4% (TUIK, 2020). In order to reduce the carbon emissions, it is necessary to increase the use of renewable energy sources. Greenhouse gas emission reduction support is provided by various countries around the world. For example, 29.6 $ / ton CO2 support is provided for greenhouse gas emission reduction in the UK. In Turkey, there is no support yet (Büyükzeren et al., 2015).

Despite the lack of support for Turkey in greenhouse gas emission reduction, there is a guarantee of purchase. Electricity produced from renewable energy sources specified in the Law No. 5346 on the Use of Renewable Energy Resources for the Purpose of Generating Electrical Energy, which entered into force on 18/05/2005, is secured by the government with a 10-year purchase guarantee. In addition, according to the law numbered 5346, if the local materials specified in the law are used in renewable energy plants, additional incentives per kWh are given in addition to the purchase guarantee. According to the law, electricity generated in solar power plants will be purchased at a price of $ 0.133 / kWh for 10 years. If all the materials specified in the law in solar power plants are local, this price is $ 0.2 / kWh (Use of Renewable Energy Sources for Electric Energy Production [Yenilenebilir Enerji Kaynaklarının Elektrik Enerjisi Üretimi Amaçlı Kullanımı], 2005).

Generating electricity without harming nature is possible with the use of renewable energy sources. Turkey, despite its lack of fossil energy sources, has renewable energy sources and is a country particularly rich in solar energy (Dinçer, 2011; Evrendilek & Ertekin, 2003). Turkey has over 2,500 hours of annual sunshine. Although this varies according to province, it has an average of 7.5 hours of daily sunshine. Turkey has almost 53% more sunshine than Germany, which is the world leader in terms of installed solar power.[1] Accordingly, Turkey's solar energy resources cannot be adequately assessed. In this respect, our study aims to contribute to the evaluation of Turkey's solar energy potential.

Fossil energy sources are not distributed equally and fairly on the Earth. Countries with these resources export them and make great gains to their economies. However, this mechanism works opposite for countries without fossil fuels. These countries are energy-importing and energy-dependent countries. Countries with high energy imports cause their foreign trade to deteriorate.

Turkey, having a production structure based on fossil fuels, is an energy-importing country. Turkey's increased dependence on foreign energy, ongoing for many years, has made its current deficit problem even worse (Ari & Cergibozan, 2018). Thus, sudden changes in energy prices deepen the sensitivity of the country to external shocks and increase its fragility. By reducing the share of fossil fuels imported from abroad for its total energy use, the country's current deficit problem could become largely solvable and the sensitivity to the external shocks could be reduced. This situation can be seen more clearly in Figure 1. The graph shows Turkey's current account balance and current account balance excluding energy for the period of 1996-2019.

Figure 1. Turkey's current account balance
(Source: Central Bank of the Republic of Turkey [CBRT], 2019).

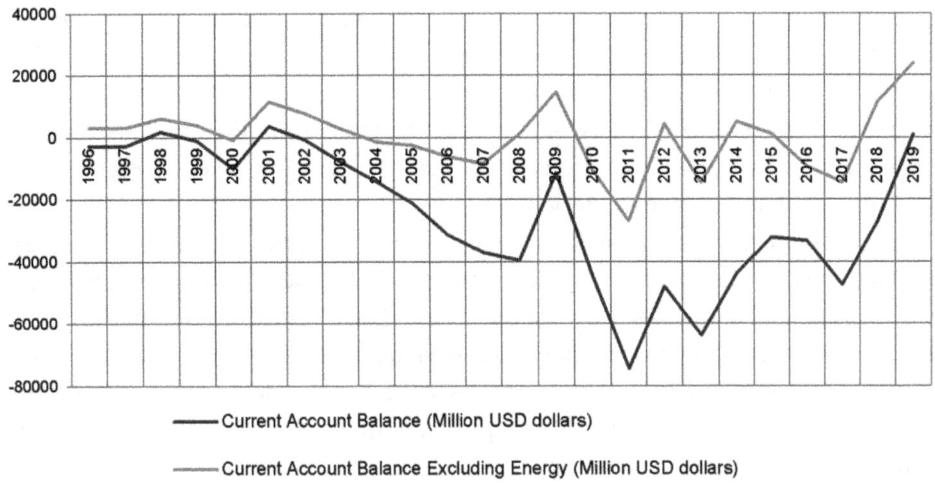

As seen in Figure 1, energy imports deepen the current account deficit. In the period of 1996-2019 which covers 24 years, Turkey's current account is in deficit for 21 years. When we subtract the energy imports, the current account deficit lessens, showing a surplus in 14 of the 24 years. This situation illustrates the vital importance of energy imports for Turkey's economy. In fact, government officials, considering the significant potential for renewable energy in Turkey's National Renewable Energy Action Plan, have ambitious targets for 2023[2] and plan to increase the percentage of renewable energy sources by 23% in generating electricity. To achieve this goal, the aim is to increase the installed power capacity to 34,000 MW for hydraulics, 20,000 MW for wind, 1,000 MW for geothermal, 5,000 MW for solar, and 1,000 MW for biomass (Ministry of Energy and Natural Resources, 2014).

In our study, all provinces of Turkey have been analyzed in terms of their global radiation value. 5 MW installed capacity solar power plants have been placed in and economic analysis is carried out with the RETScreen program for the provinces whose global radiation lower limit (1620 kWh / m2 per year) value remains above the value specified in the licensed solar power plant regulation. On this basis, 45 provinces that comply with the licensed solar power plant regulation are included in our study. This study is based on three scenarios. The first scenario is based only on the government's guarantee of purchase. In the second scenario, greenhouse gas reduction incentives were added in addition to the first scenario. Our aim here is; to investigate the effect of possible greenhouse gas emission reduction incentives on investment costs and to raise awareness about greenhouse gas emission reduction support. In the third scenario, domestic production incentives have been added in addition to the second scenario. The study first identified the provinces where solar power plants can be installed in Turkey. In addition, feasibility studies on solar power plants have been conducted for 45 provinces, and the payback period and economic efficiency of each power plant were determined accordingly. This study also includes an assessment of Turkey's solar energy potential and thus aims to reduce the use of fossil fuels. As a result, the increase in the use of renewable energy sources is seen as a solution for reducing the dependency on foreign energy as well as the current account deficit.

Energy is an important input for a country's economy. Because renewable energy production is the method for providing this input effectively to countries, national and international studies on this issue increase daily. Turkey has a high potential for solar energy, and thus solar energy has been put forward in comparison to other renewable energy sources. Several studies have been undertaken in Turkey for analyzing the feasibility of solar power plants. In this regard, feasibility studies for Turkey have been listed below.

Dağtekin et al. (2014) conducted a techno-economic feasibility study of a 100 kWe solar power plant connected to the grid for installation in Adana. This study, which has a theoretical and simulated design, determined the life of the plant at 20 years. Since the payback period for solar power plants is calculated as 7.8 years, this power plant planned to be established in Adana is expected to make profit in 12.2 years.

Özcan and Ersöz (2019) investigated the effectiveness of Turkey's cities using a simulation with certain cities in the PVSyst program. The study designed plant capacity at 8865 kWp and covers 3 provinces: Izmir, Ankara, and Istanbul. According to the outputs from the PVSyst simulation program, the payback duration for the plants established in Izmir, Ankara, and Istanbul are determined as 7.03, 6.7, and 13.6 years, respectively.

Gürtürk (2019) included a feasibility study conducted by various companies for a 1 MW solar power plant planned for establishment in Elazığ. By considering the data from various companies, the economic life for a 1 MW solar power plant was determined as 25 years and the payback period at an average of 6.6 years.

Duman and Güler (2020) analyzed the feasibility study on 5 kW capacity rooftop PV systems connected to the grid using the Homer renewable energy simulation program. According to the study, the payback periods of the simulated systems are calculated as 14.43 years for Artvin, 13.13 years for Istanbul, 11.65 years for Çanakkale, 10.85 years for Eskişehir, 9.27 years for Yozgat, 8.59 years for Denizli, 8.33 years for Van, 8.04 years for Adana, and 7.75 years for Antalya.

In addition to the studies written for Turkey, a considerable amount of international studies are found. These studies and their results are given and summarized in Table 1 for ease of comparison.

Table 1. PV power systems-related studies for countries other than Turkey

Study	Location	Capacity	Type of System	System Life	Type of Study	Objectives	Findings
Asumadu-Sarkodie & Owusu, 2016	Ghana	5 MW	Grid-connected	25	Simulations and theoretical	Feasibility analysis of the solar power plant in Ghana	Tafo has the longest simple payback and equity payback periods of 12.5 and 19.2 yrs, respectively. Navrongo has the shortest simple payback and equity payback periods of 7.3 and 5.0 yrs, respectively.
Aristizába & Páez, 2018	Colombia	6 kW	Grid-connected building-integrated	20	Simulations and theoretical	Economic feasibility analysis of a 6 kW building-integrated photovoltaic system	The system's payback period is 5.8 yrs
Samu & Fahrioglu, 2017	Zimbabwe	10 MW	Grid-connected	25	Simulations and theoretical	Economic feasibility evaluation of solar power plants for 28 locations in Zimbabwe	The place with the lowest simple payback period is Chinhoyi (6.7 yrs) and the highest is Beitbridge (7.6 yrs). The place with the lowest equity payback period is Chegutu (3.0 yrs) and the highest is Chiredzi (3.8 yrs).
Choi et al., 2018	United States	4.59 kWp	Grid-connected	25	Simulations and theoretical	Economic feasibility analysis of fixed and dual-axis tracking photovoltaic systems installed in the US Upper Midwest	The lifetime average incremental costs of energy generated by the stationary and dual-axis tracking systems are estimated to be $0.31 and $0.37 per kWh generated, respectively.
Rehman et al., 2007	Saudi Arabia	5 MW	Grid-connected	25	Simulations and theoretical	Economic feasibility evaluation of solar power plants for 41 locations	The simple payback period varied between 7.6 and 11.8 yrs, while YPCF varied between 6.7 and 10 yrs.
Rafique & Bahaidarah, 2019	Pakistan	5 MW	Grid-connected	25	Simulations and theoretical	Feasibility analysis of a 5 MW solar photovoltaic power for 10 different locations in Pakistan.	The average values of internal rate of return equity, payback period, net present value, benefit–cost ratio, and capacity factor are 21.75%, 10.58 years, 3,704,588 US$, 1.99, 24.13%, respectively.
Yadav & Bajpai, 2019	India	5 kWp/ 198 kWp/ 75 kWp/ 50 kWp	Grid-connected rooftop	25-30	Simulations and theoretical	Economic evaluation and environmental performance of 4 rooftop solar photovoltaic systems	The payback period of rooftop PV system is 13.36 years, which is higher than the other systems due to the higher system cost. Payback period of 6.71 years was calculated for rooftop PV system III, which is lower than 6.87 and 7.12 years for rooftop PV Systems II and IV, respectively.

Continued on following page

Table 1. Continued

Study	Location	Capacity	Type of System	System Life	Type of Study	Objectives	Findings
Hou et al., 2016	China	-	Grid-connected	25	Simulations and theoretical	Environmental impacts of grid-connected photovoltaic power generation from crystalline silicon solar modules in China	The energy payback times varied from 1.6 to 2.3 years.
Numbi & Malinga, 2017	South Africa	3 kW	Residential grid-interactive	30	Simulations and theoretical	Investigation of an optimal energy control of a 3 kW residential grid-interactive solar PV system	Case I, when the feed-in tariff is USD the total investment cost will be recovered over 19 years. Case II, when the feed-in tariff is increased equal to the flat tariff (USD 0.11/ kWh), the payback period is reduced from 19 to 14 years. This is further reduced to 8.6 years. Case III when the feed-in tariff is increased to be equal to the peak TOU tariff (USD 0.16/kW h)
Omar & Mahmoud, 2018	Palestine	5 kW	Grid-connected rooftop	20	Theoretical	Economic feasibility of a 5kW capacity PV system in 3 different houses located in different conflict zones in Palestine.	Payback period was calculated on average at 4.9 years.
Sagani et al., 2017	Greece	Between 2–10 kWp	Grid-connected rooftop	25	Simulations and theoretical	Economic and environmental analysis of relatively small rooftop PV-grid-interconnected energy systems of 2–10 kWp rated power, located in Athens, Greece.	The payback duration of the systems vary according to their capacities.
Sow et al., 2019	Canada	Between 100 kW-1 MW	Grid-connected and Grid-connected rooftop	30	Simulations and theoretical	Comparative economic analysis of residential solar photovoltaic systems throughout the provinces of Canada in 2013 and 2016.	The return period of the solar power plant to be built in Toronto City decreased from 12.2 in 2013 to 5.9 years in 2016. This development is also seen in other cities.
Bakhshi & Sadeh, 2018	Iran	5 kWp	Grid-connected	20	Simulations and theoretical	The viability of Grid–connected photovoltaic technology under a new 11 dynamic feed–in tariff strategy.	The return periods of the power plants vary between 2.518 and 3.763 yrs.

Continued on following page

Table 1. Continued

Study	Location	Capacity	Type of System	System Life	Type of Study	Objectives	Findings
Al-Soud & Hrayshat, 2009	Jordan	50 MW	Grid-connected-Concentrating solar power plant (CSPP)	More than 20 years	Simulations and theoretical	Economic feasibility of a prototype of a 50 MW CSPP for electricity generation in Jordan.	-Full-load operation is 2,345 (h/year) -Total investment cost is 170 million JD -O&M cost is 3.86 (million JD/year)
Mohammadi et al., 2018	Iran	5 MW	Grid-connected-1-axis and 2-axis systems	25	Simulations	The potential of developing 5 MW gird-connected PV power plants in eight selected cities in the southern coast of Iran.	The highest and lowest values of generated electricity and capacity factor were found for Jask and Bandar Abbas, respectively. Payback periods for analyzed cities vary between 5.61 and 7.00 yrs.
Lee et al., 2016	New England	67 kW	Grid-connected rooftop	25	Theoretical	The economic feasibility of the solar PV systems at the campus under realistic constraints, by analyzing actual data from the solar array on campus.	The payback period calculated for other 26 buildings ranged between 8 and 12 yrs. The average payback period and Cenlentano Hall's payback period was calculated as 11 yrs.
Edalati et al., 2016	Iran	10 MW	Grid-connected	-	Theoretical and field studies	Investigation of the behavior of the main parts of a grid-connected PV system.	The capacity factor of the modeled PV power plants ranged from a lower value of 12% in Ramsar to a maximum value of 23.13% in Kerman. The levelized cost of energy varied from 19.92 c$/kWh in the south-eastern part of Iran to 38.38 c$/kWh in northern part.
Mirzahosseini & Taheri, 2012	Iran	12 kW	Off-grid	20	Simulations	In case of energy subsidies in Iran, the feasibility study of solar power plants was carried out with the RETScreen program. The study is based on three scenarios.	- In the first case, the equity payback has been 12.1 year. - In the second case, the equity payback was 8 year. - In the last scenario, equity payback reached within 6 years.
Zhang et al., 2016	China	629 kW / 2.17 kW	Grid-connected rooftop	20	Simulations and theoretical	Techno-economic feasibility study of the Building Added PV for commercial and residential building hybrid energy systems.	The Building Added PV-based hybrid energy systems for commercial building reduces the energy costs from 1.2893 ¥/kW h to 0.6982 ¥/kW h, giving approximately 46% cost savings for customers.

Continued on following page

Table 1. Continued

Study	Location	Capacity	Type of System	System Life	Type of Study	Objectives	Findings
Xu et al., 2019	Sindh, Pakistan	Solar panel: 200 W Battery: 140 Ah/12 V	Off-Grid rooftop	Solar panel: 25 Battery: 5	Theoretical	To investigate whether solar energy is a solution for the electricity problem in rural areas.	The off-grid solar PV power generation system provides electricity at the cost of Pakistani Rupees 6.87/ kWh and is regarded as much cheaper than conventional energy sources (i.e., around Pakistani Rupees 20.79/ kWh).
Obeng et al., 2020	UENR Nsoatre Campus, Sunyani, Ghana	50 MW	Grid-connected	25	Simulations and theoretical	Technical and economic feasibility of a 50 MW grid-tied solar photovoltaic plant at UENR Nsoatre Campus.	The determined costs of energy for the systems are lower than the set feed-in tariff of 14 cents/kWh. Payback periods vary between 6.4 and 7.2 depending on the PV systems used in the power plants.
Okoye & Oranekwu-Okoye, 2018	Sub-Sahara Africa	Solar panel: 165.6 kW Battery: 43.26 kWh	Off-Grid rooftop	Solar panel: 25 Battery: 5	Theoretical	To investigate whether solar energy is a solution for the electricity problem in rural areas in Sub-Sahara Africa	The adopted 0.4 USD/ kWh electricity price is low compared to the estimated unit price of 0.62 USD/kWh for the widely utilized diesel generating set in the country.
Imam & Al-Turki, 2019	Jeddah, Saudi Arabia	12.25 kW	Grid-connected rooftop	25	Simulations and theoretical	Techno-economic feasibility evaluation for a grid-connected photovoltaic energy conversion system on the rooftop of a typical residential building in Jeddah.	The estimated NPV of the system over its lifetime is $4,378; its PBP is 14.6 years.

The contribution of this study is threefold. First, in this study, in contrast to previous literature is not a single city or region is analyzing all provinces of Turkey. Second, The study includes three different scenarios. Third, in the feasibility study, global radiation value, internal rate of return, simple payback duration, years to positive cash flow, net present value, annual life cycle savings, profitability index, greenhouse gas reduction amount (tCO2 / year), electricity to the grid (MWh), electricity export income ($) and efficiency per square meter (kWh / m2) are used. In this way, both renewable energy production and economical evaluation of the solar power plant can be made.

RETScreen Simulation

The program RETScreen is one of the world's leading clean and renewable energy simulation programs and was developed by the Canadian government. The Excel-based program determines the most optimal conditions by drawing the climate data of the selected region from the NASA database and the materials to be used in the power plant to be installed from its own pool. It also allows the user to enter this data manually. It calculates the payback duration of the facility by determining many variables such as the amount of energy to be produced, unit energy sales price, and credit and depreciation rates while simulating the plant. The RETScreen program also reveals the environmental aspect of the facility by calculating the CO2 emission amounts and emission-reduction incentives.

The values used in the RETScreen program for our study are summarized in Table 2.

Table 2. Summary of variables used in the economic feasibility study

Item Description	Value
Energy cost escalation rate	2.5%
Inflation rate	15%
Discount rate	15%
Project life	25
Greenhouse Gas Reduction Credit Rate	15 $
Initial Investment Costs	1100 $/kWh
Operation and Maintenance Costs	11 $/kWh

CLIMATE DATA OF CITIES

Turkey's compliance with all provinces in terms of solar power plant installations are analyzed using the program RETScreen and are conducted by considering the global radiation values specified in the licensed solar power plant installation. As a result of the analysis, 45 provinces are found suitable for solar power plant installation, and Table 3 shows the latitude, longitude, altitude, radiation value per square meter, and sunshine duration of these cities.

Table 3. Geographical coordinates and the daily solar radiation of 45 locations in Turkey.

Province	Latitude*	Longitude*	Altitude (m)*	kWh/m2 day*	Sunshine Duration (h)**
Muğla	36.920	28.510	579	5.28	8.313
Aydın	37.841	27.834	54	5.11	8.254
Adana	36.999	35.343	29	5.05	8.940
Bayburt	40.258	40.226	1,554	4.60	6.570
Erzurum	39.907	41.277	1,920	4.58	6.858
Hakkâri	37.572	43.697	2,161	4.77	9.617
Bingöl	38.940	40.544	1,519	4.76	7.453
Van	38.496	43.359	1,673	4.67	8.408
Elazığ	38.672	39.204	1,059	4.80	7.755
Ağrı	39.784	43.136	1,717	4.50	7.615
Denizli	37.781	29.069	423	4.76	8.028
Bitlis	38.336	42.084	1,523	4.71	7.373
Diyarbakır	37.919	40.227	674	4.77	7.155
Siirt	37.930	41.933	873	4.71	7.750
Batman	37.885	41.123	570	4.78	7.865
Şırnak	37.527	42.407	1,082	4.80	8.152
Muş	38.745	41.498	1,333	4.67	7.363
Sivas	39.746	37.009	1,286	4.52	7.270
Mardin	37.320	40.735	1,014	4.78	8.313
Konya	37.882	32.533	1,010	4.65	7.940
Kahramanmaraş	37.561	36.932	492	4.65	7.983
Gaziantep	37.070	37.388	832	4.83	8.155
Malatya	38.350	38.292	947	4.70	7.877
Kilis	36.766	37.040	509	4.83	8.154
Uşak	38.667	29.406	918	4.60	7.640
Adıyaman	37.763	38.273	683	4.70	8.114
Isparta	37.770	30.549	1,056	4.59	7.832
Kayseri	38.731	35.444	1,056	4.55	7.788
Burdur	37.733	30.281	918	4.59	8.108

Continued on following page

Table 3. Continued

Province	Latitude*	Longitude*	Altitude (m)*	kWh/m2 day*	Sunshine Duration (h)**
Nevşehir	38.578	34.635	1,368	4.56	7.767
Erzincan	39.760	39.476	1,232	4.52	7.109
Mersin	36.798	34.612	18	4.78	8.257
Aksaray	38.370	34.024	977	4.56	7.909
Karaman	37.165	33.241	1,088	4.64	8.239
Manisa	38.625	27.416	45	4.69	7.784
Niğde	38.086	34.847	1,365	4.56	8.028
Yozgat	39.750	34.763	1,183	4.46	7.353
İzmir	38.413	27.135	80	4.69	8.179
Tunceli	39.132	39.552	1,388	4.52	7.444
Osmaniye	37.075	36.253	134	4.65	8.096
Afyon	38.777	30.533	1,009	4.50	7.409
Şanlıurfa	37.147	38.801	496	4.71	8.311
Hatay	36.422	36.188	834	4.73	8.209
Kırşehir	39.174	33.987	1,545	4.45	7.587
Antalya	36.903	30.720	65	4.55	8.247
* RETScreen (Calculated for the locations to install the facility.)					
** GEPA					

Table 3 shows that the province with the lowest radiation value per square meter is Antalya with 4.55 kWh/m2 per day and the province with the highest is Muğla with 5.28 kWh/m2 per day. As one can observe, the provinces of Northern Turkey are not included in the Table, since the solar energy in Northern Turkey shows that plants are not suitable for installation.

The distribution of the global radiation values for the 45 selected provinces over a year is shown in three-dimensions by month in Figure 2. The vertical axis of the graph shows the radiation values, the horizontal axis shows the selected provinces, and the third axis shows the months. These values decrease in winter, bottom out in December-January, increase in summer, and peak in June-July. While the average radiation value for these provinces in December-January is 1.812 kWh / m2 per day, the average radiation value for June-July is 6.677 kWh / m2 per day.

Figure 2. Seasonal variation of global solar radiation for the selected provinces

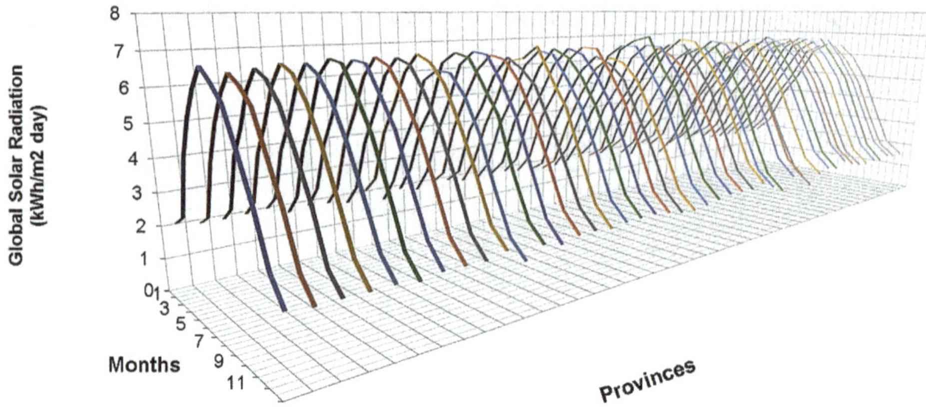

The comparison of the global radiation values for the selected 45 provinces with all of Turkey are presented in Figure 3. In the graph, the global average radiation values are given for the 45 selected provinces and for all Turkey. The selected provinces' having higher global radiation values than Turkey overall is an expected situation in our study because the solar power plants in Turkey's overall average include the provinces with values insufficient for power plant installation. However, all provinces determined for our study have radiation values that meet the requirement for solar power plant installation. Northern provinces, especially those with insufficient radiation values for the solar power plant installation, have been excluded from the study.

Figure 3. Seasonal variation of Global Solar Radiation for Turkey and selected provinces

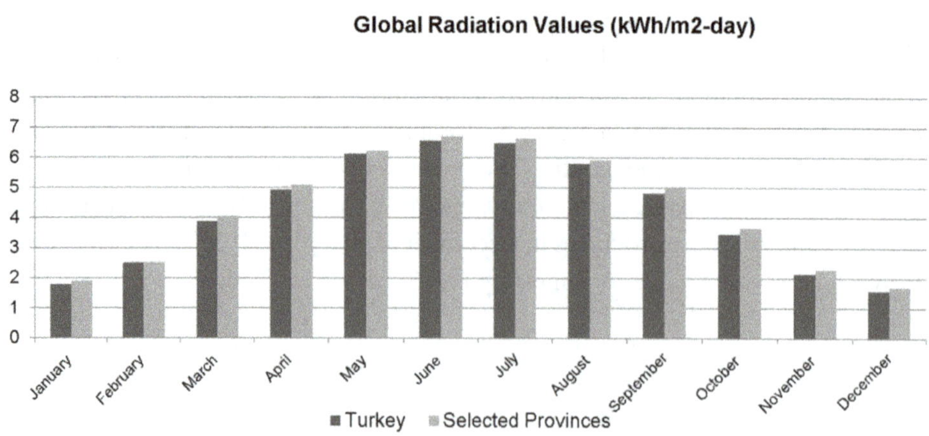

The distribution of sunshine duration for the selected 45 provinces is shown by month in Figure 4 in three dimensions. The vertical axis of the graph shows the daily sunshine duration, the horizontal axis shows the selected provinces, and the third axis shows the months. Just like the global radiation values, daily sunshine values decrease in the winter and increase in the summer months. Sunshine duration bottoms out in December-January, dropping to an average of 4.204 hours per day. It peaks in June-July and rises to an average of 11.474 hours daily. Among the selected provinces, the place with the lowest daily sunshine duration is Bayburt (6.570 hours), while the highest is Hakkari (9.612 hours).

Figure 4. Seasonal variation of sunshine duration for selected provinces

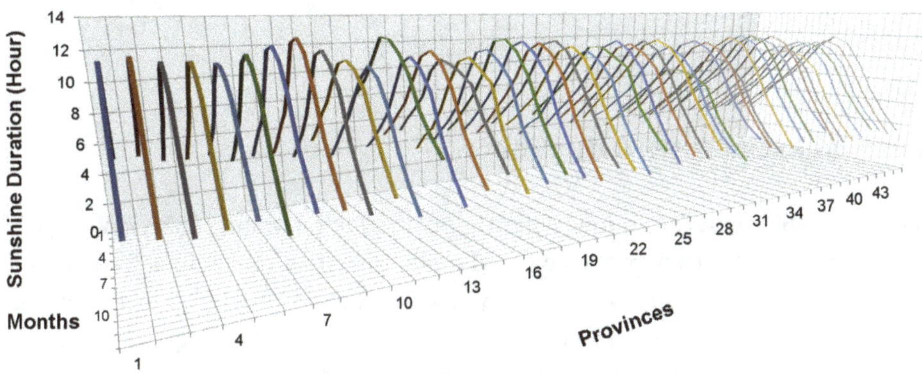

Figure 5 shows a comparison of daily sunshine duration for Turkey with the selected 45 provinces. The graph provides the global average radiation values for all of Turkey's 81 provinces, while showing the provincial averages for the 45 selected cities. For both the 45 provinces used in the study as well as Turkey overall, the daily duration of sunshine decreases in the winter months and increases in the summer. Just as with the value of global radiation, sunshine duration of the selected provinces are also above Turkey's average, as expected.

Figure 5. Seasonal variation of sunshine duration for Turkey and the selected provinces

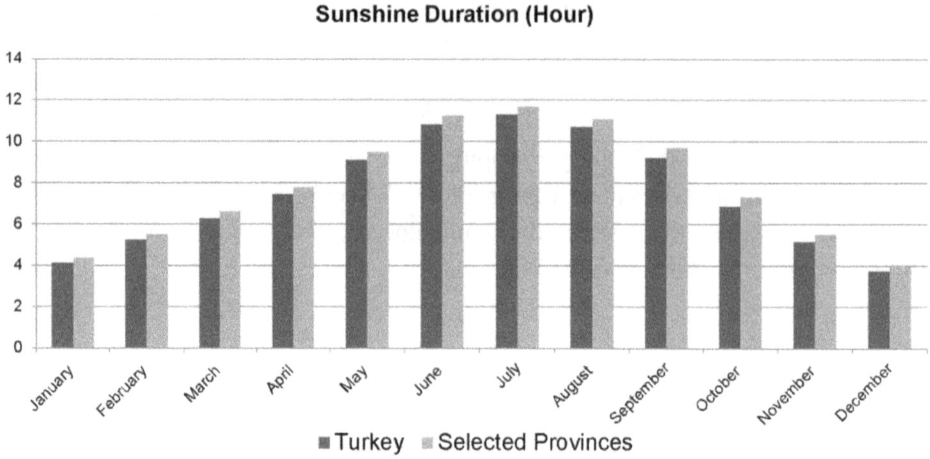

Renewable Energy Production

This study will analyze the lower-limit global radiation value (1620 kWh / m2-year) determined for each province based on the specified regulations for licensed solar power plants in Turkey. The study involves the 45 provinces that meet the criterion, and a 5 MW solar energy system is installed and analyzed using the program RETScreen. In addition, the global radiation value (*GSR*), internal rate of return (*IRR*), simple payback period (*SPP*), years to positive cash flow (*YPCF*), net present value (*NPV*), Annual Life Cycle Savings (*ALCS*), Cost-Benefit Ratio (*PI*), greenhouse gas reduction amount (tCO_2 / year), electricity to the grid (MWh), Electricity Export Revenue ($) and efficiency per square meter (kWh / m2) have also been output by RETScreen.

Figure 6 provides the efficiency values per square meter obtained as a result of installing a solar power plant when dividing the installation area by the electricity supplied to the grid for the selected provinces. The efficiency per square meter takes different values depending on the sunshine duration of the region, the radiation values it receives, and the angle at which the sun's rays fall. When comparing the provinces in our study, the province with the highest efficiency is Muğla with 0.249, while the province with the lowest efficiency is Antalya with 0.212. The fact that Muğla and Antalya, the best and worst places for solar power plant installation respectively, are very close to each other is an indication that these two provinces and the provinces with values between theirs are very suitable for solar power plant installation.

Figure 6. Variations in the specific yields of the 45 provinces

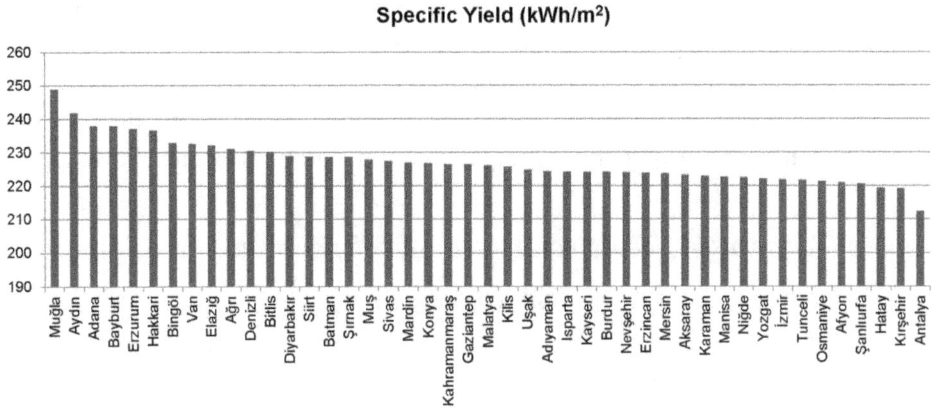

Figure 7 provides the amount of electricity exported to the grid as a result of installing 5 MW solar power plants in the 45 selected provinces. As observable in the Figure, the amount of electricity obtained varies provincially. This situation varies according to the radiation values received by the region, daily sunshine duration, and angle with which the sun rays strike the earth. According to the Figure, the highest energy is obtained from Muğla (7.901 MWh), while the lowest energy is obtained from Antalya (6.742 MWh).

Figure 7. Variations in electricity exported to the grid for the 45 provinces

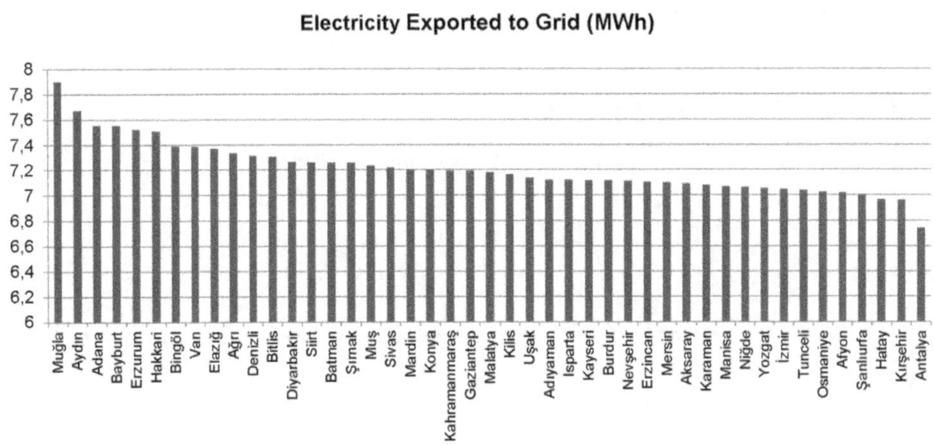

Figure 8 provides the average annual global radiation values of the selected provinces. The province with the highest radiation value is Muğla (5.28 kWh / m2-day), while the city with the lowest radiation value is Kırşehir (4.45 kWh / m2-day). Considering the average radiation value for Turkey's 81 provinces (4.425 kWh / m2-day), even Kırşehir with the lowest radiation values among the selected provinces appears to be quite high compared to the Turkey's provincial average.

Figure 8. Variations in global solar radiation for 45 provinces

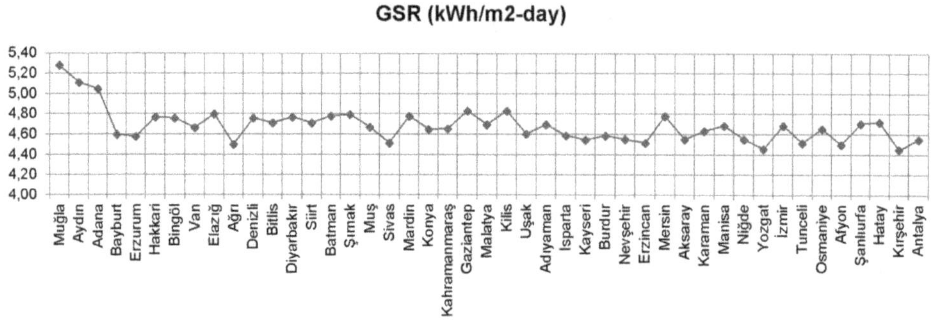

Greenhouse Gas Reduction

Figure 9 shows how much reduction in greenhouse gas emissions will occur from installing 5 MW solar power plants in the selected provinces. While the province with the highest greenhouse gas reduction is Şırnak (6,555 tons annually), Antalya has the least (3,179 tons annually).

Figure 9. Greenhouse gas reduction

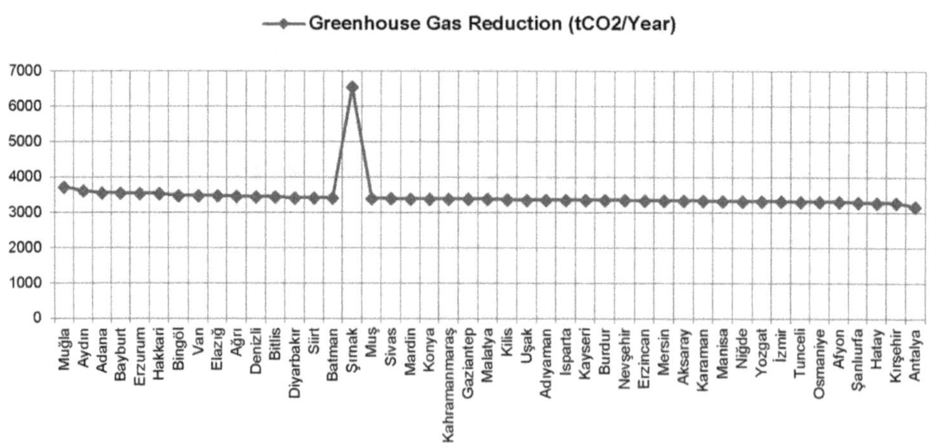

This analysis shows that when installing a 5 MW solar energy power plant, an annual average reduction of 3,470 tons of greenhouse gas emissions occurs. Since the life of a project is assumed to be 25 years in the analysis, an average reduction of 86,753 tons of greenhouse gases occurs from a 5 MW solar power plant over 25 years.

Economic Feasibility Analysis

The program RETScreen provides the opportunity to analyze power plants by determining rates such as internal rate of return (*IRR*), simple payback period (*SPP*), years to positive cash flow (*YPCF*), annual life cycle savings (*ALCS*), net present value (*NPV*), and cost-benefit ratio (*PI*).

IRR expresses the real interest yield calculated nominally during the total period of the investment and indicates the pre-tax internal rate of return on the assets. This rate also expresses how much added value the investment will create. *SPP* calculates how long a facility pays total investment costs using the revenue and savings it generates. This determines how long a facility needs to repay the investment cost. Instead of making a comparison between the profitability of the investments, *SPP* provides the opportunity to compare the investment costs in terms of payback period. *YPCF* shows how long until an investor gets compensated for their initial facility-installation investment, namely, the resources they use. *ALCS* is another term used for project suitability. *NPV* is calculated by considering the discount rate and project life. The model computes the annual life cycle savings, which is the leveled nominal yearly savings that have exactly the same life and net present value as the project. *ALCS* is calculated using *NPV*, the discount rate, and the project life. *NPV*, which is associated with the *IRR*, represents the present value minus the discount rate of all future cash flows from the facility. *PI* represents the net benefit, or the ratio of income earned to total cost. A *PI* ratio greater than 1 indicates that the facility is profitable. Thanks to this rate, comparing facilities' profitability rates is possible.

The study has three scenarios, and the *IRR*, *SPP*, *YPCF*, *ALCS*, *NPV*, and *PI* rates differ according to these scenarios. These scenarios and results are included in the continuation of our study.

Our study's first scenario assumes the domestic machinery and equipment to not be used in solar power plants; greenhouse gas reduction incentives are also not used. In other words, the first scenario is based only on the government's guarantee of purchase; one kWh of produced electricity is assumed to be sold at $ 0.133. The *IRR*, *SPP*, *YPCF*, *ALCS*, *NPV*, and *PI* values for Scenario 1 are shown in Figure 10.

Figure 10. The results from Scenario 1

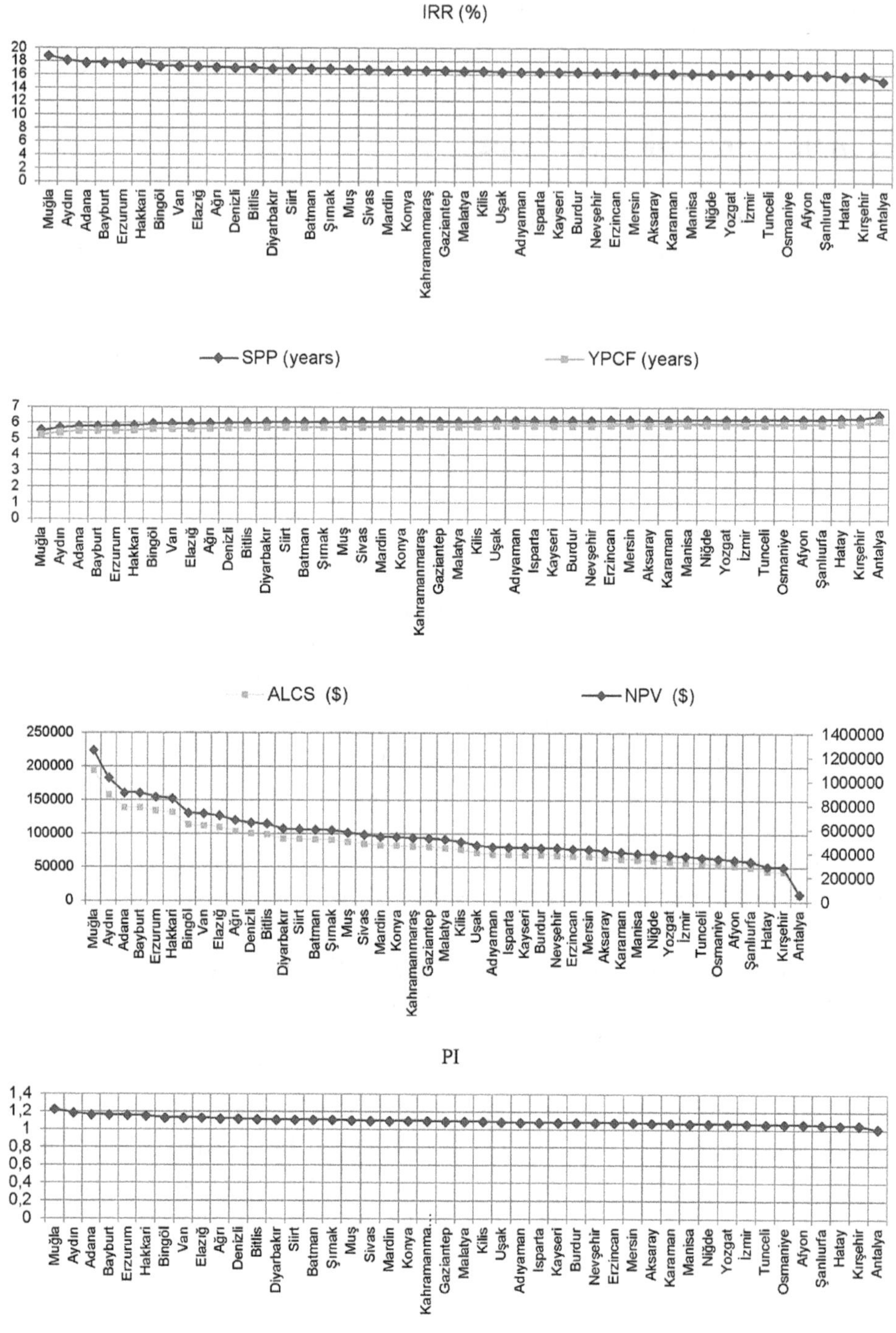

The results from our study's first scenario are as follows. When the solar power plants installed in the 45 provinces are evaluated in terms of *IRR*, the province with the highest *IRR* is Muğla (18.86) and the lowest IRR is Antalya (15.21). When considering that the average *IRR* for the 45 provinces is 16.732, 27 of 45 provinces are above this average. The *SPP* of solar power plants varies between 6.535 and 5.523 years. The province that pays the investment the earliest is Muğla (5.523 yrs), and the province that pays last is Antalya (6.535 yrs). The average *SPP* for Scenario 1 is 6.088. *YPCF* values are lower than *SPP* values due to assumption of no borrowing during power plant installation. In this case, *YPCF* values vary between 5.243 and 6.199 years. While the solar energy power plant with the lowest *YPCF* was installed in Muğla, the solar energy power plant with the highest *YPCF* was installed in Antalya. *ALCS* and *NPV* are parallel to each other as they concern project life, discount rate, and cash flows. The province with the lowest *NPV* is Antalya ($63,458), while the highest is Muğla ($1,256,681). While the province with the lowest *ALCS* is Antalya ($9,817), the highest is Muğla ($194,408). The average *NPV* and *ALCS* values for the solar power plants installed in these 45 provinces are $547,384 and $84,680, respectively. *PI* values for the solar power plants in the 45 provinces analyzed in the study are all greater than 1. While the lowest *PI* value belongs to Antalya (1.012), the highest *PI* value belongs to Muğla (1.229).

Our study's second scenario adds the greenhouse gas reduction incentive to the first scenario. As a result of reducing one ton of greenhouse gas, the additional scenario of $15 per kWh electricity export price is $ 0.133, just as in the first scenario. The *IRR*, *SPP*, *YPCF*, *ALCS*, *NPV*, and *PI* values for Scenario 2 are shown in Figure 11.

Figure 11. The results from Scenario 2

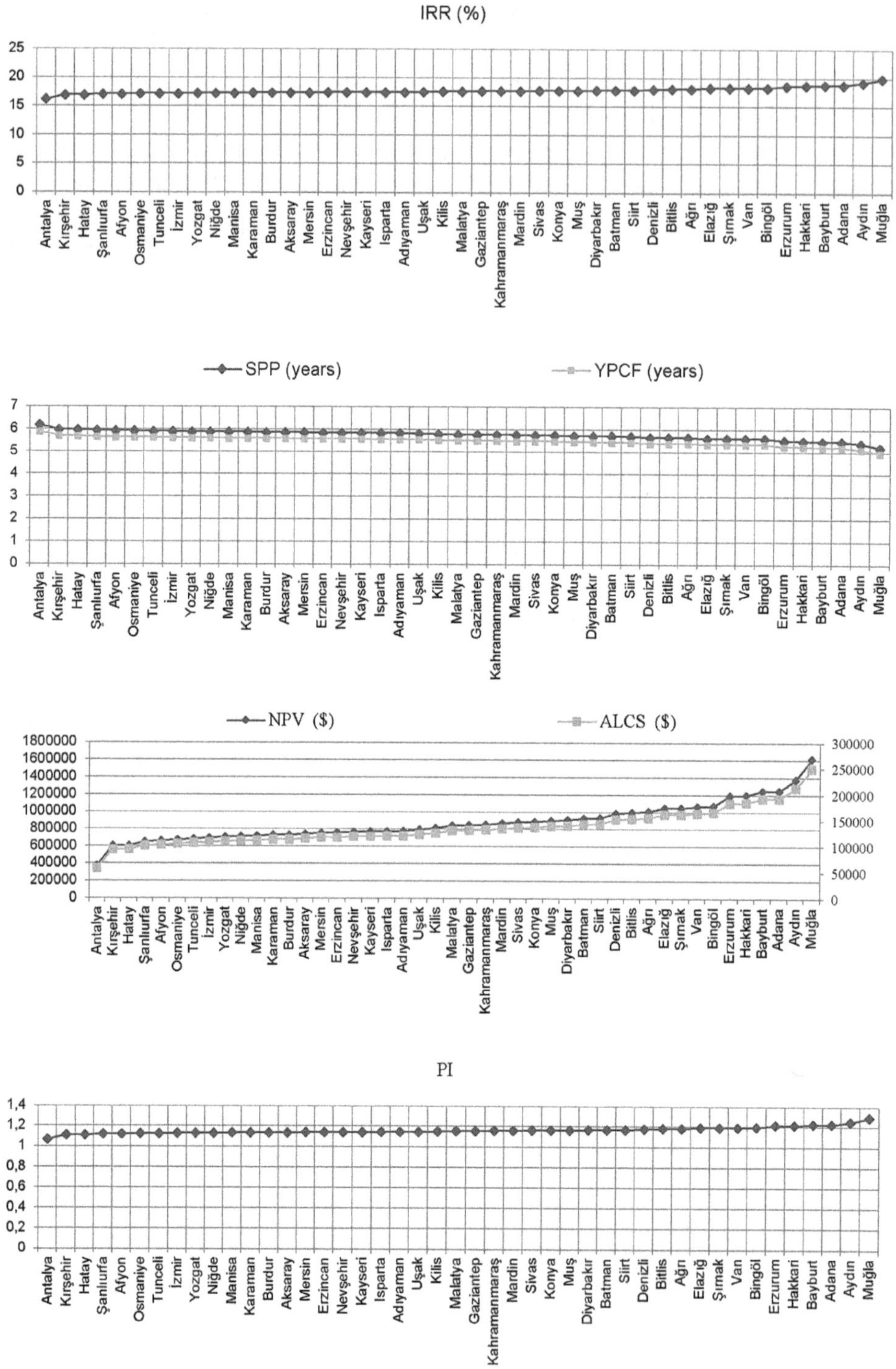

Scenario II includes the government purchase guarantee and greenhouse gas reduction incentive. In the case of Scenario 2, the results where the *IRR*, *NPV*, *ALCS*, and *PI* values are higher and the *SPP* and *YPCF* values are lower are expected. Among the 45 solar power plants installed, the province with the lowest *IRR* is Antalya with 16.20, while the highest province is Muğla (*IRR* = 19.94); the average *IRR* value is 17.757. In the study, the lowest *SPP* value is observed in Muğla (5.230) and the highest in Antalya (6.184). The lowest *YPCF* value is seen in Muğla (4.987) and the highest in Antalya (5.893). The average *SPP* and *YPCF* values are found to be 5.762 and 5.490, respectively. The province with the lowest *NPV* value is Antalya ($371,679), while the highest value is in Muğla ($1,617,908). The lowest *ALCS* value is seen in Antalya ($57,498) and the highest in Muğla ($250,289). The average *NPV* and *ALCS* values are found as $878,609 and $135,920, respectively. In Scenario II, all *PI* ratios are greater than 1, and the average *PI* ratio in Scenario II is 1.159; 26 of the 45 selected provinces are below average, while the remaining provinces are above average. While the solar power plant with the highest *PI* rate is the Muğla installation (1.294), the lowest is the Antalya installation (1.068). In general, the solar power plant in Turkey is evaluated when the investment advantages than disadvantages installation is becoming more attractive with each passing day. It will become even more attractive to invest in solar power plant if given support reduction of greenhouse gas emissions in Turkey are considered.

In our study's Scenario 3, the domestic production incentive is added to the second scenario. In Scenario 3, all the machinery and equipment used in solar power plants are assumed to be domestic products, and by adding the incentives, one kWh of electricity is sold at over $ 0.2. The *IRR*, *SPP*, *YPCF*, *ALCS*, *NPV*, and *PI* values for Scenario 3 are shown in Figure 12.

Figure 12. The results from Scenario 3

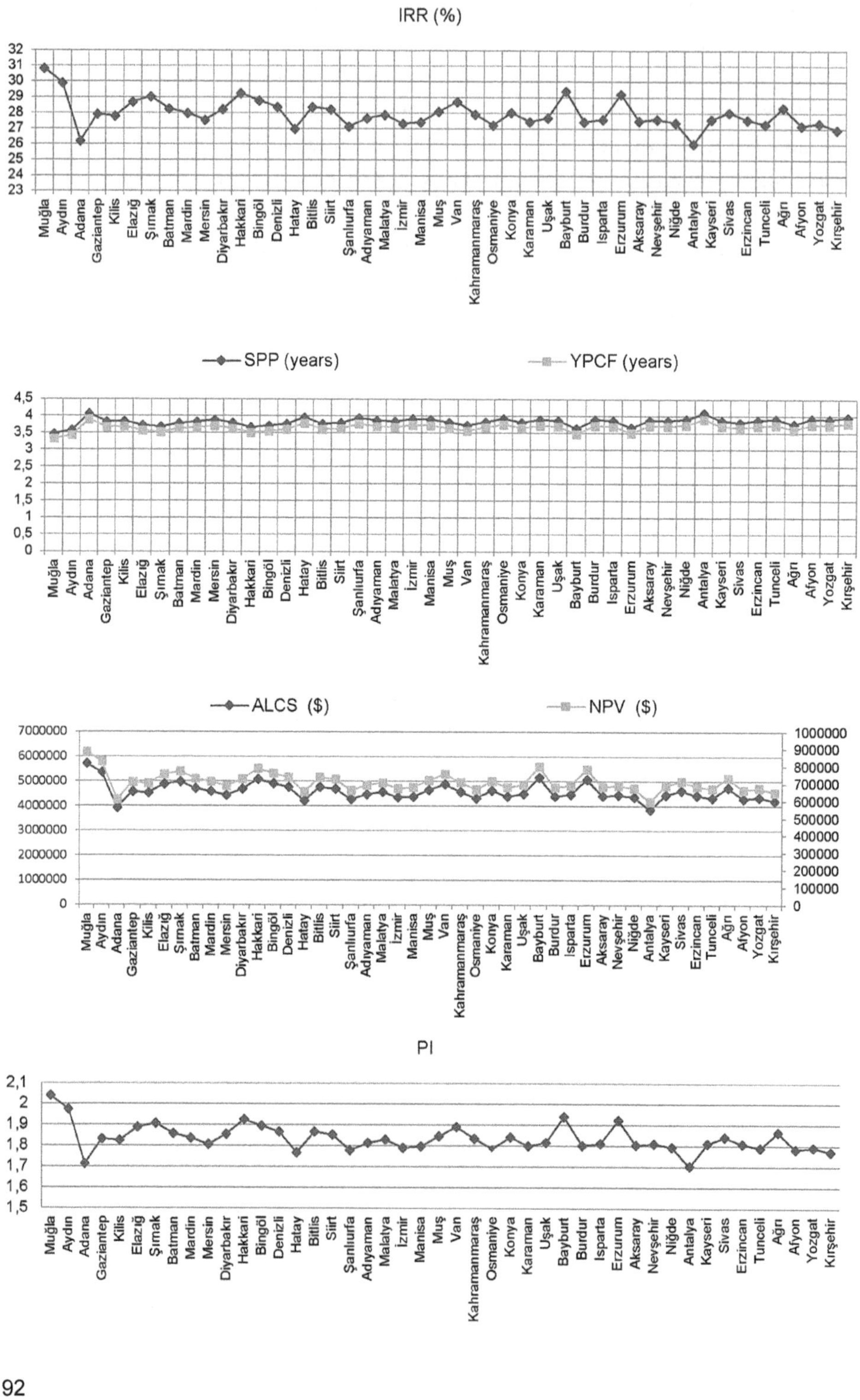

Scenario 3 includes the government purchase guarantee, greenhouse gas reduction incentive, and domestic production incentive. In this case, the *IRR*, *NPV*, *ALCS*, and *PI* values in Scenario III are the highest among the scenarios. The lowest *SPP* and *YPCF* values being found in Scenario 3 are an expected result in our study. When evaluating the selected provinces in terms of *IRR*, Antalya has the lowest *IRR* (26.06) and Muğla has the highest (30.84). The study shows the lowest *SPP* value to be in the solar power plant installed in Muğla (3.479), while the highest *SPP* value is seen in Antalya (4.101). Having great importance in Scenario III, the incentives decrease the *SPP* rate considerably, with an average of 3.835. Similarly, the *YPCF* value in Scenario 3 reached the lowest average value from all three scenarios. While the lowest *YPCF* value is seen in Muğla (3.335), the highest *YPCF* value is seen in Antalya (3.921). The average *YPCF* value has been found as 3.669. In terms of *ALCS*, the lowest province is Antalya ($598,221) while the highest province is Muğla ($884,002). Similarly, the lowest province with respect to *NPV* is Antalya ($3,866,992) while the highest province is Muğla ($5,714,319). The scenario with the highest *PI* values is Scenario 3ö wıth *PI* varying between 1.703 and 2.039. According to this scenario, the solar power plant with the lowest *PI* is the Antalya installation and the solar power plant with the highest *PI* is the Muğla installation. Scenario 3 finds the average *PI* to be 1.836, and 19 of the selected provinces are above this average.

CONCLUSION

Our study has reached the following results:

- The total installation cost for power plants installed in selected provinces has been determined as $5.5 million and the installation cost per unit is $1,100 / kWh.
- As expected, parallelism was found among the sunshine duration, global radiation values, electricity supplied to the grid, efficiency per square meter, and the simple and equity repayment values for the selected 45 provinces.
- The global radiation values for the selected provinces vary between 5.28 kWh / m2 and 4.45 kWh / m2. While the province with the highest global radiation value is Muğla, the one with the lowest is Kırşehir.
- When evaluating the selected provinces in terms of sunshine duration, Hakkari has the highest (average of 9.62 hours / day), while Bayburt has the lowest (average of 6.57 hours / day). Although sunshine duration fluctuates over the seasons, electricity can be produced efficiently throughout the year.

- The amount of electricity obtained from 5 MW installed solar power plants varies between 6,742 kWh and 7,901 kWh with an average generation of 7,212 kWh. In this case, 17 of the selected provinces are above average and the others below average.
- The average internal rate of return for the provinces selected for Scenarios 1, 2, and 3 is 16.73%, 17.76%, and 27.96%, respectively; 17 of the selected provinces are above average and the others are below average. The province with the highest internal rate of return is Muğla while the lowest is Antalya.
- The province with the lowest *SPP* and *YPCF* is Muğla for all three scenarios, while the highest is Antalya. In terms of the Scenario 1, these rates are 5.553 and 5.243 years for Muğla, and 6.535 and 6.199 years for Antalya, respectively. In terms of Scenario 2, the *SPP* and *YPCF* values are 5.230 and 4.987 years for Muğla, and 6.184 and 5.893 years for Antalya, respectively. In terms of Scenario 3, these rates are 3.479 years and 3.335 years for Muğla, and 4.101 and 3.921 years for Antalya, respectively.
- According to Scenarios 1, 2, and 3, the province with the highest cost-benefit ratio is Muğla, while Antalya has the lowest. In terms of Scenario 1, the benefit cost ratios for Muğla and Antalya are 1.229 and 1.011, respectively; Scenario 2 shows 1.294 and 1.068, respectively; and Scenario 3, 2.039 and 1.703, respectively.
- The study shows 45 of Turkey's 81 provinces to be eligible for solar power plant installations. Looking at the economic indicators, the most suitable place among the selected cities is Muğla, while Antalya has been determined as the most unsuitable.
- As a result of installing solar power plants with a 5 MW capacity, the power plant will prevent an average of 3,470 tons per year of greenhouse gases from being emitted to the atmosphere, with an average of 86,753 tons at the end of 25 years.
- Distributing solar power plants in the selected provinces will also contribute to the country's industry and employment. In an environment where Scenario 3 is valid, the payback times for the plants in particular will be shortened and prevent product imports from being required for the plant installation.

REFERENCES

Al-Soud, M. S., & Hrayshat, E. S. (2009). A 50 MW concentrating solar power plant for Jordan. *Journal of Cleaner Production*, *17*(6), 625–635. doi:10.1016/j.jclepro.2008.11.002

Ari, A., & Cergibozan, R. (2018). Sustainable growth in Turkey: The role of trade openness, financial development, and renewable energy use. *Industrial Policy and Sustainable Growth*, 1-21.

Aristizábal, A. J., & Páez, C. A. (2018). Power quality analysis of 6 kW building-integrated photovoltaic system and economic evaluation using RETScreen™. *Intl J Ambient Energy*, *39*(8), 813–822. doi:10.1080/01430750.2017.1354319

Asumadu-Sarkodie, S., & Owusu, P. A. (2016). The potential and economic viability of solar photovoltaic power in Ghana. *Energ Source Part A*, *38*(5), 709–716. doi:10.1080/15567036.2015.1122682

Bakhshi, R., & Sadeh, J. (2018). Economic evaluation of grid–connected photovoltaic systems viability under a new dynamic feed–in tariff scheme: A case study in Iran. *Renewable Energy*, *119*, 354–364. doi:10.1016/j.renene.2017.11.093

Büyükzeren, R., Altintaş, H. B., Martin, K., & Kahraman, A. (2015). Binalardaki Fotovoltaik Uygulamasının Teknik, Çevresel ve Ekonomik İncelenmesi: Meram Tıp Fakültesi Hastanesi Örneği. *EMO Bilimsel Dergi*, *5*(10), 9–14.

Choi, W., Pate, M. B., Warren, R. D., & Nelson, R. M. (2018). An economic analysis comparison of stationary and dual-axis tracking grid-connected photovoltaic systems in the US Upper Midwest. *International Journal of Sustainable Energy*, *37*(5), 455–478. doi:10.1080/14786451.2017.1294178

Dağtekin, M., Kaya, D., Öztürk, H. H., & Kiliç, F. Ç. (2014). A study of techno-economic feasibility analysis of solar photovoltaic (PV) power generation in the province of Adana in Turkey. *Energ Explor Exploit*, *32*(4), 719–735. doi:10.1260/0144-5987.32.4.719

Dinçer, F. (2011). Overview of the photovoltaic technology status and perspective in Turkey. *Renewable & Sustainable Energy Reviews*, *15*(8), 3768–3779. doi:10.1016/j.rser.2011.06.005

Duman, A. C., & Güler, Ö. (2020). Economic analysis of grid-connected residential rooftop PV systems in Turkey. *Renewable Energy*, *148*, 697–711. doi:10.1016/j.renene.2019.10.157

Edalati, S., Ameri, M., Iranmanesh, M., Tarmahi, H., & Gholampour, M. (2016). Technical and economic assessments of grid-connected photovoltaic power plants: Iran case study. *Energy*, *114*, 923–934. doi:10.1016/j.energy.2016.08.041

Evrendilek, F., & Ertekin, C. (2003). Assessing the potential of renewable energy sources in Turkey. *Renewable Energy*, *28*(15), 2303–2315. doi:10.1016/S0960-1481(03)00138-1

Gürtürk, M. (2019). Economic feasibility of solar power plants based on PV module with levelized cost analysis. *Energy*, *171*, 866–878. doi:10.1016/j.energy.2019.01.090

Hou, G., Sun, H., Jiang, Z., Pan, Z., Wang, Y., Zhang, X., Zhao, Y., & Yao, Q. (2016). Life cycle assessment of grid-connected photovoltaic power generation from crystalline silicon solar modules in China. *Applied Energy*, *164*, 882–890. doi:10.1016/j.apenergy.2015.11.023

Imam, A. A., Al-Turki, Y. A., & R, S. K. (2019). Techno-Economic Feasibility Assessment of Grid-Connected PV Systems for Residential Buildings in Saudi Arabia—A Case Study. *Sustainability*, *12*(1), 1–25. doi:10.3390u12010262

Lee, J., Chang, B., Aktas, C., & Gorthala, R. (2016). Economic feasibility of campus-wide photovoltaic systems in New England. *Renewable Energy*, *99*, 452–464. doi:10.1016/j.renene.2016.07.009

Melikoglu, M. (2013). Vision 2023: Feasibility analysis of Turkey's renewable energy projection. *Renewable Energy*, *50*, 570–575. doi:10.1016/j.renene.2012.07.032

Ministry of Energy and Natural Resources. (2014). National Renewable Energy action plan for Turkey. Author.

Mirzahosseini, A. H., & Taheri, T. (2012). Environmental, technical and financial feasibility study of solar power plants by RETScreen, according to the targeting of energy subsidies in Iran. *Renewable & Sustainable Energy Reviews*, *16*(5), 2806–2811. doi:10.1016/j.rser.2012.01.066

Mohammadi, K., Naderi, M., & Saghafifar, M. (2018). Economic feasibility of developing grid-connected photovoltaic plants in the southern coast of Iran. *Energy*, *156*, 17–31. doi:10.1016/j.energy.2018.05.065

Numbi, B. P., & Malinga, S. J. (2017). Optimal energy cost and economic analysis of a residential grid-interactive solar PV system-case of eThekwini municipality in South Africa. *Applied Energy*, *186*, 28–45. doi:10.1016/j.apenergy.2016.10.048

Obeng, M., Gyamfi, S., Derkyi, N. S., Kabo-bah, A. T., & Peprah, F. (2020). Technical and economic feasibility of a 50 MW grid-connected solar PV at UENR Nsoatre Campus. *Journal of Cleaner Production*, *247*, 119159. doi:10.1016/j.jclepro.2019.119159

Okoye, C. O., & Oranekwu-Okoye, B. C. (2018). Economic feasibility of solar PV system for rural electrification in Sub-Sahara Africa. *Renewable & Sustainable Energy Reviews*, *82*, 2537–2547. doi:10.1016/j.rser.2017.09.054

Omar, M. A., & Mahmoud, M. M. (2018). Grid connected PV-home systems in Palestine: A review on technical performance, effects and economic feasibility. *Renewable & Sustainable Energy Reviews*, *82*, 2490–2497. doi:10.1016/j.rser.2017.09.008

Özcan, O., & Ersöz, F. (2019). Project and cost-based evaluation of solar energy performance in three different geographical regions of Turkey: Investment analysis application. *Eng Sci Technol Int J*, *22*(4), 1098–1106. doi:10.1016/j.jestch.2019.04.001

Rafique, M. M., & Bahaidarah, H. M. (2019). Thermo-economic and environmental feasibility of a solar power plant as a renewable and green source of electrification. *International Journal of Green Energy*, *16*(15), 1–14. doi:10.1080/15435075.2019.1677237

Rehman, S., Bader, M. A., & Al-Moallem, S. A. (2007). Cost of solar energy generated using PV panels. *Renewable & Sustainable Energy Reviews*, *11*(8), 1843–1857. doi:10.1016/j.rser.2006.03.005

Sagani, A., Mihelis, J., & Dedoussis, V. (2017). Techno-economic analysis and life-cycle environmental impacts of small-scale building-integrated PV systems in Greece. *Energy and Building*, *139*, 277–290. doi:10.1016/j.enbuild.2017.01.022

Samu, R., & Fahrioglu, M. (2017). An analysis on the potential of solar photovoltaic power. *Energy Source Part B*, *12*(10), 883–889. doi:10.1080/15567249.2017.1319437

Sow, A., Mehrtash, M., Rousse, D. R., & Haillot, D. (2019). Economic analysis of residential solar photovoltaic electricity production in Canada. *Sustain Energy Techn*, *33*, 83–94. doi:10.1016/j.seta.2019.03.003

TUIK. (2020). *Sera Gazı Emisyon İstatistikleri: 1990-2018*. https://data.tuik.gov.tr/Bulten/Index?p=Sera-Gazi-Emisyon-Istatistikleri-1990-2018-33624

Xu, L., Wang, Y., Solangi, Y. A., Zameer, H., & Shah, S. A. A. (2019). Off-grid solar PV power generation system in Sindh, Pakistan: A techno-economic feasibility analysis. *Processes (Basel, Switzerland)*, *7*(5), 308. doi:10.3390/pr7050308

Yadav, S. K., & Bajpai, U. (2019). Energy, economic and environmental performance of a solar rooftop photovoltaic system in India. *International Journal of Sustainable Energy*, 1–16.

Yenilenebilir Enerji Kaynaklarının Elektrik Enerjisi Üretimi Amaçlı Kullanımına İlişkin Kanun. (2005). *Kanun No. 5346*. Resmi Gazete (Sayı: 25819). Accessed October 2020, https://www.resmigazete.gov.tr/eskiler/2005/05/20050518-1.htm

Zhang, X., Li, M., Ge, Y., & Li, G. (2016). Techno-economic feasibility analysis of solar photovoltaic power generation for buildings. *Applied Thermal Engineering*, *108*, 1362–1371. doi:10.1016/j.applthermaleng.2016.07.199

ENDNOTES

1 The data for Turkey is from GEPA; Germany's data were taken from https://www.climatedata.eu/

2 More information about the 2023 vision regarding Turkey's renewable energy targets can be found in Melikoglu (2013).

Chapter 5
Lessons From COVID-19 Conferring Environmental Re-Engineering Opportunity

Mir Sayed Shah Danish
https://orcid.org/0000-0002-5196-2280
University of the Ryukyus, Japan

Tomonobu Senjyu
https://orcid.org/0000-0003-4494-6773
University of the Ryukyus, Japan

Najib Rahman Sabory
University of the Ryukyus, Japan

Alexey Mikhaylov
https://orcid.org/0000-0003-2478-0307
Financial University Under the Government of the Russian Federation, Russia

ABSTRACT

This chapter presents re-engineering environmental sustainability opportunities within constrained pandemic situations that suggest solutions. Climate change disaster and COVID-19 socio-economic impacts are leading nations to dramatic tragedy. Simultaneously, increasing demand for energy are due to population, political competition, and industrial growth globally. At present, the pandemic attracts more attention due to its immediate effect. Ignoring climate change can have a worse consequence not only on the present but for the future generation in the long run. Therefore, re-engineering the current pandemic situation with a futurism outlook for saving the world will enable nations to transform and rethink strategies, policies, procedures, processes, and any actions that can cope with present and possible future pandemics and climate change tragedies.

DOI: 10.4018/978-1-7998-9502-2.ch005

INTRODUCTION

Referring to the global coronavirus report released on 20 January 2022, 332,617,707 confirmed COVID-19 cases, including approximately five million counts of deaths, are reported (World Health Organization (WHO), 2022). Climatic changes affect many of the environmental determinants of health, death, and infection. This phenomenon is likely to arise when environmental changes such as water pollution, global warming, air pollution, extreme heat, and natural disasters cause approximately 250,000 deaths per year (Word Vision, 2021). COVID-19 epidemic is still ongoing from the first reported outbreak in 2019 to the world experiencing more than two waves of infections and death cases leading to a health crisis that needed an emergency response and action plan. COVID-19 has been a global pandemic that has created medical emergencies worldwide. When the pandemic overwhelmed the existing health facilities, COVID-19 became a disaster because of its high spread and mortality rate. Covid-19 Pandemic threw into question our level of preparedness when it comes to first aid response that would facilitate holistic recovery and well-being from patients to healthcare workers.

The screening was determined as the most appropriate method to establish the accurate trend and number of confirmed positive COVID-19 cases across countries. However, screening was limited due to the unavailability of requisite testing tools, including swabs for sample collection and laboratory reagents for actual testing of samples. The unavailability of equipment is attributed to the high global demand against a low supply chain that could not match up to the urgent demand in affected countries. Besides, the cost and performance of available diagnostic services are insufficient and inadequate There is a need for rapid testing, tracing possible contacts, and improvement of preparedness for pandemics (Zhu & Wong, 2020).

Dealing with global pandemic on top of the present climate-related crises is a challenge because climate change is known to have created conditions that favor the spread of numerous communicable diseases, including tuberculosis and malaria, which are still a part of eradication target the United Nation's Sustainable Development Goals. Globally, COVID-19 restrictions have threatened the ease of vaccines transportation, medical aid supplies, food supply, and rescue mission to vulnerable populations. In order to maintain hope for life on earth, dismantling toxic systems and climate change need to be treated as an emergency to prevent another infectious pandemic from happening.

Human survival is under threat from global infections, particularly those transmitted from animals to humans as pathogens causing epidemics and pandemics over the years. Contagious zoonoses are particularly dreadful to the public health system. The ravaging viral COVID-19 pandemic is primarily attributed to wild bats. Its impact has been felt across various economic sectors resulting in job losses that have largely destabilized the globe. Other reported issues associated with COVID-19 among the masses include anxiety in response to economic stress, panic due to future uncertainty, and depression, among other psychosocial challenges (Khanna et al., 2020). It is important to note there are other threats to global health security, which are closely related to other dimensions of human activities, such as food production, environmental conservation, biodiversity, water and sanitation, trade, education, technology innovation, and economic growth. All these dimensions are dependent on each other to promote good health embedded in livelihood and human dignity for a healthy population. Pandemics are unpredictable and events likely to trigger such pandemics cannot be easily foreseen. Hence, strategizing how we can prevent such occurrence is key from the lessons we have learned from the COVID-19 pandemic to apply a multifaceted approach. A zoonotic disease is a contagious animal disease that can be transmitted to and infect humans. Most of the transmission occurs due to consumption of wild meat and interaction with the infected animals. The unintended transmission happens with increased human to animal contact when people practice encroaching activities in wild habitats that interfere with the nature paradigms.

The transmission of infectious diseases can lead to a worst-case scenario of a global pandemic. During the pandemic, human movement through the globe while traveling or trading animals was immediately considered a geographical spread of pathogens, increasing the transmission rate of contagious diseases and extending spread across the world. Nowadays, ease of international travel and enhanced trading between countries in animals and animal products, labor mobility, the emergence of urban centers, and growth of towns to cater for farming and thriving zones for pathogens, disease hosts, and vectors that raise chances of transmission to humans. The transmission of pathogens can also be influenced by varied climatic patterns that hasten or limit their spread.

ENVIRONMENTAL RE-ENGINEERING OPPORTUNITY

Sustainability can be introducing the requirements balancing of socioeconomic growth and prosperity with the reservation of resources with an intimate use of these resources in a way that does not compromise the ability of future generations to meet their needs (Danish et al., 2017).

Intentional separation of environmental policy from health is risky since there is evidence of co-dependency between the two. Human healthy living is dependent on climate and the living organisms we co-exist with. Meanwhile, deforestation practiced mostly for agricultural activities causes animal habitat loss worldwide. Such a loss causes migration of wildlife, thus causing them to come into contact with other animals or people and end up sharing pathogens. According to the UN

Environment Programme Organization, changes in climate and global warming are currently interfering with the planet's sources of food, natural resources, and health/ immunity of man. The future is still likely to be threatened with the co-occurrence of climate change and a pandemic. Global climate change mitigation efforts in 2015 under the Paris Climate Summit "to strengthen the global response to the threat of climate change by keeping a global temperature rise this century well below 2 degrees Celsius above pre-industrial levels and to pursue efforts to limit the temperature increase even further to 1.5 degrees Celsius." has demonstrates with no significant outcomes that followed by the 2021 United Nations Climate Change Conference (COP26) (United Nations (UN), 2022).

Therea are many studies on combatting climate change, which the 4E (energy, exergy, environment, and economic) concept can be the best option (Danish et al., 2019c), with considering suitability pillars: technical, technological, institutional etc. (Danish et al., 2020), in aligning with reduction in cost of emissions pollution and leveling cost of energy production.

Human activities affecting climate change and loss of biodiversity have partly driven the risk of global pandemics through their significant influence on environmental changes. Land-use diversity through expansion of agriculture, especially in production and consumption rates, changes nature and adds to the contact between human beings, wildlife, microbes, and livestock, relying on fossil fuel energy production. Maintaining a sustainable and clean environment has been an economical large investment, especially in the energy sector compared to traditional fossil fuels (Danish et al., 2014).

Ecological sustainability ensuring a peaceful and sustainable relation between humankind and nature that retaining biodiversity over time to supplies essential functions for a routine life (Danish et al., 2021c). The exploitation of natural resources by humankind (air, soil, marine, forests, aquatic resources) threatens biodiversity. Other global risks include overgrowth of the population and urbanization, changes in climate and rise in temperature, conversion of the habitat, over-exploitation of natural resources, and environmental degradation and pollution. Meaning there is an urgent need to tackle climate changes, and more should be done to guard the life on earth considering its diversity. This diversity is being lost at a rate that could make extinct half of the life on earth.

COVID-19 pandemic added to the issues such as conflicts, literacy, perception of risks emanating from natural disasters, insecurities, poor infrastructure, and environmental degradation has increased vulnerability. To reduce the direct negative impact on the economy and the associated reduction in global Gross Domestic Product (GDP), investment in preparedness offers a greater emphasis on building capacities and reducing vulnerabilities, allowing communities to reduce risks and recover themselves, thereby reducing the high levels of aid and dependency that is systematic.

Occurrence of the COVID -19 pandemic has also had a profound effect on the environment when it comes to the loss of biodiversity that is an essential stimulator of newly occurring diseases caused by pathogens. The global unemployment rate has increased during the pandemic, with still over 200 million people expected to be unemployed next year (Rume & Islam, 2020). More people had to depend on biodiversity for farming, domesticating animals, exploring land use and wildlife to meet economic challenges. Containment measures and a reduction in revenue from tourism have created wildlife protection challenges. Government and investors should put more effort into protecting biodiversity from risks such as extinction and exploitation that disrupt the ecological processes that keep infections in check. However, an ignorable and temporary impact of COVID-19 on climate change will not be a solution, but by re-engineering natural resources utilization, animal species cut-off prevention, forest and nature protection can contribute to environmental distress. Environmental-friendly projects can be initiated from a small rural to a large scale at urban by re-engineering and resources optimization. Real life examples in action are revision of inadequate policies, communities scale clean energy supply, fishery, aquaculture, and many more (Danish et al., 2016a, 2020, 2021d). Such projects are small in scale, but large in consequence in terms of socioeconomic development at the constrained pandemic situation, and environmental sustainability.

There has been a global shift to prioritize green investment in COVID-19 recovery plans driven by UNEP. The majority of the key causes of changes in climate propagate chances of pandemic emergence (United Nations Environment Programme (UNEP), 2021). UN Environment Programme (UNEP) publishes a Year Annual Book focusing on global problems of environment (2016-2019). In addition, the UNEP published annual Frontier reports identifying emerging issues of environmental concern in the last years. Analytical details of upcoming issues related to environment and innovative policy interventions, emerging solutions, and adapting practices already in existence are provided in the scientific reports. The influence main factors on environmental sustainability, e.g., geographical factors, theater conditions, nature parameters, etc. are characterized in terms of indicators (details are given at the end of chapter) such as (Danish et al., 2021b, 2021c):

- Gross Domestic Product (GDP)
- Human Development Index (HDI)
- Environmental Sustainability Index (ESI)
- Dashboard of Sustainability (DS)
- Well-being Index (WI)
- Ecological Footprint (EF)

- Living Planet Index (LPI)
- Direct Material Consumption (DMC)

In a capitalist world, among all other sustainability pillars, the economic and production performance indicators and variables are focused on in different studies, which are provided in the below table.

Table 1. Indicators and variables of economic and production performances (Danish et al., 2021a)

Indicator		Abbreviation	Formula / Value	Remark
Economic	Gross income	GI	$Gross\ pay$	
	Gross margin	GM	$GM = GR - VC$	VC - Variable costs
	Gross margin index	GMI	$GMI = \frac{GR - VC}{GR} \times 100$	Percentage of contribution over gross income
	Profitability index	PI	$PI = \frac{PV\ of\ CF}{Initial\ investment}$	
	Present value	PV	$PV = \frac{CF}{\left(1+r\right)^{t}}$	r - Discount rate t - Number of years
	Profit	P	$P = GR - TC$	
	Gross revenue	GR	$GR = q \times p$	Revenue that resulted from sales q - Quantity sold p - Selling price
	Total cost	TC	$VC - FC$	FC - Fixed costs
	Average variable cost	AVC	$\frac{Monetary\ unit\left(USD\right)}{kg}$	Ratio between cost and quantity produced
	Equilibrium point	EP	$\frac{FC}{p - AVC}\left(\left(ha.yr\right)^{-1}\right)$	Minimum production level needed to offset total costs, at a given selling price (p)

Continued on following page

Table 1. Continued

Indicator		Abbreviation	Formula / Value	Remark
Production performance	Production cycle	PC	$length\ of\ production\ cycle$	Month
	Apparent feed conversion	FC	$\frac{Feed\ consumed\left(kg\right)}{Weight\ gained\left(kg\right)}$	At a PC
	Average biomass storage	ABS	$\frac{Live\ weight}{m^2}$	
	Survival index	SI	$\frac{Number\ of\ harvested\ fish}{Number\ of\ stocked\ fish} \times 100$	Percentage (%)
	Final average weight	FAW	$\frac{Total\ harvested\ weight}{Number\ of\ harvested\ fish}$	kg

According to (Danish et al., 2019b), criteria for setting sustainability measures in defending indicators are reported as follows:

- Representativeness and interpretation capability
- Simple and easy to interpret
- Scientifically valid
- Able to show trends over time
- Give an early warning, and influence about irreversible trends where possible
- Sensitive to change in the environment, society, or economy
- Based on readily available and adequately documented data
- Capable of being updated at regular intervals
- Have a target level or guideline against which to compare

SOLID WASTE MANAGEMENT

Waste management involves preventing and minimizing waste production. The most paramount medical waste management tread is cutting back on waste creation with conventional segregation of waste sources (Singh et al., 2021). The world's challenging experience with coronavirus transmission has been felt of a broad impact on our socioeconomic systems to health care systems, transport, water, and sanitation infrastructure vital to limit its spread around the globe. The challenges we have experienced have taught us how critical rapid preparation and planning are to prevent another pandemic that will bring severe consequences to public health and the environment (Truelove et al., 2020). Solid and hazardous waste has been on the rise during this pandemic, with decreasing human capacity and the use of infrastructure to enable its management in our communities and health facilities.

Globally, solid waste contributes around 5 percent of global greenhouse gas emissions (Danish et al., 2019c). Global management of waste during the COVID-19 outbreak is crucial for human health. A negative impact on the environment arises from potentially hazardous waste like used gloves, masks and face shields. In the presence of COVID-19 pandemic fighting against environmental pollution is challenging since waste volumes have increased in our public locations and households. This is escalated by mishandling and irresponsible disposal aside from other hazardous waste generated from industrial activities, including processing, agriculture, building and construction, and many more.

Raising awareness, training and education to protect persons in contact with waste should involve handling waste safely, storage and transportation, pathways of pathological infections and associated risks of hazardous waste to prevent further exposure to COVID-19 and other zoonotic diseases. Proximity of people at risk in hazardous medical waste, including those who handle waste, those generating the waste, patients or residents outside the establishment. Education and training to promote adequate waste management should target at risk groups (Ilyas et al., 2020). To properly manage solid waste, a standardized paradigm to enforce standardizations aligning interchangeability and converge diversities in terms of qualitative and quantitative values optimization and economic integration in adopting the ISO 14001 or BS 7750 standards is required (Danish et al., 2019d). Solid waste management best practices shall be applied to fulfill this purpose, aligning with smart energy consumption standards (ISO-37120, ISO-37122, ITU-4901, ITU-4903) within smart cities application (Sabory et al., 2021).

Failing to manage COVID-19 wastes causes an increase in transmission and number of infections, especially in health facilities that do not have a precautious approach or guidelines for dealing with health care waste. The application of waste management guidelines varies globally; the United Nations Environment Programme (UNEP) plays a role in helping low and middle-income states create and implement policies and procedures that are favorable to the environment. UN bodies and associates are insistent that when sound policies are implemented, a considerable amount of toxicity waste disposed of yearly can be reduced to a large extent or even destroyed. Multiplying waste management challenges during the pandemic has created a learning opportunity for future, long-term enhancement, and maintenance of waste collection and disposal systems. Healthcare waste regulation municipalities often come across different challenges that hinder the process of giving key solutions of clinical waste collection and management and similarly counter growth of infectious waste prospects while preserving the lives of staff at risk. Reprocessing waste to an extent conceivable through salvaging reusable objects that are papery, glassy, plastic, and metallic items that are malleable (*Source and Types of Waste*, 2022) and treating waste by using cautious and ecologically thoughtful procedures. Disposing waste scraps into a dumping ground. Landfills are designed to operate with less threat to health subjects and our surroundings (Nathanson, 2020).

In low and middle-income countries with 97% of cities with over 100,000 inhabitants, solid waste management remains a challenge (Danish et al., 2021b). Recycling and composting solid waste will reduce greenhouse emissions, minimize transportation, and collect organic and nonorganic waste. It also enhances workforce immunity by reducing workload and being in contact with others.

Senior Homes-Population of persons considered elderly has increased over the years. The zeal to follow proper waste management guidelines among older people is lacking. In addition, the senior population experiences health issues growing demand for healthcare services and necessities that produce more waste (Brunn, 2019). Senior citizens are more vulnerable to COVID-19 infection than young people, so their health and hygiene through proper waste management are critical to improve their health (Ian Guenn, 2020). Most senior homes frequently offer psychosocial support, safe haven, and medical care; therefore, these spaces generate considerable solid and hazardous waste. Formal alliances with elderly home facilities and waste management companies can provide reliable waste collection and disposal services to maintain health safety for workers and the elderly. Cautious handling of household waste apart from medical waste is pivotal during the COVID-19 emergency. Medical waste should be treated as hazardous waste and not mixed with domestic waste but should be treated as hazardous waste and discarded separately for collection by public municipalities (United Nations (UN), 2020, p. 19). Effective waste management steps such as segregation and secure disposal after treatment during the pandemic are exigency reactions. Schools' formal settings like schools and higher learning institutions present vulnerable settings where COVID-19 will enter. Social distancing and constant hand hygiene is not enough precaution in schools because some schools have safe water shortage, poor sanitation, inferior communal health solutions, and waste management Programme (*Preparedness, Prevention and Control of Coronavirus Disease (COVID-19) for Refugees and Migrants in Non-Camp Settings*, 2020). COVID -19 outbreak had refugee camps put on lockdown that halted solid waste management program. Still, with the strict health regulation, camp employees needed supplies such as facemasks, gloves, hand sanitizers, disinfecting gears, safety boots, reusable water bottles and thermometer guns on site (European Commission, 2022). Landfills are mostly preferred for waste disposal to help maintain safer and healthier environment. Migrants and refugees should be remembered as states tussle with pandemic consequences (Sonja, 2020). Human shelters provide temporary housing services for homeless people and families. Individuals infected with COVID-19 in such households have to be put under investigation (PUI) or under monitoring (PUM) with the available waste management resources that encourage segregation of all medical waste (face masks, wipes, tissues). Since human shelters depend on informal workers who handle waste multiple times, the threat of infection in such surroundings needs to be circumvented.

GOVERNING THE PANDEMIC WITH MINIMUM ADVANTAGES OF CLIMATE CHANGE MITIGATION

Governments have the responsibility to protect their subjects. The commitment involves disaster preparedness aimed at shielding their citizens from disasters. Therefore, local communities are aware of their local surrounding like topography, socioeconomic trends, emergency response activities when they experience disasters. The governments must ensure they have sufficient capacity to manage the COVID-19 pandemic. The governments ought to be committed to developing intervention mechanisms. Besides, community participation and thorough assessment of available facilities boost awareness and handling of disasters. The aim is to ensure a self-reliant community through economic and strategic awareness. This responsibility does not fall under only pandemic situation management. The governments must also take care of citizens' lives, which were threatened long ago due to environmental pollution and climate change. However, it does not have a surge impact in the short-term likewise COVID-19, but its impact will be more severe in the long run. This distress will demonstrate challenges for mankind and impact climate change, ozone depletion, ecosystem, biodiversity, wildlife conservation, natural and anthropogenic disturbances, deforestation, genetically engineering, life expectancy, lifestyle change, urbanizations, etc. (Danish & Senjyu, 2021). We must protect nature and curb the exploitation of wild animals to reduce disease risk. Making a shift to sustainable agriculture production practices and consumption of healthier foods can reduce pressure on forests and wildlife food market operation. Coherent policy reforms are required to push this transformation with more strategic investment in forest conservation.

The COVID-19 pandemic exposed multiple issues in the surveillance and monitoring of health crises. Extensive research by scientists and monitoring of these zoonotic viruses to prevent future pandemics could highlight which regions are most vulnerable to future infectious diseases and warn governments ahead of time. Around a quarter of reported global deaths are linked to infectious diseases, a tragic inclination that correlates with increased exploitation and misuse of natural resources. These cause sharp drops in biodiversity, increasing the risk of pathogen transmission (Carroll, 2021). Pioneering organizations in providing digital solutions in the health care sector should be provided with incentives for further research and development (Min ReadTime, 2021). The COVID-19 crisis reminds the interdependence and interconnection existing among communities. On the other hand, it should be perceived as an opportunity for reevaluation and reassessment to improve understanding of pathogens causing a pandemic, the importance of forestry, health care systems (telemedicine) and health care workers, and biodiversity to respond diligently on time. All these outbreaks experienced over time have been serving as a reminder that human survival depends on the relationship with nature. Therefore, Sustainable Development Goals aimed at supporting the green economy will help detect, prevent, and battle imminent future pandemics paged on the experience attained through the handling of COVID-19 (World Health Organization (WHO), 2021).

PROPOSED SOLUTIONS

The fundamental solid waste management is pre-determined by human, technical and financial resources assigned to curb its impact. There is a substantial difference in the amount of solid and hazardous waste generated by private manufacturing companies (Gakungu et al., 2012). The anticipation of Private establishments is to introduce improved technologies because of abundant funding resources private businesses have. For example, it would be encouraging to see a public-private partnership that seeks to explore the energy potential of reusable waste developed to function as energy conservation to reduce high energy costs. On this subject, many case studies can be counted as a roadmap for proposing waste to energy production in a sustainable manner.

Policy commitment on healthcare waste management needs funds designation consideration relevant at distinctive level of administration. There has to be a commitment when developing a federal approach before executing proposals and delegating responsibility to the health and local environment department. A direction from executive subsidiary intended for topmost capability in utilizing at hand measurers from healthcare setups partnering with auxiliary agencies, the independent, professional, and non-governmental federations are inevitable to guarantee plan execution. Ample health and welfare amenities for laborers at all instances of handling waste (Mishra et al., 2014). In addition, deploying the best practices, tools, and techniques will enable the waste management sector to properly handle this issue. Among these tools and techniques, some of them are listed herein that contribute to administrative and management affairs of well-managing waste collection, recycling, reuse, and so on (Danish et al., 2019d):

- Alternative Generation
- Analogous Estimation Technique
- Analytical Techniques
- Benchmarking
- Bottom-up Estimation Technique
- Change Control
- Context Diagram
- Contingent Response
- Cost Aggregation Technique
- Critical Path Method
- Decomposition
- Dependency Determination Technique
- Diagramming Technique
- Direct and Indirect Observation
- Document Analysis
- Expert Judgment
- Facilitated Workshops
- Facilitation Technique
- Focus Group
- Group decision-making Techniques
- Historical Relationship Method
- Inspection
- Interview
- Leads and Lags Technique
- Make-or-Buy Analysis
- Meetings
- Modeling Technique
- Network Analysis
- Networking
- Parametric Estimation Technique
- Precedence Diagramming Method
- Probability and Impact Matrix
- Prototype
- Quantitative Risk Analysis and Modeling Techniques
- Questionnaires and Surveys
- Reserve Analysis Technique
- Resource Optimization Technique

- Schedule Compression Techniques
- Statistical Sampling
- SWOT Analysis
- Three-point Estimation Technique
- To-complete Performance Index
- Variance Analysis Technique

In the pandemic era, health and other community services works are the center of focus, in which their role in risky environments is more significant. Advanced systems in managing healthcare waste are essential to Sustainable Development Goals, including hygiene, water supply, health security, respectable work, and fiscal growth, climate change, accountability in utilization, and manufacturing commodities. To prioritize global public health, eliminating challenges like false information intended to mislead the public that similarly hampered COVID-19 efforts from the start must be waded through by getting science-based facts before undertaking public health campaigns. Implementing public health measures serves as a temporary solution to slow down infection rates and protect vulnerable persons at greater risk of severe effects before an effective vaccine is discovered. Such measures prevent overwhelming of the available medical facilities. These measures are important in preventing the next pandemic before it starts. Assessment restoration recovery is a post-disaster recovery planning phase of developing strategies to assist communities in rebuilding after a disaster. Communities can engage in several activities post-Covid, such as practicing future capitalism, diversifying the economy, remote work transition, addressing structural problems in public, and private sectors, revamping healthcare systems, which is critical to facilitate human survival.

Shaping environmental sustainability by optimally utilizing available opportunities with the help of viable strategies is a profound concern not only for a particular territory but also globally. This provision follows the policy cycle, which includes a systematic process of setting objectives, program costing, implementation and monitoring measures, guiding well understanding of variables considered by policymakers (Danish et al., 2021b). Key strategies include accessing disaster effects and impacts to determine the value of recovery needs and assigning clear roles and responsibilities of the particular government departments to ensure assessment is undertaken jointly. The government will have to collaborate with different agencies to undertake assessment responsibility. The assessment should have a wealth of data and analysis of losses, damages, and priority recovery cost and action. An appropriate community-based approach in promoting development and controlling disaster risks needs all stakeholders' commitment and active participation. There is a need for coordination in the SDG among all the stakeholders (Ramalingam & Prabh, 2020).

CONCLUSION

Pandemic recovery plans should focus on our natural environment and the implementation of acquired knowledge on climate change and health agendas. It is important to identify and adopt methods to bring both short term and long term to ensure a healthy and reliable future for mankind. This is done through scientific findings supported by evidence-based policymaking promoted by health and environment activists and researchers in media and publishing platforms, fostering communication between researchers, policymakers, the public, and experts.

Global organizations, research and learning bodies availed guidelines on how to manage infectious waste that target countries on different developmental spectrums. The treatment, together with disposal method of waste from COVID-19 medical response equipment and kits should be implemented by adhering to guidelines standards at national level.

Abundance of vaccines after their discovery has not been experienced by low-income countries. According to some research, only high-income countries have enough to sustain immunity of their population against COVID-19 according to some studies. Vaccine donations have been made to low-income countries to support its widespread distribution to persons at risk of infection, especially medical workers and medical waste employees. Vaccination debate topics (consent) have been about whether it should embody free will or be mandatory like made into a law. Still, medical workers and biomedical researchers, waste workers and teachers had to be vaccinated as a prevention measure against COVID-19.

REFERENCES

Brunn, M. (2019, July 3). Waste management and the elderly: A neglected problem? *Rycycling Magazine*. https://www.recycling-magazine.com/2019/07/03/waste-management-and-the-elderly-a-neglected-problem/

Carroll, D. (2021, May 17). Opinion: We need a new global surveillance system to detect and prevent the next pandemic. *Devex*. https://www.devex.com/news/sponsored/opinion-we-need-a-new-global-surveillance-system-to-detect-and-prevent-the-next-pandemic-99900

Danish, M. S. S., Elsayed, M. E. L., Ahmadi, M., Senjyu, T., Karimy, H., & Zaheb, H. (2020). A strategic-integrated approach for sustainable energy deployment. *Energy Reports*, *6*, 40–44. doi:10.1016/j.egyr.2019.11.039

Danish, M. S. S., Ibrahimi, A. M., Yaqobi, M. A., Udagawa, S., Mikhaylov, A., Faisal, N., & Senjyu, T. (2021a). Afghanistan aquaculture and fishery sectors—A foresight outlooks. *Journal of Ecoscience and Plant Revolution*, *2*(1), 17–37. doi:10.37357/1068/jepr.2.1.03

Danish, M. S. S., Matayoshi, H., Howlader, H. O. R., Chakraborty, S., Mandal, P., & Senjyu, T. (2019a). Microgrid Planning and Design: Resilience to Sustainability. *2019 IEEE PES GTD Grand International Conference and Exposition Asia (GTD Asia)*, 253–258. 10.1109/GTDAsia.2019.8716010

Danish, M. S. S., Sabory, N. R., Danish, S. M. S., Ludin, G. A., Yona, A., & Senjyu, T. (2016a). An open-door immature policy for rural electrification: A case study of Afghanistan. *International Journal of Sustainable and Green Energy*, *6*(3), 8–13. doi:10.11648/j.ijrse.s.2017060301.12

Danish, M. S. S., Sabory, N. R., Ershad, A. M., Danish, S. M. S., Ohta, R., Sediqi, M. M., Ahmadi, M., & Senjyu, T. (2017). The least developed countries need for changing the passive trend of renewable energy exploitation to a proactive trend. *International Journal of Energy and Power Engineering*, *5*(6), 215–221. doi:10.11648/j.ijepe.20160506.17

Danish, M. S. S., Sabory, N. R., Ershad, A. M., Danish, S. M. S., Yona, A., & Senjyu, T. (2016b). Sustainable Architecture and Urban Planning trough Exploitation of Renewable Energy. *International Journal of Sustainable and Green Energy*, *6*(3), 1. doi:10.11648/j.ijrse.s.2017060301.11

Danish, M. S. S., Senjyu, T., Ahmadi, M., Ludin, G. A., Ahadi, M. H., Karimy, H., & Khosravy, M. (2021b). A review on energy efficiency for pathetic environmental trends mitigation. *Journal of Sustainability Outreach*, *2*(1), 1–8. doi:10.37357/1068/jso.2.1.01

Danish, M. S. S., Senjyu, T., Faisal, N., Stannikzai, M. Z., Nazari, A. M., & Vargas-Hernández, J. G. (2021c). A review on environmental-friendly energy multidisciplinary exposition from goals to action. *Journal of Environmental Science Revolution*, *2*(1), 1–9. doi:10.37357/1068/jesr.2.1.01

Danish, M. S. S., Senjyu, T., Ibrahimi, A. M., Ahmadi, M., & Howlader, A. M. (2019b). A managed framework for energy-efficient building. *Journal of Building Engineering*, *21*, 120–128. doi:10.1016/j.jobe.2018.10.013

Danish, M. S. S., Senjyu, T., Sabory, N. R., Khosravy, M., Grilli, M. L., Mikhaylov, A., & Majidi, H. (2021d). A forefront framework for sustainable aquaponics modeling and design. *Sustainability*, *13*(16), 9313. doi:10.3390u13169313

Danish, M. S. S., Senjyu, T., Zaheb, H., Sabory, N. R., Ibrahimi, A. M., & Matayoshi, H. (2019c). A novel transdisciplinary paradigm for municipal solid waste to energy. *Journal of Cleaner Production*, *233*, 880–892. doi:10.1016/j.jclepro.2019.05.402

Danish, M. S. S., & Senjyu, T. S. (Eds.). (2021). System of green resilience eco-oriented land uses in urban socio-ecosystems. In Y. Wang (Ed.), Eco-Friendly Energy Processes and Technologies for Achieving Sustainable Development: (pp. 1–23). IGI Global. doi:10.4018/978-1-7998-4915-5

Danish, M. S. S., Yona, A., & Senjyu, T. (2014). Pre-design and life cycle cost analysis of a hybrid power system for rural and remote communities in Afghanistan. *The Journal of Engineering-IET*, *2014*(8), 438–444. doi:10.1049/joe.2014.0172

Danish, M. S. S., Zaheb, H., Sabory, N. R., Karimy, H., Faiq, A. B., Fedayi, H., & Senjyu, T. (2019d). The Road Ahead for Municipal Solid Waste Management in the 21st Century: A Novel-standardized Simulated Paradigm. *IOP Conference Series. Earth and Environmental Science*, *291*(1), 1–5. doi:10.1088/1755-1315/291/1/012009

European Commission. (2022, January 20). *Strict health regulations to ensure workers' protection in Azraq camp*. https://ec.europa.eu/trustfund-syria-region/strict-health-regulations-ensure-workers-protection-azraq-camp_en

Gakungu, N. K., Gitau, A., Njoroge, B. K., & Kimani, M. (2012). Solid waste management in Kenya: A case study of public technical training institutions. *ICASTOR Journal of Engineering*, *5*(3), 127–138.

Ian Guenn, T. (2020, July 6). Waste management in aged care facilities—Taking care of Australia's elderly. *Waster*. https://waster.com.au/waste-management-aged-care-facilities/

Ilyas, S., Srivastava, R. R., & Kim, H. (2020). Disinfection technology and strategies for COVID-19 hospital and bio-medical waste management. *The Science of the Total Environment*, *749*, 141652. doi:10.1016/j.scitotenv.2020.141652 PMID:32822917

Khanna, R. C., Cicinelli, M. V., Gilbert, S. S., Honavar, S. G., & Murthy, G. S. V. (2020). COVID-19 pandemic: Lessons learned and future directions. *Indian Journal of Ophthalmology*, *68*(5), 703–710. doi:10.4103/ijo.IJO_843_20 PMID:32317432

Min ReadTime. (2021, July 21). *Learning from Covid-19 and preparing for the future: Reflections from a collaborative partnership*. The Rockefeller Foundation. https://www.rockefellerfoundation.org/case-study/learning-from-covid-19-and-preparing-for-the-future-reflections-from-a-collaborative-partnership/

Mishra, A. R., Mishra, S. A., & Tiwari, A. V. (2014). Solid waste management—Case study. *International Journal of Research in Advent Technology*, *2*(1), 1–4.

Nathanson, J. A. (2020, November 10). *Solid waste management* [Britannica]. https://www.britannica.com/technology/solid-waste-management

Preparedness, prevention and control of coronavirus disease (COVID-19) for refugees and migrants in non-camp settings. (2020). *World Health Organization (WHO)*. https://apps.who.int/iris/bitstream/handle/10665/331777/WHO-2019-nCoV-Refugees_Migrants-2020.1-eng.pdf?sequence=1&isAllowed=y

Ramalingam, B., & Prabh, J. (2020). Innovation, development and COVID-19: Challenges, opportunities and ways forward. *OECD Policy Responses to Coronavirus (COVID-19)*, 1–14.

RobinsonJ. (2020, April 2). *Pandemics*. WebMD. https://www.webmd.com/cold-and-flu/what-are-epidemics-pandemics-outbreaks

Rume, T., & Islam, S. M. D.-U. (2020). Environmental effects of COVID-19 pandemic and potential strategies of sustainability. *Heliyon*, *6*(9), e04965. doi:10.1016/j.heliyon.2020.e04965 PMID:32964165

Sabory, N. R., Senjyu, T., Danish, M. S. S., Hosham, A., Noorzada, A., Amiri, A. S., & Muhammdi, Z. (2021). Applicable Smart City Strategies to Ensure Energy Efficiency and Renewable Energy Integration in Poor Cities: Kabul Case Study. *Sustainability*, *13*(21), 11984. doi:10.3390u132111984

Singh, N., Ogunseitan, O. A., & Tang, Y. (2021). Medical waste: Current challenges and future opportunities for sustainable management. *Critical Reviews in Environmental Science and Technology*, 1–23. doi:10.1080/10643389.2021.1885325

Sonja, A.-K. (2020). *How can we protect refugees during the COVID-19 crisis? Our world*. United Nations University. https://ourworld.unu.edu/en/how-can-we-protect-refugees-during-the-covid-19-crisis

Source and types of waste. (2022, January 20). *BYJUS*. https://byjus.com/chemistry/waste/

Truelove, S., Abrahim, O., Altare, C., Lauer, S. A., Grantz, K. H., Azman, A. S., & Spiegel, P. (2020). The potential impact of COVID-19 in refugee camps in Bangladesh and beyond: A modeling study. *PLoS Medicine*, *17*(6), e1003144. doi:10.1371/journal.pmed.1003144 PMID:32544156

United Nations Environment Programme (UNEP). (2021, August 9). *Time to get serious about climate change. On a warming planet, no one is safe*. https://www.unep.org/news-and-stories/speech/time-get-serious-about-climate-change-warming-planet-no-one-safe

United Nations (UN). (2020, March 24). *Waste management an essential public service in the fight to beat COVID-19*. UN Environment. https://www.unep.org/news-and-stories/press-release/waste-management-essential-public-service-fight-beat-covid-19

United Nations (UN). (2022, January 22). *COP26: Together for our planet*. https://www.un.org/en/climatechange/cop26

Word Vision. (2021, June 21). *How climate change impacts poverty*. https://www.worldvision.ca/stories/climate-change/how-climate-change-impacts-poverty

World Health Organization (WHO). (2021, September 13). *Lessons learned from the COVID-19 pandemic – achieving resilience for the next health emergency*. https://www.euro.who.int/en/health-topics/health-emergencies/coronavirus-covid-19/news/news/2021/9/lessons-learned-from-the-covid-19-pandemic-achieving-resilience-for-the-next-health-emergency

World Health Organization (WHO). (2022, January 20). *Coronavirus (COVID-19) dashboard*. https://covid19.who.int

Zhu, N., & Wong, P. K. (2020). Advances in viral diagnostic technologies for combating COVID-19 and future pandemics. *SLAS Technology*, *25*(6), 513–521. doi:10.1177/2472630320953798 PMID:32833548

KEY TERMS AND DEFINITIONS

Dashboard of Sustainability (DS): Is software developed by the European Union's Joint Research Centre at Ispra and represents the complicated relationship between environmental, economic, and social issues. Moreover, it provides information in a way palatable to decision-makers and the general public.

Direct Material Consumption (DMC): Measures the total sum of the domestic extraction flows, including imported but excludes exported.

Disaster: Is an unexpected event whose occurrence results in extensive damage, including properties, the environment, animals, and humans inhabiting the affected region. Since disasters occur unexpectedly, there is a need to understand and determine strategies to minimize the effects caused by such disasters.

Ecological Footprint (EF): Is the area of productive land needed to maintain its current consumption ratio while using the prevailing technology to absorb wastes and calculated for a specific population.

Environmental Sustainability Index (ESI): Measures overall progress towards environmental sustainability for 142 countries measured by the World Economic Forum. It provides a more analytical approach to environmental decision-making and allows comparison of the progress among the nations.

Gross Domestic Product (GDP): Gives the total value of goods and services produced in a specific time frame, usually a year. It is a popular indicator in the economic sphere but does not apply in the social sphere.

Hazard: Is the potential for a sudden event to occur and whose occurrence can lead to unforeseen consequences and emergencies.

Human Development Index (HDI): Measures the average achievement in the three primary aspects of human development; knowledge, longevity, and improved living standards. Life expectancy defines longevity, enrollment in schools, and adult literacy provides awareness, and the GDP per capita defines the living standard. The UN development program published it.

Living Planet Index (LPI): Assesses the overall global state of the ecosystem using national and international data on the impact of human activities on the environment.

Mitigation: Is measures taken before a disaster happens to neutralize or decrease its impact on society or the environment.

Pandemic: Is a disease outbreak that spreads across countries or continents that affects more people and takes more lives than an epidemic (Robinson, 2020).

Risk: Is the probability, depending on how high or low a hazard will cause harm. Risk is determined by vulnerable conditions associated with physical surroundings, social setup, environmental factors, or economic factors.

Salience, Credibility, and Legitimacy of GDP and HDI: Salience means that the indicators are useful, applicable, and attractive to the user. Credibility means that the pointers are valid and make scientific sense. Finally, legitimacy touches pointers' perception from the perspectives of users, stakeholders, businesses, trade unions, and environmental non-governmental organizations.

Sustainability in Terms of Energy and Environment: Defines developing affordable and sustainable protocols for various applications of clean and green energy aligned with environmental requirements with minimum greenhouse gas emission over time with contemplating some indicators such as deployment diversity, policy developments, technology costs, and investment in renewable energy (Danish et al., 2016).

Sustainability Pillars: Have counted the parameters to be analyzed to balance the proposed system or solution in accordance with resiliency and sustainability pillars within these criteria: technical sustainability, economic sustainability, institutional sustainability, environmental sustainability, social sustainability, etc. (Danish et al., 2019a).

Well-Being Index (WI): Is called stress index, combines two types of indicators then combines them. The first type has thirty-six pointers for health, population, wealth, freedom, peace, crime, equity, communication, and education. The other type has fifty-one land, water, air, and energy.

Chapter 6
BODIPY Dyes in Solar Energy

Seda Cetindere
Gebze Technical University, Turkey

ABSTRACT

Solar energy shows great potential in the future energy market due to its essentially unlimited supply and renewable, cheap, and clean nature. Among the third-generation solar technologies that are able to harvest light and convert it to electricity are dye-sensitized solar cells (DSSCs), organic solar cells (OSCs), and inorganic-organic hybrid solar cells also known as perovskite solar cells (PSCs). Among the various organic dyes, BODIPY (4,4-difluoro-4-bora-3a,4a-diaza-s-indacene) dyes have been recognized as promising candidates for solar cells due to their intrinsic advantages such as sharp and strong absorption near 500 nm, which is leading to efficient light-harvesting capability, diverse modification on the core structure at all positions and with any desired functionality, long excited state lifetimes, excellent photo-stability, and good solubility in organic solvents. This chapter will focuse on studies of BODIPY dyes in solar energy technology.

INTRODUCTION

World's energy demand is growing fast because of increasing population and technological advancements. It is therefore important to go for reliable, cost effective and everlasting renewable energy source for energy demand arising in future. Solar energy, among other renewable sources of energy, is a promising and freely available energy source for managing long-term issues in energy crisis. Solar energy is the conversion of sunlight into usable energy forms. It is an essential source of renewable energy that has sufficient capacity for the global energy need, and it is the most important that can address the issues of energy problems (Lewis & Nocera,

DOI: 10.4018/978-1-7998-9502-2.ch006

2006). Solar energy is free, clean and abundant in most places throughout the year and is important especially at the time of high fossil fuel costs and degradation of the atmosphere by the use of these fossil fuels. It can be converted by natural and technological processes into other useful forms of energy. Such as the chemical process through photosynthesis or the electric process using photovoltaic (PV) equipment to produce electricity or the thermal process to produce heat or the conversion of solar radiation into mechanical energy. Relatively well-spread over the globe since solar energy comes from the sun and it represents a limitless source of power (Ladomenou et al., 2017). Solar photovoltaics (PV), solar thermal electricity and solar heating and cooling are well established solar technologies. These technologies are depending on how they capture and distribute solar energy or convert it into solar power. The conversion of solar energy into electricity is done through photovoltaic (PV) solar cells. Solar cells represent the building block and main component of PV systems. A solar cell is defined as an electrical device that directly converts the energy of photons into direct current (DC) electricity through a chemical/physical phenomenon called the photovoltaic effect (Shubbak, 2019). Solar PV combines two advantages: module manufacturing can be done in large plants, which allows for economies of scale, and it is also a very modular technology and can be deployed in very small quantities at a time. This allows for a wide range of applications. Systems can be very small, from personal electronics or off-grid applications, up to utility-scale power generation facilities. However, solar radiation that is absorbed but not converted into electricity promotes the temperature increase of the solar cells, thus reducing their conversion efficiency. To counteract this phenomenon, the cells may be cooled by a working fluid (thermal fluid) in order to maintain a high level of conversion efficiency. In this setup, the thermal fluid - water or air - extracts the heat from the cells to be used elsewhere, resulting in a hybrid solar equipment with simultaneous generation of electric energy and thermal energy: this is called a Photovoltaic-Thermal collector or simply PVT. Thus, a PVT hybrid solar collector is a device formed by a PV module with an attached thermal unit on its back, being thus considered a cogeneration equipment. The importance of this equipment is that it can generate energy, electric and thermal simultaneously, with a good level of efficiency and in a smaller area than current thermal collectors and photovoltaic panels operating separately (Ramos et al., 2019). In the 21st century solar energy is expected to become increasingly attractive as a renewable energy source because of its inexhaustible supply and its nonpolluting character, in stark contrast to the finite fossil fuels coal, petroleum, and natural gas. Solar energy could be a best option for the future world because of several reasons: First, solar energy is the most abundant energy source of renewable energy (Panwar et al., 2011). Solar energy reaches the earth in various forms like heat and light. Studies revealed that global energy demand could be fulfilled by using solar energy satisfactorily as it is abundant in nature and freely available source of energy with no cost (Lewis, 2007). Second, it is a promising source of energy in the world because it is not exhaustible, giving solid and increasing output efficiencies than other sources of energy (Nozik, 1978).

It is clear that optimal solar cell performance of a photosensitizer is dependent on a large number of parameters; however, absorption range, anchoring groups and the direction of electronic reorganization on excitation should be among the most important ones (Kolemen et al., 2011). Donor–acceptor π-conjugated (D-π-A) dyes possessing both electron-donating (D) and electron-accepting (A) groups linked by π-conjugated bridges are expected to be one of the most promising classes of organic dyes due to their effective photo-induced intramolecular charge transfer. (Nazeeruddin et al., 2001; Gao et al., 2008; Cao et al., 2009, Chen et al., 2006). Boron-dipyrromethene (BODIPY) (Figure 1) is one class of organic dyes that has been used widely in bioimaging, sensing, and logic gate design (Klfout et al., 2017). BODIPY dyes possess intense absorption profiles that can be exploited in various light harvesting applications. However, redox stability and optimization of frontier molecular orbital energies in these dyes are critical for their successful incorporation into new solar cell materials (Hussein et al., 2018). BODIPY dyes have certain unique features that could make them highly advantageous compared to most other organic dyes, and thus, they are highly promising in this regard. (i) BODIPY dyes have high extinction coefficients (70000-80000 $M^{-1}cm^{-1}$) and can easily be modified with any desired functionalities. (ii) Absorption peak can be moved to longer wavelengths through simple modifications, keeping strong absorption cross sections. (iii) BODIPY dyes have inherent asymmetry in charge redistribution when they undergo S_0-S_1 transition upon excitation, increasing the charge density on the meso-carbon (C-8), while decreasing it in most other positions in the boradiazaindacene system. This inherent directionality of charge redistribution pinpoints C-8 as the optimal position of charge injection. (iv) The directionality observed in excitation can be further enhanced with strategically placed electronwithdrawing and electron-donating groups, and on the basis of earlier work, cyanoacrylic acid and 4-N,N¢-diphenylaminophenyl groups are to be of tremendous utility in this regard (Hagberg et al., 2006; Erten-Ela et al., 2008;). Most BODIPY dyes for solar application have a methyl group in the C3 and C5 position due to the availability of pyrrole on the market. Owing to these unique features, BODIPY dyes are among the most popular class of functional dyes which find applications in solar energy conversion (Qin et al., 2014; Mao et al., 2012; Kolemen et al., 2014).

Figure 1. Chemical structure and numbering of BODIPY core.

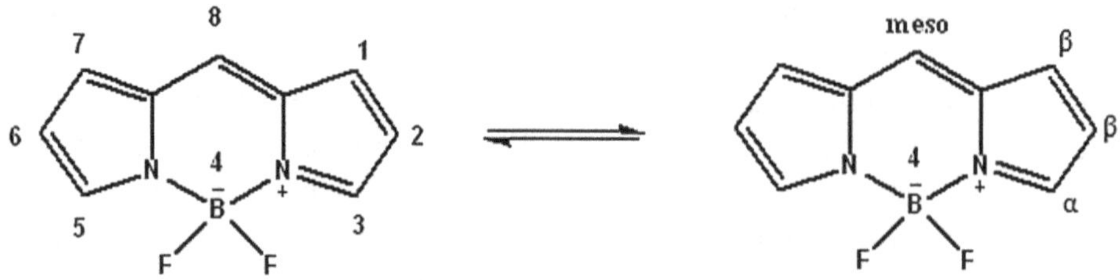

Recently, BODIPY dyes have been explored for solar cells such as dye-sensitized solar cells (DSSCs), organic solar cells (OSCs) and inorganic-organic hybrid solar cells also known as perovskite solar cells (PSCs). These studies demonstrate their potential as light absorbers for the conversion of solar energy to electricity. However, their photovoltaic performance is inferior to many other dyes, including porphyrin dyes.

DYE-SENSITIZED SOLAR CELLS (DSSCS)

Dye-sensitized solar cells (DSCs) are lightweight devices made from inexpensive materials and can be adapted for a variety of indoor and outdoor applications with minimal environmental impact (Vlachopoulos et al., 1988; O'Regan & Gratzel, 1991; Hagfeldt & Graetzel, 1995; Kay & Gratzel, 1996; Kalyanasundaram & Gratzel, 1998; Hagfeldt & Gratzel, 2000). In these types of devices the dye plays an important role whenever it meets the following requirements: (a) radiation absorption in the visible–ultraviolet (UV–vis) and nearinfrared (NIR) region, (b) large molar absorption coefficients, (c) energy of its lowest unoccupied molecular orbital (LUMO) above the conduction band edge of the semiconductor, with electrons located near the molecular group of the dye anchored to the semiconductor for an efficient electron injection, (d)energy of its highest occupied molecular orbital (HOMO) below the redox mediator for an efficient electron reduction in the regeneration of the dye, and (e) absence of aggregation on the semiconductor surface (Borges-Martínez et al., 2019). As such, they are a promising technology for the cost-effective conversion of solar energy to electricity (Shrestha et al., 2012). DSSC made of four primary parts, mesoporous layer of oxide having nanostructured particles embedded, photoelectrode (PE); the substrate of glass, having Transparent Conducting Oxide (TCO) layer which is transparent and act as conducting layer, TCO is used as counter electrode (CE) covered with platinum, photosensitive dye and the solution of electrolyte (Iqbal et al., 2019). DSSCs based on dye photosensitizers (organic dyes or metal complex dyes) are of considerable practical interest as third-generation photovoltaic cells (Mao et al., 2017). To develop high-performance DSSCs, it is essential to create efficient dye sensitizers. Generally, organic sensitizers can be divided into two mainstreams: metal complexes (ruthenium polypyridyl, zinc porphyrin) and metal free dyes. Among the various metal free organic dyes, BODIPYs have been recognized as promising candidates for DSSCs due to the following intrinsic advantages: i) sharp and strong absorption near 500 nm (4-10 x 10^4 M^{-1} cm^{-1}), leading to efficient light-harvesting capability; ii) diverse modification on the core structure, at all positions and with any desired functionality; iii) long excited state lifetimes, excellent photo-stability and good solubility in organic solvents (Loudet & Burgess, 2007; Ulrich et al., 2008; Kolemen et al., 2010; Wang et al., 2012; Mao et al., 2014; Mao et al., 2015, Graf et al., 2013; Kubo et al., 2014; Cakmak et al., 2015; Lin et al., 2015; Mao & Song, 2016; Kaneza et al., 2016; Summers et al, 2016; Cheema et al., 2016; Liu et al., 2016; Zhang et al., 2016; Lu et al., 2016; Watson et al., 2017; Hattori et al., 2005). Therefore, design and development of effective photosensitizers are currently an important topic in DSSCs.

Due to the high absorption coefficients and versatile structural modification capability, BODIPY dyes have attracted attention in recent years for DSSCs. The first BODIPY photosensitizer for DSSCs was reported by Fukuzumi and co-workers (Kumaresan et al., 2009). They modified the BODIPY molecule as a donor–acceptor dyad by introducing a 2,4,5-trimethoxyphenyl group on C-8 as an electron donor, with the BODIPY core itself acting as an electron acceptor. On photoexcitation, this dyad exhibited electron transfer from the donor to the acceptor moiety, which was observed by femtosecond laser flash photolysis measurements. The lifetime of the charge separated state of the dyad was found to be 59 ps at 298 K. Motivated by the results, they extended their study of BODIPY dyes as photosensitizers for DSSCs. Akkaya and co-workers synthesized a BODIPY dye with cyanoacrylic acid and 4-(diphenylamino) phenyl groups as electron-withdrawing/anchoring and electron-donating groups (Erten-Ela et al., 2008). This is the largest out of the reported organic dyes with monochromatic incident photon-to-current conversion efficiency (IPCE) in the near-IR region. In another study, they reported a library of distyryl BODIPY chromophores, optimising their parameters with various structural modifications (Kolemen et al, 2011). All the sensitizers exhibited slight broader absorption in the 650–760 nm region. The DSSC performance parameters of all sensitizers were observed under similar conditions. Also, Kolemen et al. designed BODIPY sensitizers for solid-state DSSCs with use of 2,2',7,7'-tetrakis[di-(p-methoxyphenyl)amino]-9,9'-spiro-bifluorene (spiro-OMeTAD) as hole transport material (Kolemen et al., 2010). The presence of methoxy groups on the donor moiety might result in enhanced excited-state charge transfer, with the carboxylic acid group acting as an anchoring/electron-withdrawing group. Thummel and co-workers designed some BODIPY sensitizers for DSSCs, by exchanging fluoride on boron with side chains attached through ethynyl functions; these not only enhance the solubilities of the dyes in organic solvents but also prevent aggregation of the dyes by virtue of steric bulk around the boron. Carboxylic acid groups serve as anchors, placed at the pseudocentral meso position on all three sensitizers (Warnan et al., 2011). They pointed that it is not necessary to substitute donor–acceptor groups on the dye for it to function as a sensitizer, as observed in earlier systems. Thummel's group also mentioned a few requirements for a dye to be an efficient sensitizer; they include (i) matching of excited state redox potential with the energy level of the TiO_2 conduction band, (ii) directed electron flow from the dye to the surface, and (iii) a potential anchoring group. Odobel and co-workers synthesized trichromophoric supramolecular sensitizer T, based on the antenna effect, for DSSCs (Geiger et al., 2009). This trichromoric dye consists of BODIPY, zinc porphyrin (ZnP) and squaraine moieties linked covalently. In this trichromophore BODIPY and ZnP function as light harvesters, whereas the squaraine unit acts as sensitizer for electron injection into TiO_2. The squaraine unit was chosen as sensitizer because it exhibits a few characteristic properties, such as high molar extinction coefficients and low lying excited state, that favour it both as final energy acceptor and as electron injector. Ooyama et al. designed a D-π-A-type BODIPY based sensitizer containing a pyridine system as an anchoring group for DSSCs (Ooyama et al., 2013). The dye was synthesized with carbazole-diphenylamine as electron donor and pyridine units as electron-withdrawing/anchoring groups. Thiophene spacers at the C-3, C-5 and C-8 positions of the BODIPY core were introduced to red-shift the absorption spectrum. The synthesized dye had good light-harvesting efficiency along with adsorption ability on TiO_2. It showed a good IPCE of 10% over the 500–700 nm range, with an onset at 800 nm. A series of donor-acceptor- π -linker-acceptor (D-A-π-A) featured organic dyes containing BODIPY derivative as the auxiliary acceptor, furan unit as the π conjugated spacer, and 2-cyanoacrylic acid as the anchor group were synthesized and applied in DSSCs by Mao et al. (Mao et al., 2017). The incorporated electron-withdrawing unit of BODIPY enhances light harvesting by decreasing the molecular energy gap and red-shifting absorption spectra. For comparison, three different arylamine chromophores namely 3-methoxy-9H-carbazole (CBZ-B), triphenylamine (TPA-B) and phenoxazine (POZ-B) were separately appended onto the 6-position of BODIPY unit to study the effect of the electron donating groups on device performance. The UV-vis absorption spectra, electrochemical properties, density functional theory calculation, photovoltaic properties and electrochemical impedance measurements of DSSCs with these three dyes were systematically investigated. They reported that their research demonstrated that BODIPY based D-A-π-A molecular architecture is a highly promising class for the improvement of the performance of DSSCs in the future.

According to these reports, there are two BODIPY strategies that have been used to design BODIPY dyes for DSSC applications (Figure 2). These strategies are based on the position of the anchoring group. The first one is the vertical design, in which an anchoring group such as benzoic acid or phenylcyanoacetic acid group is located on the meso-position of the BODIPY core structure. In this design, the phenyl or other anchoring units align almost vertically to the BODIPY core. The second strategy is the horizontal design, in which an anchoring group such as cyanoacetic acid group is linked to the BODIPY core through its C2 position. The donor group such as a triphenylamine can also link the BODIPY core through π-conjugation, forming a premium donor–π–acceptor system. In this design, the anchoring group generally falls into the BODIPY core plane.

Figure 2. Two design strategies of BODIPY dyes for DSSCs. D stands for an electron donor group, and A stands for an electron acceptor group.

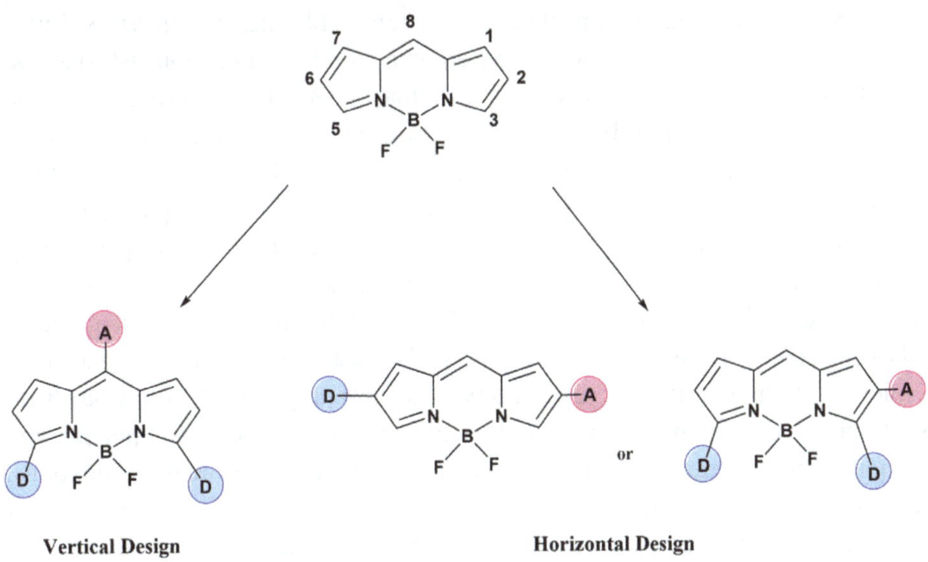

To make BODIPY dyes more promising for DSSCs, the following shortcomings need to be overcome. The first is the narrow light absorption capability, which is characteristic of BODIPY dyes. Generally, it is ideal for a dye to exhibit absorption in a broader range in order to capture more photons. The second is insufficient driving force for electron injection. Even though the LUMO energy levels of BODIPY dyes are usually higher than the conduction band of TiO_2, it may not be high enough to facilitate efficient charge separation at the dye–TiO_2 interface. The third is the significant aggregation on the surface of TiO_2 nanoparticles. BODIPY dyes with a small and flat core structure have a great tendency to form dimers, as demonstrated by absorption spectra of dyes on the TiO_2 surface and numerous single crystal structures of BODIPY dyes. A general strategy to overcome this problem is to introduce bulky groups, including long alkyl chains. Such groups can be introduced to the meso-position or some groups in the C2 and C6 positions; however, introducing groups to the B atom could be more effective (Klfout et al., 2017).

ORGANIC SOLAR CELLS (OSCs)

Organic solar cells (OSCs) are a type of photovoltaics that deals with conductive organic polymers or small organic molecules, for light absorption and charge transport to produce electricity from sunlight by the PV effect (Ameri et al., 2009). These are exhibit advantages such as low–cost fabrication, lightweight, color-tunable feature, manufacturability and the ability to manufacture large area devices, which makes them promising candidates for next generation renewable energy convertors (Liu et al., 2016; Chen & Cao, 2009). OSCs based on a bulk heterojunction (BHJ) active layer that typically consists of a blend of a photoactive polymer or a small molecule as an electron donor and fullerene derivatives as an electron acceptor. (Cheng et al., 2009; Li et al., 2012; Li, 2012; Tong et al., 2010). High-performance OSCs are usually made on the basis of mixed donor (polymer) and acceptor (fullerene and nonfullerene) materials with the formation of bi-continuous interpenetrating networks with large inter-phase regions that provide effective dissociation of excitons. Regarding the donor materials for OSCs, extensive research has centered on polymeric donors. Though polymers represent the most important class of donor-type semiconducting materials, there exist several disadvantages, such as low batch-to-batch reproducibility, polydispersity, difficult synthetic protocols and purification methods. These obstacles lead to multiple processing and performance limitations. In contrast, small molecules (SMs) possess a specific molecular weight, distinct molecular structure and high purity without batch-to-batch variation. Absorption profiles and orbital energy levels of small molecules can be fine-tuned with suitable structural modifications so as to match the device characteristics to a specific given acceptor, which is highly difficult in polymer solar cells (PSCs). Moreover, the reproducible fabrication protocols of SM-based OSCs provide a better understanding of molecular structure–property relationships and facile commercialization (Coffin et al., 2009; Kan et al., 2015). Power conversion efficiencies (PCEs) of over 10% have been achieved by various solution-processed SMs, (Liu et al., 2013; Liu et al., 2017; You et al., 2013) matching those of polymer-based OSCs (Chen et al., 2015; Knag et al., 2015; Hagfeldt et al., 2010).

The unique properties of BODIPY dyes, such as facile synthesis, strong absorption in the UV-Vis region with high absorption coefficients, excellent photochemical and thermal stability, and appropriate energy levels, make them potential candidates for OSCs. In the solar spectrum ca. 50% of solar photons is located in the wavelength of 600-1000 nm and the highest photon flux of the solar spectrum is distributed around 600-800 nm (Bessette & Hanan, 2014). Therefore, designing narrow-band gap (<1.6 eV, or with onset absorption beyond 800 nm) small-molecule donors are required in the field of SM-OSCs. The relatively high extinction coefficient of the BODIPY core combined with deep HOMO energy levels and propensity to π-stack in solid-state, make BODIPY-based materials attractive candidates for organic solar cells (Rousseau et al., 2019). Rousseau et al. reported the first examples of BODIPY donors involving two styryl units along with PCBM as the electron acceptor for solution-processed BHJ-OSCs (Rousseau et al., 2009). They reported the first solution-processed BHJ-OSC consist of BODIPY donors and [6,6]- phenyl C_{61}-butyric acid methyl ester (PCBM) as electron acceptor (Liaoa et al., 2018). A novel donor-acceptor-donor (D-A-D) type chromophore, named BDP-dBDT, with a boron dipyrromethene (BODIPY) linked through alkynyl with two benzo [1,2-b:4,5-b'] dithiophene (BDT) terminal donors was designed and synthesized for solution-processed small molecule bulk heterojunction (BHJ) solar cell (Sharma et al., 2015). The thermal and optochemical properties, BHJ morphology and solar cell performance were investigated. The new molecule shows panchromatic absorption with narrow optical band gap and high molar extinction coefficient. Moreover, the molecule displays deep highest occupied molecular orbital (HOMO) and lowest unoccupied molecular orbital (LUMO) energy levels. The BHJ morphology was systematically optimized by additives, thermal annealing (TA) and solvent vapor annealing (SVA). The BDPdBDT: $PC_{71}BM$ device exhibits a PCE of 5.61%, which is one of the best values among the ever reported BODIPY based organic solar cells (Sharma et al., 2015). Bulk heterojunction (BHJ) solution processed organic solar cells have been prepared by Bucher et al. with using a mono(carboxy)porphyrin-triazine-$(BODIPY)_2$ triad $(PorCOOH)(BDP)_2$ as a donor and ([6,6]-phenyl C_{71} butyric acid methyl ester) ($PC_{71}BM$) as an acceptor (Bucher et al. 2018). This donor–acceptor system aims to increase the light capturing process efficiency of the device. The solution processed BHJ-OSC with an optimized weight ratio of 1:1 $(PorCOOH)(BDP)_2$:$PC_{71}BM$ processed with THF as a solvent showed an overall PCE of 3.48%. When the active layer of the solar cell was processed using a mixture of 4% v/v of pyridine in THF solvent, it achieved a PCE value of 5.29%. Bucher et al. have designed and synthesized two novel SMs consisting of a BODIPY central core surrounded by two diketopyrrolopyrrole (DPP) and two porphyrin units. Both SMs have similar chemical structures except that the porphyrin units are linked to the BODIPY unit through phenyl (BD-pPor) and thiophene (BD-tPor) (Li et al., 2018). They reported the optical properties of these compounds and they showed that efficient energy transfer processes were taking place within these materials from both porphyrins and DPPs toward the central BODIPY unit, and electrochemical studies further confirmed that the two SMs were suitable as donors along with a $PC_{71}BM$ acceptor for the fabrication of solution processed bulk heterojunction OSCs. The OSCs based on BD-pPor and BD-tPor exhibited overall PCEs of 6.67% and 8.98% with an energy loss of 0.63 eV and 0.50 eV. To the best of our knowledge, these results are among the highest efficiencies recorded for donor materials based on one or a combination of two of these dyes (i.e., BODIPY, DPP and porphyrin). Three furan-fused BODIPYs were synthesized with perfluorinated methyl, ethyl and n-propyl groups on the meso-carbon by Li et al. (Li et al., 2018). They were obtained with high yields by reacting the furan-fused 2-carboxylpyrrole in corresponding perfluorinated acid and anhydride. With the increase in perfluorinated alkyl chain length, the molecular packing in the single crystal is influenced, showing increasing stacking distance and decreasing slope angle. All the BODIPYs were characterized as intense absorbers in near infrared region in solid state, peaking at near 800 nm with absorption coefficient of over $2.8x10^5$ cm^{-1}. Facilitated by high thermal stability, the furan-fused BODIPYs were employed in vacuum-deposited organic solar cells as electron donors. All devices exhibit PCE over 6.0% with the external quantum efficiency (EQE) maximum reaching 70% at near 790 nm. The chemical modification of the BODIPY donors have certain influence on the active layer morphology, and the highest PCE of 6.4% was obtained. Thumuganti and co-workers reported two efficient donor materials for OCSs, namely Si-BDP and Ge-BDP, composed by the novel union of thienyl Bodipy wings and a dithienosilole/dithienogermole (DTS/DTG) core (Thumuganti et al., 2019). These novel donor molecules exhibited excellent solubility in organic solvents, extended p-conjugation and low-lying energy levels that matched with fullerene acceptors, supporting their application as donors in OSCs. Under optimized conditions and using $PC_{70}BM$ as an acceptor, desirable PCEs of 4.58% and 4.12% were observed for Si-BDP and Ge-BDP, respectively. Liu et al. reported an efficient A^1-C≡C-A^2 -C≡C-A^1 type SM-BODIPY acceptor (A^1 =BODIPY and A^2 =diketopyrrolopyrrole (DPP)) by following the A-to-A excited electron delocalization via the BODIPY meso-position, the inherent directionality for the excited electron delocalization (Liu et al. 2017). It has a dual role as wide and narrow band gap organic acceptor due to its broad and intense absorption in the near ultraviolet to near infrared wavelength region. The unprecedented broad and intense absorption is achieved through the excited electron delocalization between BODIPY (A^1) and DPP (A^2), which is across the BODIPY's meso-carbon, the ideal position for the inherent excited electron delocalization of BODIPY derivatives. The employment of the ethylnyl bridge between the BODIPY and DPP and their close LUMO-to-LUMO energies both facilitate the LUMO delocalization across over the entire BODIPY-DPP-BODIPY backbone. Preliminary PV tests with the ternary approach give 3% efficiency. They reported that their work opens a new horizon toward dual-role small molecule acceptor with not only decent HOMO/LUMO energy levels, but broad and intensive light absorption. Kraner et al. reported three aza-BODIPY dyes absorbing in the infrared (Kraner et al., 2015). The addition of side groups leads to a red shift of the optical gap from 802 to 818 nm. In optimized devices using these donors in a bulk heterojunction with C_{60}, they observed a higher charge carrier mobility and a higher power conversion efficiency for the molecules without a methyl or methoxy side group lowering the molecular reorganization energy. Surprisingly, the donor–acceptor blend with the lowest energy loss during the electron transfer to the C_{60} yields the highest short circuit current. With increasing size of the attached side chain, the devices exhibit a larger trap density, measured by impedance spectroscopy. Based on the investigation of different blend ratios, we conclude that these traps are mainly present in the donor phase. Jadhav et al. reported three D–A type carbazole substituted BODIPY (carbazole connected to the meso position of BODIPY) small molecules as donors along with $PC_{71}BM$ as an electron acceptor for the fabrication of solution processed bulk heterojunction organic solar cells (Jadhav et al., 2015). The devices based on the as cast active layer showed PCE in the range of 2.20–2.70%, with high open circuit voltage (Voc) in the range of 0.94–1.08 V. The high Voc is related to the deeper HOMO energy level of these small molecules. Zhang et al. have designed star shaped carbazole–BODIPY derivates and used them as donors in BHJ organic solar cells and reported a moderate PCE of 2.70% with a high V_{oc} of 0.85 V (Zhang et al., 2015) and Liu et al. have reported a PCE of 3.13% for BODIPY based solution processed solar cells by dimerisation via the meso position (Liu et al., 2014). Mishra et al. reported a study that introduced corrole derivatives as a new and promising donor platform for BHJ solar cells (Mishra et al., 2018). The Cor-BODIPY derivative exhibited good photovoltaic performance when used as a donor, employing $PC_{71}BM$ as an acceptor in solution-processed small-molecule OSCs. After the optimization of the donor-to-acceptor weight ratio (1:1.5) and solvent vapor annealing time (40 s), the OSC showed an overall PCE of 6.60%, which is significantly higher than that of the as-cast active layer. They suggested that corroles can be promising motifs in OPV technology at low cost and will open up new avenues in the direction of developing highly efficient solar cells. Two novel triads based on a diketopyrrolopyrrole (DPP) central core and two BODIPY units attached by thiophene rings that having high molar extinction coefficients was reported (Cortizo-Lacalle et al., 2014). These triads were characterised and used as donor materials in small molecule, solution processable OCSs. Both triads were blended with $PC_{71}BM$ as an acceptor in different ratios by wt % and their PV properties were studied. For both the triads a modest PV performance was observed, having an efficiency of 0.65%. Four novel symmetrical donor-acceptor-donor (D-A-D) type BODIPY based dyes, namely BDP1-4, featuring an electron-withdrawing BODIPY motif as the central core, and several types of electron-donating segments, including thiophene, bithiophene, fluorene and carbazole, flanked at its 2,6-positions as terminal groups were synthesized and characterized by Liao et al. (Liao et al. 2016). All of these BDPs exhibit panchromatic absorption covering the wavelength range 300-900 nm with high extinction coefficients (0.78 x 10^5 to 1.22 x 10^5 cm^{-1} M^{-1}) and relatively lower HOMO energy levels from -5.17 to -5.48 eV. Furthermore, BDP1-4 show good thermal stabilities, excellent film-forming abilities and low fluorescent quantum yields. BDP2 flanking with bithiophene units demonstrates superior charge transport property and favorable nanoscale morphology relative to BDP1, BDP3 and BDP4. PCE values of the OSCs based on the BDPs/PC_{61}BM (1:0.5, w/w) are 1.49% for BDP1, 2.15% for BDP2, 1.06% for BDP3 and 1.56% for BDP4, respectively. They reported that these results show that these series of BODIPY based dyes are potential good candidates for the efficient organic solar cell materials.

PEROVSKITE SOLAR CELLS (PSCs)

Perovskite solar cells (PSCs) are solar cells which includes a perovskite structured compound, most commonly a hybrid organic-inorganic lead or tin halide-based material, as the light-harvesting active layer. The perovskite materials are considered as one of the eminent materials for new generation PV technology because of their unique properties, such as high electron mobility (800 cm^2/Vs), high carrier diffusion length (exceeding 1 μm), ambipolar charge transport behaviour, high absorption coefficient (greater than 10^5 cm^{-1})due to s-p antibonding coupling, low exciton binding energy (less than 10 meV), high photoluminesce (PL) quantum efficiency (as high as 70%), high carrier lifetime (exceeding 300 ns), optimum bandgap, low surface recombination velocity, tunable bandgap, great structural defect tolerance, and amiable grain boundary effect (Sahoo et al., 2018; Valverde-Chávez et al., 2015; Miyata et al., 2015). In addition, flexibility, lightweight, and semitransparency are some of the important properties of perovskites (Roy et al., 2020). A perovskite material incorporated in a solar cell can serve as both, an absorber layer and an efficient charge transport layer (Mei et al., 2014). These materials are gaining huge interest among the researchers because of their brilliant PV performance, low-cost raw material, and requirement of easy processing conditions.

A PSC generally consists of: a transparent conductive oxide, a compact TiO_2 layer, a mesoporous scaffold of TiO_2, an organohalide lead halide perovskite and hole-transporting material (HTM), and a vi) metal cathode (Ag or Au). PSCs are the most emerging area of research among different new generation PV technologies due to its super PCE (Shi & Ahalapitiya, 2018). PSCs are the future of the PV technology as they are capable of generating power with performance being comparable with the leading silicon solar cells (SSCs), with the cost being lower than SSCs. The enormous potential of PSCs is evident from the fact that the efficiency of these cells has risen from 3.8% to 25.2% up to date. It is a very attractive option for commercial applications because this type of cells is very cheap during the scale-up process. PSC has emerged as one of the most standout cell in terms of efficiency. Therefore the researchers have shown tremendous interest in PSC. The first PCS device reported by Kim et al. (Kim et al., 2012). They used a solid-state HTL (spiro-OMeTAD) with remarkable efficiency (PCE=9.7%). After that Li et al. reported a PSC device with a solid-state HTL based on thiophene cores with efficiency of 15.4%. (Li et al., 2014). Bi and co-workers used 2,2,7,7-tetrakis(N,N-di-p-methoxyphenyl-amine)-9,9-spirobifluorene (Spiro-OMeTAD) as a representative HTM, PSC device stability and performance could be improved by the use of poly(methyl methacrylate) or α-bis-[6,6]-phenyl-C_{61}-butyric acid methyl ester (PCBM) as a template (Bi et al., 2016). Kyeong and co-workers developed novel conjugate polymers of benzo[1,2-b:4,5-b¢]-dithiophene (BDT) and 4,4-difluoro-4-bora-3a,4a-diaza-s-indacene (BODIPY) for use as HTMs without dopants (Kyeong, et al., 2018). They reported that the pBDT−BODIPY polymer allows individual "dialing" of HOMO or LUMO levels with small modifications to the molecular structure, enabling study of the impact of the frontier molecular orbital on PSC performance. Different alkyl chains on BDT can minutely adjust the HOMO level, and mesosubstituents on BODIPYs can selectively set the LUMO level of the resulting polymers. Application of BODIPY-containing polymer into the perovskite solar cell as an HTM leads to a high PCE value (16.02%) and exceptional solar cell stability shown by the fact that over 80% of its original PCE value was maintained after 10 days under ambient air conditions. In 2019, the syntheses and characterization of four new triphenylamine-BODIPY (TPA-BODIPY) dyads as HTLs are reported (Ortiz, 2019). The photophysical and electrochemical properties of these dyads were investigated and compared with 4,4¢-dimethoxytriphenylamine-BODIPY reference dyads. It was shown that the TPA-BODIPY derivatives strongly absorb visible light and upon photoexcitation at the BODIPY unit and these dyads undergo photoinduced electron transfer to form the corresponding charge-separated species, which was confirmed by solvatochromic measures of fluorescence. It was suggested that their redox potentials and photophysical properties makes them promising candidates to use as HTLs.

CONCLUSION

In this chapter, the third-generation solar technologies like DSSC, OSC and PSC were mentioned and the use and importance of BODIPY dyes in DSSC, OSC and PSC devices were discussed. According to the reported studies it was shown that BODIPY dyes have great potentials as light absorbers for the conversion of solar energy to electricity. Also, BODIPY based dyes are good candidates for the efficient organic solar cell materials. Solar PV technology is a promising energy source for a sustainable future and BODIPYs are an important part of this.

REFERENCES

Ameri, T., Dennler, G., Lungenschmied, C., & Brabec, C. J. (2009). Organic tandem solar cells: A review. *Energy & Environmental Science*, *2*(4), 347–363. doi:10.1039/b817952b

Bessette, A., & Hanan, G. S. (2014). Design, synthesis and photophysical studies of dipyrromethene-based materials: Insights into their applications in organic photovoltaic devices. *Chemical Society Reviews*, *43*(10), 3342–3405. doi:10.1039/C3CS60411J PMID:24577078

Bi, D., Yi, C., Luo, J., Décoppet, J. D., Zhang, F., Zakeeruddin, S. M., Li, X., Hagfeldt, A., & Grätzel, M. (2016). Polymer-templated nucleation and crystal growth of perovskite films for solar cells with efficiency greater than 21%. *Nature Energy*, *1*(10), 16142. doi:10.1038/nenergy.2016.142

Borges-Martínez, M., Alvarez, D., Montenegro-Pohlhammer, N., Menendez, M. I., Lopez, R., & Cardenas-Jiron, G. (2019). Assessment of bodipy−oxasmaragdyrin dyads for dye-sensitized solar cells: Aromaticity, photosensitization capability, and charge transport. *The Journal of Physical Chemistry C*, *123*(32), 19362–19375. doi:10.1021/acs.jpcc.9b05136

Bucher, L., Desbois, N., Koukaras, E. N., Devillers, C. H., Biswas, S., Sharma, G. D., & Gros, C. P. (2018). BODIPY–diketopyrrolopyrrole–porphyrin conjugate small molecules for use in bulk heterojunction solar cells. *Journal of Materials Chemistry. A, Materials for Energy and Sustainability*, *6*(18), 8449–8461. doi:10.1039/C8TA01291A

Cakmak, Y., Kolemen, S., Buyuktemiz, M., Dede, Y., & Erten-Ela, S. (2015). Synthesis and dye sensitized solar cell applications of bodipy derivatives with bisdimethylfluorenyl amine donor groups. *New Journal of Chemistry, 39*(5), 4086–4092. doi:10.1039/C4NJ02393E

Cao, Y., Bai, Y., Yu, Q., Cheng, Y., Liu, S., Shi, D., Gao, F., & Wang, P. (2009). Dye-sensitized solar cells with a high absorptivity ruthenium sensitizer featuring a 2-(hexylthio) thiophene conjugated bipyridine. *The Journal of Physical Chemistry C, 113*(15), 6290–6297. doi:10.1021/jp9006872

Cheema, H., Younts, R., Gautam, B., Gundogdu, K., & El-Shafei, A. (2016). Design and synthesis of BODIY sensitizers with long alkyl chains tethered to N-carbazole and their application for dye sensitized solar cells. *Materials Chemistry and Physics, 184*, 57–63. doi:10.1016/j.matchemphys.2016.09.024

Chen, C. Y., Wu, S. J., Wu, C. G., Chen, J. G., & Ho, K. C. (2006). A ruthenium complex with superhigh light-harvesting capacity for dye-sensitized solar cells. *Angewandte Chemie International Edition, 45*(35), 5822–5825. doi:10.1002/anie.200601463 PMID:16865765

Chen, J., & Cao, Y. (2009). Development of novel conjugated donor polymers for high-efficiency bulk-heterojunction photovoltaic devices. *Accounts of Chemical Research, 42*(11), 1709–1718. doi:10.1021/ar900061z PMID:19572607

Chen, J. D., Cui, C., Li, Y. Q., Zhou, L., Ou, Q. D., Li, C., Li, Y., & Tang, J. X. (2015). Single-junction polymer solar cells exceeding 10% power conversion efficiency. *Advanced Materials, 27*(6), 1035–1041. doi:10.1002/adma.201404535 PMID:25408480

Cheng, Y. J., Yang, S. H., & Hsu, C. S. (2009). Synthesis of conjugated polymers for organic solar cell applications. *Chemical Reviews, 109*(11), 5868–5923. doi:10.1021/cr900182s PMID:19785455

Coffin, R. C., Peet, J., Rogers, J., & Bazan, G. C. (2009). Streamlined microwave-assisted preparation of narrow-bandgap conjugated polymers for high-performance bulk heterojunction solar cells. *Nature Chemistry, 1*(8), 657–661. doi:10.1038/nchem.403 PMID:21378958

Cortizo-Lacalle, D., Howells, C. T., Pandey, U. K., Cameron, J., Findlay, N. J., Inigo, A. R., Tuttle, T., Skabara, P. J., & Samuel, I. D. W. (2014). Solution processable diketopyrrolopyrrole (DPP) cored small molecules with BODIPY end groups as novel donors for organic solar cells. *Beilstein Journal of Organic Chemistry, 10*, 2683–2695. doi:10.3762/bjoc.10.283 PMID:25550732

Dai, S. Y., & Song, Q. H. (2012). New 2,6-modified BODIPY sensitizers for dye-sensitized solar cells. *Chemistry, an Asian Journal*, *7*(4), 696–700. doi:10.1002/asia.201100779 PMID:22250129

Erten-Ela, S., Yilmaz, M. D., Icli, B., Dede, Y., Icli, S., & Akkaya, E. U. (2008). A panchromatic boradiazaindacene (bodipy) sensitizer for dye-sensitized solar cells. *Organic Letters*, *10*(15), 3299–3302. doi:10.1021/ol8010612 PMID:18588306

Gao, F., Wang, Y., Shi, D., Zhang, J., Wang, M., Jing, X., Humphry-Baker, R., Wang, P., Zakeeruddin, S. M., & Grtzel, M. (2008). Enhance the optical absorptivity of nanocrystalline TiO2 film with high molar extinction coefficient ruthenium sensitizers for high performance dye-sensitized solar cells. *Journal of the American Chemical Society*, *130*(32), 10720–10728. doi:10.1021/ja801942j PMID:18642907

Geiger, T., Kuster, S., Yum, J. H., Moon, S. J., Nazeeruddin, M. K., Grätzel, M., & Nuesch, F. (2009). Molecular Design of Unsymmetrical Squaraine Dyes for High Efficiency Conversion of Low Energy Photons into Electrons Using TiO2 Nanocrystalline Films. *Advanced Functional Materials*, *21*(17), 2720–2727. doi:10.1002/adfm.200900231

Graf, K., Körzdörfer, T., Kümmel, S., & Thelakkat, M. (2013). Synthesis of donor-substituted meso-phenyl and meso-ethynylphenyl BODIPYs with broad absorption. *New Journal of Chemistry*, *37*(5), 1417–1426. doi:10.1039/c3nj00157a

Hagberg, D. P., Edvinson, T., Marinado, T., Boschloo, G., Hagfeldt, A., & Sun, L. (2006). A novel organic chromophore for dye-sensitized nanostructured solar cells. *Chemical Communications*, (21), 2245–2247. doi:10.1039/b603002e PMID:16718317

Hagfeldt, A., Boschloo, G., Sun, L., Kloo, L., & Pettersson, H. (2010). Dye-sensitized solar cells. *Chemical Reviews*, *110*(11), 6595–6663. doi:10.1021/cr900356p PMID:20831177

Hagfeldt, A., & Graetzel, M. (1995). Light-Induced Redox Reactions in Nanocrystalline Systems. *Chemical Reviews*, *95*(1), 49–68. doi:10.1021/cr00033a003

Hagfeldt, A., & Gratzel, M. (2000). Molecular Photovoltaics. *Accounts of Chemical Research*, *33*(5), 269–277. doi:10.1021/ar980112j PMID:10813871

Hattori, S., Ohkubo, K., Urano, Y., Sunahara, H., Nagano, T., Wada, Y., Tkachenko, N. V., Lemmetyinen, H., & Fukuzumi, S. (2005). Charge separation in a nonfluorescent donor–acceptor dyad derived from boron dipyrromethene dye, leading to photocurrent generation. *The Journal of Physical Chemistry B*, *109*(32), 15638–15375. doi:10.1021/jp050952x PMID:16852949

He, Z., Xiao, B., Liu, F., Wu, F., We, H., Yang, Y., Xiao, S., Wang, C., Russell, T. P., & Cao, Y. (2015). Single-junction polymer solar cells with high efficiency and photovoltage. *Nature Photonics*, *9*(3), 174–179. doi:10.1038/nphoton.2015.6

Hussein, B. A., Huynh, J. T., Prieto, P. L., Barran, C. P., Arnold, A. E., Sarycheva, O. V., Lough, A. J., & Koivisto, B. D. (2018). Molecular lemmings: Strategies to avoid when designing BODIPY ferrocene dyads for dye-sensitized solar cell applications. *Dalton Transactions (Cambridge, England)*, *47*(14), 4916–4920. doi:10.1039/C8DT00174J PMID:29546908

Iqbal, M. Z., Ali, S. R., & Khan, S. (2019). Progress in dye sensitized solar cell by incorporating natural photosensitizers. *Solar Energy*, *181*, 490–509. doi:10.1016/j.solener.2019.02.023

Jadhav, T., Misra, R., Biswas, S., & Sharma, G. D. (2015). Bulk heterojunction organic solar cells based on carbazole–BODIPY conjugate small molecules as donors with high open circuit voltage. *Physical Chemistry Chemical Physics*, *17*(40), 26580–26588. doi:10.1039/C5CP04807A PMID:26394373

Kalyanasundaram, K., & Gratzel, M. (1998). Applications of functionalized transition metal complexes in photonic and optoelectronic devices. *Coordination Chemistry Reviews*, *177*(1), 347–414. doi:10.1016/S0010-8545(98)00189-1

Kan, B., Li, M., Zhang, Q., Liu, F., Wan, X., Wang, Y., Ni, W., Long, G., Yang, X., Feng, H., Zuo, Y., Zhang, M., Huang, F., Cao, Y., Russell, T. P., & Chen, Y. (2015). A Series of Simple Oligomer-like Small Molecules Based on Oligothiophenes for Solution-Processed Solar Cells with High Efficiency. *Journal of the American Chemical Society*, *137*(11), 3886–3893. doi:10.1021/jacs.5b00305 PMID:25736989

Kaneza, N., Zhang, J., Liu, H., Archana, S., Shan, Z., Vasiliu, M., Polansky, S. H., Dixon, D. A., Adams, R. E., Schmehl, R. H., Gupta, A., & Pan, S. (2016). Electrochemical and spectroscopic properties of boron dipyrromethene-thiophenetriphenylamine-based dyes for dye-sensitized solar cells. *The Journal of Physical Chemistry C*, *120*(17), 9068–9080. doi:10.1021/acs.jpcc.6b01611

Kay, A., & Gratzel, M. (1996). Low cost photovoltaic modules based on dye sensitized nanocrystalline titanium dioxide and carbon powder. *Solar Energy Materials and Solar Cells*, *44*(1), 99–117. doi:10.1016/0927-0248(96)00063-3

Kim, H. S., Lee, C. R., Im, J. H., Lee, K. B., Moehl, T., Marchioro, A., Moon, S. J., Humphry-Baker, R., Yum, J. H., Moser, J. E., Grätzel, M., & Park, N. G. (2012). Lead iodide perovskite sensitized all-solidstate submicron thin film mesoscopic solar cell with efficiency exceeding 9%. *Scientific Reports*, *2*(1), 591. doi:10.1038rep00591 PMID:22912919

Klfout, H., Stewart, A., Elkhalifa, M., & He, H. (2017). BODIPYs for dye-sensitized solar cells. *ACS Applied Materials & Interfaces*, *9*(46), 39873–39889. doi:10.1021/acsami.7b07688 PMID:29072443

Knag, H., Kee, S., Yu, K., Lee, J., Kim, G., Kim, J., Kim, J. R., Kong, J., & Lee, K. (2015). Simplified tandem polymer solar cells with an ıdeal self-organized recombination layer. *Advanced Materials*, *27*(8), 1408–1413. doi:10.1002/adma.201404765 PMID:25449142

Kolemen, S., Bozdemir, O. A., Cakmak, Y., Barin, G., Erten-Ela, S., Marszalek, M., Yum, J. H., Zakeeruddin, S. M., Nazeeruddin, M. K., Grätzel, M., & Akkaya, E. U. (2011). Optimization of distyryl-Bodipy chromophores for efficient panchromatic sensitization in dye sensitized solar cells. *Chemical Science (Cambridge)*, *2*(5), 949–954. doi:10.1039/c0sc00649a

Kolemen, S., Cakmak, Y., Erten-Ela, S., Altay, Y., Brendel, J., Thelakkat, M., & Akkaya, E. U. (2010). Solid state dye-sensitized solar cells using red and near-IR absorbing Bodipy sensitizers. *Organic Letters*, *12*(17), 3812–3815. doi:10.1021/ol1014762 PMID:20704314

Kolemen, S., Cakmak, Y., Ozdemir, T., Erten-Ela, S., Buyuktemiz, M., Dede, Y., & Akkaya, E. U. (2014). Design and characterization of Bodipy derivatives for bulk heterojunction solar cells. *Tetrahedron*, *70*(36), 6229–6234. doi:10.1016/j.tet.2014.03.049

Kraner, S., Widmer, J., Benduhn, J., Hieckmann, E., Jageler-Hoheisel, T., Ullbrich, S., Schütze, D., Radke, K. S., Cuniberti, G., Ortmann, F., Lorenz-Rothe, M., Meerheim, R., Spoltore, D., Vandewal, K., Koerner, C., & Leo, K. (2015). Influence of side groups on the performance of infrared absorbing aza-BODIPY organic solar cells. *Physica Status Solidi. A, Applications and Materials Science*, *212*(12), 2747–2753. doi:10.1002/pssa.201532385

Kubo, Y., Eguchi, D., Matsumoto, A., Nishiyabu, R., Yakushiji, H., Shigaki, K., & Kaneko, M. (2014). Boron-dibenzopyrromethene-based organic dyes for application in dyesensitized solar cells. *Journal of Materials Chemistry. A, Materials for Energy and Sustainability*, *2*(15), 5204–5211. doi:10.1039/C3TA15340A

Kumaresan, D., Thummel, R.P., Bura, T., Ulrich, G., & Ziessel, R. (2009). Color tuning in new journal. metal-free organic sensitizers (bodipys) for dye-sensitized solar cells. *Chemistry A Europian Journal, 15*, 6335–6339.

Kyeong, M., Lee, J., Lee, K., & Hong, S. (2018). BODIPY-based conjugated polymers for use as dopant-free hole transporting materials for durable perovskite solar cells: Selective tuning of HOMO/LUMO levels. *ACS Applied Materials & Interfaces*, *10*(27), 23254–23262. doi:10.1021/acsami.8b05956 PMID:29916246

Lewis, N. S. (2007). Toward cost-effective solar energy use. *Science*, *315*(5813), 798–801. doi:10.1126cience.1137014 PMID:17289986

Lewis, N. S., & Nocera, D. G. (2006). Powering the planet: Chemical challenges in solar energy utilization. *Proceedings of the National Academy of Sciences of the United States of America*, *103*(43), 15729–15735. doi:10.1073/pnas.0603395103 PMID:17043226

Li, G., Zhu, R., & Yang, Y. (2012). Polymer solar cells. *Nature Photonics*, *6*(3), 153–161. doi:10.1038/nphoton.2012.11

Li, H., Fu, K., Boix, P. P., Wong, L. H., Hagfeldt, A., Grätzel, M., Mhaisalkar, S. G., & Grimsdale, A. C. (2014). Hole-transporting small molecules based on thiophene cores for high efficiency perovskite solar cells. *Chemistry-Sustainability-Energy-Materials*, *7*(12), 3420–3425. doi:10.1002/cssc.201402587 PMID:25233841

Li, Q. S., Zhang, X. L., Wu, G. H., Dai, C. G., Ding, Y., Dai, S. Y., & Song, Q. H. (2017). Effects of donors of bodipy dyes on the performance of dye-sensitized solar cells. *Dyes and Pigments*, *141*, 148–160. doi:10.1016/j.dyepig.2017.02.017

Li, T. Y., Benduhn, J., Li, Y., Jaiser, F., Spoltore, D., Zeika, O., Ma, Z., Neher, D., Vandewal, K., & Leo, K. (2018). Boron dipyrromethene (BODIPY) with meso-perfluorinated alkyl substituents as near infrared donors in organic solar cells. *Journal of Materials Chemistry. A, Materials for Energy and Sustainability*, *6*(38), 18583–18591. doi:10.1039/C8TA06261G

Li, Y. (2012). Molecular design of photovoltaic materials for polymer solar cells: Toward suitable electronic energy levels and broad absorption. *Accounts of Chemical Research*, *45*(5), 723–733. doi:10.1021/ar2002446 PMID:22288572

Liao, J., Zhao, H., Xu, Y., Cai, Z., Peng, Z., Zhang, W., Zhou, W., Li, B., Zong, Q., & Yang, X. (2016). Novel D–A–D type dyes based on BODIPY platform for solution processed organic solar cells. *Dyes and Pigments*, *128*, 131–140. doi:10.1016/j.dyepig.2016.01.020

Liaoa, J., Zhao, H., Cai, Z., Xu, Y., Qin, F. G. F., Zong, Q., Peng, F., & Fang, Y. (2018). BODIPY-based panchromatic chromophore for efficient organic solar cell. *Organic Electronics*, *61*, 215–222. doi:10.1016/j.orgel.2018.05.049

Lin, Y. D., Ke, B. Y., Chang, Y. J., Chou, P. T., Liau, K. L., Liu, C. Y., & Chow, T. J. (2015). Pyridomethene-BF2 complex/phenothiazine hybrid sensitizer with high molar extinction coefficient for efficient sensitized solar cells. *Journal of Materials Chemistry. A, Materials for Energy and Sustainability*, *3*(32), 16831–16842. doi:10.1039/C5TA03807C

Liu, C., Wang, K., Gong, X., & Heeger, A. J. (2016). Low bandgap semiconducting polymers for polymeric photovoltaics. *Chemical Society Reviews*, *45*(17), 4825–4846. doi:10.1039/C5CS00650C PMID:26548402

Liu, J. Y., Hou, X. N., Tian, Y., Jiang, L., Deng, S., Röder, B., & Ermilov, E. A. (2016). Photoinduced energy and charge transfer in a bis(triphenylamine)-BODIPY-C_{60} artificial photosynthetic system. *RSC Advances*, *6*(62), 57293–57305. doi:10.1039/C6RA06841C

Liu, W., Tang, A., Chen, J., Wu, Y., Zhan, C., & Yao, J. (2014). Photocurrent enhancement of bodipy-based solution-processed small-molecule solar cells by dimerization via the meso position. *ACS Applied Materials & Interfaces*, *6*(24), 22496–22505. doi:10.1021/am506585u PMID:25427221

Liu, W., Yao, J., & Zhan, C. (2017). A novel bodipy-based low-band-gap small-molecule acceptor for efficient non-fullerene polymer solar cells. *Chinese Journal of Chemistry*, *35*(12), 1813–1823. doi:10.1002/cjoc.201700542

Liu, X., Wang, H. Q., Li, Y., Gui, Z., Ming, S., Usman, K., Zhang, W., & Fang, J. (2017). Regular organic solar cells with efficiency over 10% and promoted stability by ligand- and thermal annealing-free al-doped ZnO cathode interlayer. *Advancement of Science*, *4*(8), 1700053. doi:10.1002/advs.201700053 PMID:28852624

Liu, Y., Chen, C. C., Hong, Z., Gao, J., Yang, Y., Zhou, H., Dou, L., Li, G., & Yang, Y. (2013). Solution-processed small-molecule solar cells: Breaking the 10% power conversion efficiency. *Scientific Reports*, *3*(1), 3356. doi:10.1038rep03356 PMID:24285006

Loudet, A., & Burgess, K. (2007). BODIPY dyes and their derivatives: Syntheses and spectroscopic properties. *Chemical Reviews*, *107*(11), 4891–4932. doi:10.1021/cr078381n PMID:17924696

Lu, Z., Liang, M., Dai, P., Miao, K., Zhang, C., Sun, Z., & Xue, S. (2016). A strategy for enhancing the performance of Borondipyrromethene dye-sensitized solar cells. *The Journal of Physical Chemistry C*, *120*(45), 25657–25667. doi:10.1021/acs.jpcc.6b07356

Mao, M., & Song, Q. H. (2016). The structure-property relationships of D-π-A BODIPY dyes for dye-sensitized solar cells. *Chemical Record (New York, N.Y.)*, *16*(2), 719–733. doi:10.1002/tcr.201500251 PMID:26846846

Mao, M., Wang, J., Xiao, Z., Dai, S., & Song, Q. (2012). New 2,6-modified BODIPY sensitizers for dye-sensitized solar cells. *Dyes and Pigments*, *94*(2), 224–232. doi:10.1016/j.dyepig.2012.01.011

Mao, M., Zhang, X. L., Cao, L., Tong, Y., & Wu, G. H. (2015). Design of bodipy based organic dyes for high-efficient dye-sensitized solar cells employing double electron acceptors. *Dyes and Pigments*, *117*, 28–36. doi:10.1016/j.dyepig.2015.02.001

Mao, M., Zhang, X. L., Fang, X. Q., Wu, G. H., Dai, S. Y., Song, Q. H., & Zhang, X. X. (2014). Highly efficient light-harvesting boradiazaindacene sensitizers for dye-sensitized solar cells featuring phenothiazine donor antenna. *Journal of Power Sources*, *268*, 965–976. doi:10.1016/j.jpowsour.2014.05.079

Mao, M., Zhang, X. L., Fang, X. Q., Wu, G. H., Ding, Y., Liu, X. L., Dai, S. Y., & Song, Q. H. (2014). 2,6-Conjugated Bodipy sensitizers for high-performance dye-sensitized solar cells. *Organic Electronics*, *15*(9), 2079–2090. doi:10.1016/j.orgel.2014.05.024

Mishra, R., Basumatary, B., Singhal, R., Sharma, G. D., & Sankar, J. (2018). Corrole-BODIPY Dyad as Small-Molecule Donor for Bulk Heterojunction Solar Cells. *ACS Applied Materials & Interfaces*, *10*(37), 31462–31471. doi:10.1021/acsami.8b08519 PMID:30136584

Miyata, A., Mitioglu, A., Plochocka, P., Portugall, O., Wang, J. T. W., Stranks, S. D., Snaith, H. J., & Nicholas, R. J. (2015). Direct measurement of the exciton binding energy and effective masses for charge carriers in organic–inorganic tri-halide perovskites. *Nature Physics*, *11*(7), 582–587. doi:10.1038/nphys3357

Nazeeruddin, M. K., Pchy, P., Renouard, T., Zakeeruddin, S. M., Humphry-Baker, R., Comte, P., Liska, P., Cevey, L., Costa, E., Shklover, V., Spiccia, L., Deacon, G. B., Bignozzi, C. A., & Grtzel, M. (2001). Engineering of efficient panchromatic sensitizers for nanocrystalline TiO2-based solar cells. *Journal of the American Chemical Society*, *123*(8), 1613–1624. doi:10.1021/ja003299u PMID:11456760

Nikolaou, V., Charalambidis, G., Charisiadis, A., & Coutsolelos, A. G. (2017). Porphyrin–BODIPY-based hybrid model compounds for artificial photosynthetic reaction centers. *Comptes Rendus. Chimie*, *20*(3), 314–322. doi:10.1016/j.crci.2016.07.009

Nozik, A. J. (1978). Photoelectrochemistry: Applications to solar energy conversion. *Annual Review of Physical Chemistry*, *29*(1), 189–222. doi:10.1146/annurev.pc.29.100178.001201

O'Regan, B., & Gratzel, M. (1991). A low-cost, high-efficiency solar cell based on dye-sensitized colloidal TiO2 films. *Nature*, *353*(6346), 737–740. doi:10.1038/353737a0

Ooyama, Y., Hagiwara, Y., Mizumo, T., Harima, Y., & Oshita, J. (2013). Photovoltaic performance of dye-sensitized solar cells based on D–π–A type BODIPY dye with two pyridyl groups. *New Journal of Chemistry*, *37*(8), 2479–2485. doi:10.1039/c3nj00456b

Ortiz, A. (2019). Triarylamine-BODIPY derivatives: A promising building block as hole transporting materials for efficient perovskite solar cells. *Dyes and Pigments*, *171*, 107690. doi:10.1016/j.dyepig.2019.107690

Panwar, N., Kaushik, S., & Kothari, S. (2011). Role of renewable energy sources in environmental protection: A review. *Renewable & Sustainable Energy Reviews*, *15*(3), 1513–1524. doi:10.1016/j.rser.2010.11.037

Qin, C., Mirloup, A., Leclerc, N., Islam, A., El-Shafei, A., Han, L., & Ziessel, R. (2014). Molecular engineering of new thienyl-bodipy dyes for highly efficient panchromatic sensitized solar cells. *Advanced Energy Materials*, *4*(11), 1400085. doi:10.1002/aenm.201400085

Ramos, C. A. F., Alcaso, A. N., & Cardoso, A. J. M. (2019). Photovoltaic-thermal (PVT) technology: Review and case study. *IOP Conference Series. Earth and Environmental Science*, *354*(1), 012048. doi:10.1088/1755-1315/354/1/012048

Rousseau, T., Cravino, A., Bura, T., Ulrich, G., Ziessel, R., & Roncali, J. (2009). BODIPY derivatives as donor materials for bulk heterojunction solar cells. *Chemical Communications*, *1*(13), 673–675. doi:10.1039/b822770e PMID:19294258

Rousseau, T., Cravino, A., Bura, T., Ulrich, G., Ziessel, R., & Roncali, J. (2009). Multi-donor molecular bulk heterojunction solar cells: Improving conversion efficiency by synergistic dye combinations. *Journal of Materials Chemistry*, *19*(16), 2298–2300. doi:10.1039/b903189h

Roy, P., Sinha, N. K., Tiwari, S., & Khare, A. (2020). A review on perovskite solar cells: Evolution of architecture, fabrication techniques, commercialization issues and status. *Solar Energy*, *198*, 665–688. doi:10.1016/j.solener.2020.01.080

Sahoo, S. K., Manoharan, B., & Sivakumar, N. (2018). *Introduction: Why Perovskite and Perovskite Solar Cells? Perovskite Photovoltaics Basic to Advanced Concepts and Implementation*. Academic Press.

Sharma, G. D., Siddiqui, S. A., Nikiforou, A., Zervaki, G. E., Georgakaki, I., Ladomenou, K., & Coutsolelos, A. G. (2015). A mono(carboxy)porphyrin-triazine-(bodipy)$_2$ triad as a donor for bulk heterojunction organic solar cells†. *Journal of Materials Chemistry. C, Materials for Optical and Electronic Devices*, *3*(24), 6209–6217. doi:10.1039/C5TC01076D

Shi, Z., & Ahalapitiya, J. H. (2018). Perovskites-based solar cells: A review of recent progress, materials and processing methods. *Materials (Basel)*, *11*(5), 729. doi:10.3390/ma11050729 PMID:29734667

Shubbak, M. H. (2019). Advances in solar photovoltaics: Technology review and patent trends. *Renewable & Sustainable Energy Reviews*, *115*, 109383. doi:10.1016/j.rser.2019.109383

Si, L., Chang, C. W., He, H., Sykes, A., Lin, C. Y., & Diau, E. W. G. (2012). Dual functionality of bodipy chromophore in porphyrin-sensitized nanocrystalline solar cells. *The Journal of Physical Chemistry C*, *116*(19), 10451–10460. doi:10.1021/jp210458j

Summers, G. H., Lefebvre, J. F., Black, F. A., Davies, E. S., Gibson, E. A., Pullerits, T., Wood, C. J., & Zidek, K. (2016). Design and characterisation of bodipy sensitizers for dye-sensitized NiO solar cells. *Physical Chemistry Chemical Physics*, *18*(2), 1059–1070. doi:10.1039/C5CP05177K PMID:26660278

Thumuganti, G., Gupta, V., & Singh, S. P. (2019). New dithienosilole- and dithienogermole-based BODIPY for solar cell applications. *New Journal of Chemistry*, *43*(22), 8735–8740. doi:10.1039/C9NJ00550A

Tong, M., Cho, S., Rogers, J. T., Schmidt, K., Hsu, B. B. Y., Moses, D., Coffin, R. C., Kramer, E. J., Bazan, G. C., & Heeger, A. J. (2010). Higher Molecular Weight Leads to Improved Photoresponsivity, Charge Transport and Interfacial Ordering in a Narrow Bandgap Semiconducting Polymer. *Advanced Functional Materials*, *20*(22), 3959–3965. doi:10.1002/adfm.201001271

Ulrich, G., Ziessel, R., & Harriman, A. (2008). The chemistry of fluorescent bodipy dyes: Versatility unsurpassed. *Angewandte Chemie International Edition*, *47*(7), 1184–1201. doi:10.1002/anie.200702070 PMID:18092309

Valverde-Chávez, D.A., Ponseca, C. S., Stoumpos, C., Yartsev, A., Kanatzidis, M. G., Sundström, V., & Cooke, D. G. (2015). *Optical pump - multi-thz probe spectroscopy of a single crystal organic hybrid lead halide perovskite.* Optical Society of America.

Vlachopoulos, N., Liska, P., Augustynski, J., & Graetzel, M. (1988). Very efficient visible light energy harvesting and conversion by spectral sensitization of high surface area polycrystalline titanium dioxide films. *Journal of the American Chemical Society, 110*(4), 1216–1220. doi:10.1021/ja00212a033

Warnan, J., Buchet, F., Pellegrin, Y., Blart, E., & Odobel, F. (2011). Panchromatic trichromophoric sensitizer for dye-sensitized solar cells using antenna effect. *Organic Letters, 13*(15), 3944–3947. doi:10.1021/ol2014686 PMID:21721515

Watson, B. L., Moore, T. A., Moore, A. L., & Gust, D. (2017). Synthesis of a novel building block for the preparation of multi-chromophoric sensitizers for panchromatic dyesensitized solar cells. *Dyes and Pigments, 136*, 893–897. doi:10.1016/j.dyepig.2016.09.037

You, J., Dou, L., Yoshimura, K., Kato, T., Ohya, K., Moriarty, T., Emery, K., Chen, C. C., Gao, J., Li, G., & Yang, Y. (2013). A polymer tandem solar cell with 10.6% power conversion efficiency. *Nature Communications, 4*(1), 1446. doi:10.1038/ncomms2411 PMID:23385590

Zhang, F., Shi, W., Luo, J., Pellet, N., Yi, C., Li, X., Zhao, X., Dennis, T. J. S., Li, X., Wang, S., Xiao, Y., Zakeeruddin, S. M., Bi, D., & Gratzel, M. (2017). Isomer-pure bis-pcbm-assisted crystal engineering of perovskite solar cells showing excellent efficiency and stability. *Advanced Materials, 29*(17), 1606806. doi:10.1002/adma.201606806 PMID:28240401

Zhang, J., Lu, F., Qi, S., Zhao, Y., Wang, K., Zhang, B., & Feng, Y. (2016). Influence of various electron-donating triarylamine groups in BODIPY sensitizers on the performance of dye-sensitized solar cells. *Dyes and Pigments, 128*, 296–303. doi:10.1016/j.dyepig.2016.02.008

Zhang, X., Zhang, Y., Chen, L., & Xiao, Y. (2015). Star-shaped carbazole-based BODIPY derivatives with improved hole transportation and near-infrared absorption for small-molecule organic solar cells with high open-circuit voltages. *RSC Advances, 5*(41), 32283–32289. doi:10.1039/C5RA02414E

Chapter 7
Do Remittance Inflows Increase Energy Security Risk in the Long Run?
Evidence From Selected MENA Countries

Alper Karasoy
https://orcid.org/0000-0001-9447-3375
Afyon Kocatepe University, Turkey

ABSTRACT

This chapter examines the long-run effect of remittances on energy security risks of Egypt, Morocco, and Tunisia in a bi-variate setting by utilizing the DOLS, FMOLS, and CCR estimation methods for the 1980-2018 period. The empirical findings of this chapter show that remittances and energy security risks are cointegrated. Additionally, the estimation results show that remittances significantly aggravate energy insecurities in selected MENA countries in the long run. Further, the results of this chapter reveal that remittances' role in increasing energy insecurity should not be overlooked. Therefore, the author proposes that remittance inflows can be utilized as a long-term integral instrument in combating energy insecurity issues and developing new energy policies.

INTRODUCTION

One of the sustainable development goals (SGDs) is "*affordable and clean energy.*" In other words, the 7th of the SDGs is to "*ensure access to affordable, reliable, sustainable and modern energy for all*" (UN, 2020, p. 12). Although some progress

DOI: 10.4018/978-1-7998-9502-2.ch007

has been made to accomplish this goal, 789 million people still lack access to electricity (UN, 2020, p. 36). One of the ways to achieve this goal, especially in developing countries, is to ensure energy security and sustainability. However, energy security that is "*the uninterrupted availability of energy sources at an affordable price*" (IEA, 2021), remains to be a problem in the Middle East and North Africa (MENA) countries especially in Egypt, Morocco, and Tunisia.

In these 3 MENA countries, energy consumption has been increasing sharply since 1980. In Egypt, energy use per capita rose by almost 2.4 folds from 1980 (348.5 kilograms of oil equivalent [kgoe]) to 2014 (827.49 kgoe). In the same period, Morocco experienced a more than two-fold increase in per capita energy consumption, from 270.47 kgoe in 1980 to 555.14 kgoe in 2014. Lastly, in Tunisia's experience, the increase was more than 1.85 folds; in other terms, it increased from 512.7 kgoe in 1980 to 950.5 kgoe in 2014. Furthermore, another implication of increasing energy use in these countries is the heavy reliance on fossil fuels. In 2014, fossil fuel utilization in total energy usage was 97.92% in Egypt, 88.47% in Morocco, and 88.87% in Tunisia. Of course, this situation additionally produced environmental ramifications for these countries. In 2018, compared to the 1980 values, carbon dioxide emissions rose by more than five folds in Egypt, more than four folds in Morocco, and more than three folds in Tunisia (the author's calculation based on the WDI (2021) data).

In addition, according to an energy security risk index (ESRI) that is discussed in more detail in the next section and constructed by *Global Energy Institute*, most of the MENA countries, exclusively Egypt, Morocco, and Tunisia, have critically high energy security risks (i.e., energy insecurities) (Global Energy Institute - International Energy Security Risk Index, 2021). Further, as shown in Figure 1, compared to the OECD average, the ESRIs of these three countries had been consistently higher since 1981. In summary, energy security seems to be a persistent problem in these countries. Achieving energy security is vital because energy security can also help maintain economic growth (Le & Nguyen, 2019; Xu et al., 2021) and socio-economic development (Naeem Nawaz & Alvi, 2018).

Figure 1. Energy security risk indices of Egypt, Morocco, Tunisia, and the OECD average.
Source: Global Energy Institute - International Energy Security Risk Index (2021)

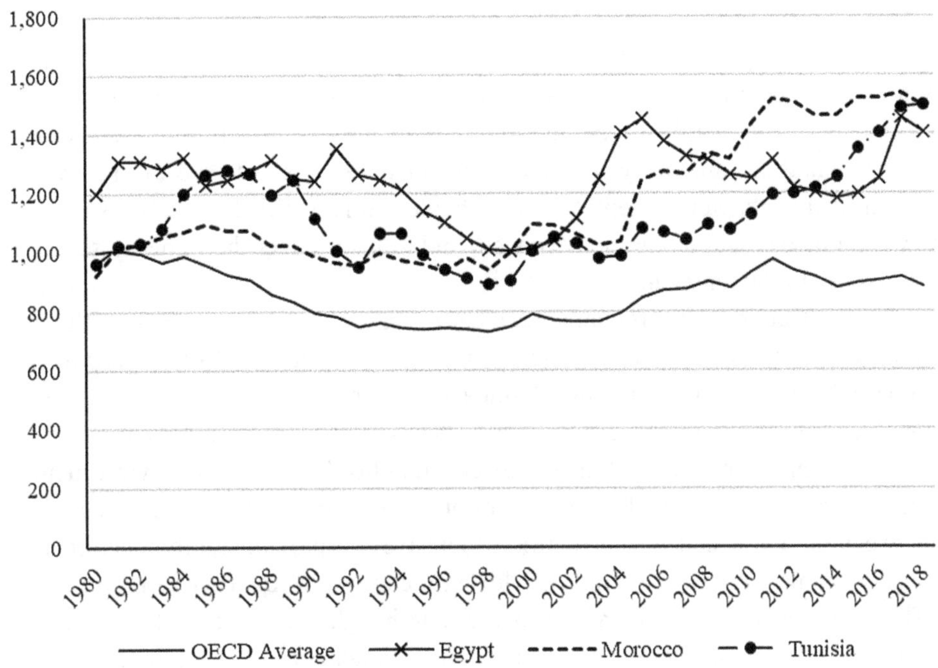

Also, besides high energy insecurity, another common feature of these three MENA countries is that they receive significant remittances, defined as "*money sent home by migrant and guest workers employed in foreign countries*" (Sayeh & Chami, 2020, p. 16). As reported by the World Bank, in 2018, remittances received equaled more than 10% of Egypt's gross domestic product (GDP), almost 6% of Morocco's GDP, and roughly 5% of Tunisia's GDP (The World Bank, 2021). Moreover, remittances are considered to be a significant external financial source and income. To the extent that, on average, these countries receive more remittances than foreign direct investments (FDI), as shown in Table 1.

Table 1. A comparison of FDI and remittance inflows

	Egypt	Morocco	Tunisia
FDI	2.4021	1.7943	2.4753
Remittances	7.1684	6.2796	4.2157

Notes: The numbers present the average inflows for the 1980-2018 period. Both indicators are presented as a percentage of gross domestic product (GDP). Data are obtained from World Bank (2020) and WDI (2021)

According to Table 1, all three countries received more remittances than FDI, implying the significance of remittances for these countries. On average, for the 1980-2018 period, remittances constituted more than 7%, 6%, and 4% of Egypt's, Morocco's, and Tunisia's GDPs, respectively.

While the impact of remittances on various macroeconomic indicators -such as growth- is empirically well-investigated (see Jushi et al., 2021; Woldemariam & Yiheyis, 2017), studies that investigate their effects on different energy and environmental sustainability indicators are relatively nascent (see the following section of this chapter). Therefore, this chapter aims to fill this gap by studying how remittances sent to Egypt, Morocco, and Tunisia affect their energy security risk indices in the long run.

The contributions of this chapter can be summarized as follows: First, although remittance inflows' impacts on energy use and environmental deterioration are empirically examined, remittances' impact on energy security has never been examined. As the energy security risk index that is utilized in this chapter is based on many different metrics, including energy consumption/intensity and environmental pollution variables (see the following section for a review of this index), the author believes this energy security index is more inclusive than previously employed energy and environment proxies. Second, exploring the long-run link between remittance inflows and energy security may provide new insights to policymakers in these countries and other developing countries. Third, implications that remittance inflows impose on energy security, in the long run, may reveal that remittances should be considered an integral part of energy policies in MENA and other developing countries that heavily rely on remittances. Finally, methodologically, the unit-root properties and the long-run association between these two variables (remittance inflows and energy security) are examined via the tests that have the Fourier components. The main advantage of this approach is that it accounts for the impacts of an unknown number of structural breaks that have unknown forms. Given that many structural shifts might have occurred in these countries through the sample period, accounting for them may provide more robust and reliable results.

This chapter is organized as follows. In the next section, the ESRI and its components are defined, a conceptual framework regarding the remittances-energy security link is developed, and a literature review is conducted. In the second section, data, model, and methods employed in this chapter are summarized. In the third section, empirical findings are discussed. In the last section, the conclusion is presented.

ENERGY SECURITY RISK INDEX, REMITTANCES-ENERGY SECURITY LINK, AND LITERATURE SURVEY

In the first sub-section of this section, the ESRI is briefly explained. In the following sub-section, the channels through which remittances may affect energy (in)security are conceptually developed with respect to the relevant literature. In the last sub-section, a literature review is conducted for the studies that -either partially or as a whole- investigated the remittances-energy-environment nexus.

Defining Energy Security and Energy Security Risk Index

Initially, energy security was (narrowly) defined as "*securing access to oil and other fossil fuels*" (von Hippel et al., 2011, p. 6719). However, this definition is rather limited and fails to identify the complexities of energy policies and security. For instance, sustainability and sustainable development are an integral part of energy security, and these dimensions should also be accounted for when energy security is examined (von Hippel et al., 2011). For this reason, in this chapter, a more comprehensive energy security risk index calculated by Global Energy Institute (2020) is utilized. This index's contents are summarized in Figure 2.

As shown in Figure 2, the ESRI is calculated by weighting eight sub-categories and taking 29 variables into account. These sub-categories consist of various dimensions of energy security.

Figure 2. Energy security risk index and its contents
Notes: Values in the parentheses show weights in percentages.
Source: Global Energy Institute - International Energy Security Risk Index (2021)

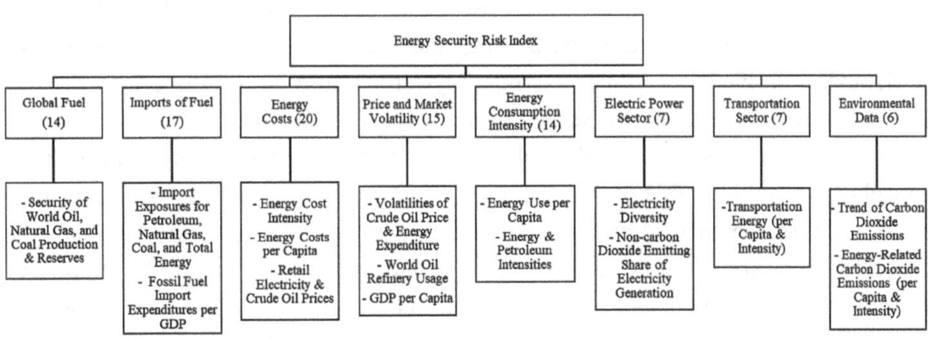

For instance, global fuel indicates the security of world fossil fuels' production and reserves. Moreover, fuel imports and the cost of energy include imports of various energy sources, expenditures of these imports per GDP, energy cost intensity and its per capita value, and retail prices of electricity and oil. Also, price and market volatilities cover volatilities in oil price and energy expenditures, world oil refinery use, and GDP per capita. Further, energy use intensity encompasses energy use per capita, besides energy and petroleum intensities. The electric power sector accommodates electric diversity and the share of electricity generation from non-carbon emitting resources. Additionally, the transportation sector with its per capita energy consumption and energy use intensity is also utilized in the ESRI. Finally, environmental pollution data such as the trend of carbon emissions and energy-related carbon emissions are also utilized to construct the ESRI. The ESRI is calculated by normalizing the aforementioned scores and taking the OECD average and the year 1980 as the main reference points. A higher index value corresponds to higher energy insecurity (i.e., higher energy security risk), while a lower value indicates lower energy insecurity or higher energy security (for a more detailed description, see Global Energy Institute - International Energy Security Risk Index (2021); Global Energy Institute (2020); United States Chamber of Commerce (2014)).

Establishing the Remittance Inflows-Energy (In)security Link

As mentioned previously, there is no empirical study that links remittances to energy security. However, the author believes that a conceptual framework can be developed based on the previous studies that examined the link between energy use and remittances or between environmental damage and remittances because the ESRI used in this chapter is closely related to energy use and environmental damage variables. Therefore, based on the previous empirical research, Figure 3 is constructed to conceptualize the remittances-energy security link.

According to Ahmad et al. (2019) and Rahman, Cai, & Ahmad (2019), remittances would initially result in increased income and savings. As savings would have to be deposited in banks, a proper financial system should be developed, leading to financial development. At this stage, higher income levels would make households more resilient against price and market shocks in the energy sector because these households would be able to spend more on energy. This situation would, in turn, improve energy security. Nevertheless, remittances would also create a "*consumer effect*," which would allow remittance recipient households to purchase energy-consuming goods such as cars, air conditioners, washing machines (Sahoo & Sethi, 2020, p. 2). Additionally, remittance inflows would also increase the demand for nonessential products that are imported from abroad. This situation would additionally boost energy usage in the transportation sector (Amuedo-Dorantes, 2014, p. 4). Given that the sample of countries examined in this chapter is heavily reliant upon fossil fuels, increased demand and a corresponding rise in investments would intensify fossil fuel demand. Accordingly, higher fossil fuel consumption, imports, and expenditures would lead to higher volatilities in the energy sector and higher energy security risk (i.e., energy insecurity). In addition, increased fossil fuel consumption would also worsen environmental damage through escalated carbon emissions and intensity. Eventually, through this environmental damage channel, energy insecurity would be worsened.

As highlighted previously, remittances can boost savings and therefore induce financial development. Sahoo & Sethi (2020, p. 2) claim that would create a "*wealth effect*," which would induce new investments and existing investments to expand further because financial development eases access to new financial instruments and credits. At this level, remittances' impact on energy security through financial development can be positive or negative. On the one hand, the proposed impact can be negative given that remittances are used -through improved financial systems- to support and promote technologically innovative investment venues (such as green energy and environmentally sustainable production methods) (Villanthenkodath & Mahalik, 2020, p. 5). Then, besides energy efficiency, alternative energy investment and production would increase, leading to a decrease in energy insecurity. On the other hand, the impact can be positive, given that new investments are conventional in the sense that they promote the usage of fossil fuels. If new investments are fossil fuel dependent, it is much more likely that financial development would induce energy insecurity.

In short, as shown in Figure 3, although it is possible to claim that remittances can improve energy security through various channels, it is much more reasonable to posit that remittances sent to Egypt, Morocco, and Tunisia diminish the energy securities of these countries, especially given that these countries are significantly dependent on fossil fuel consumption.

Figure 3. Remittances-energy security link: A conceptual framework
Source: Based on Ahmad et al. (2019), Khan, Ahmad, & Khan (2020), Neog & Yadava (2020), Rahman, Cai, & Ahmad (2019), Sahoo & Sethi (2020), Villanthenkodath & Mahalik (2020), Yang, Jahanger, & Ali (2021), Yang, Jahanger, & Khan (2020), and the author's conceptualizations.

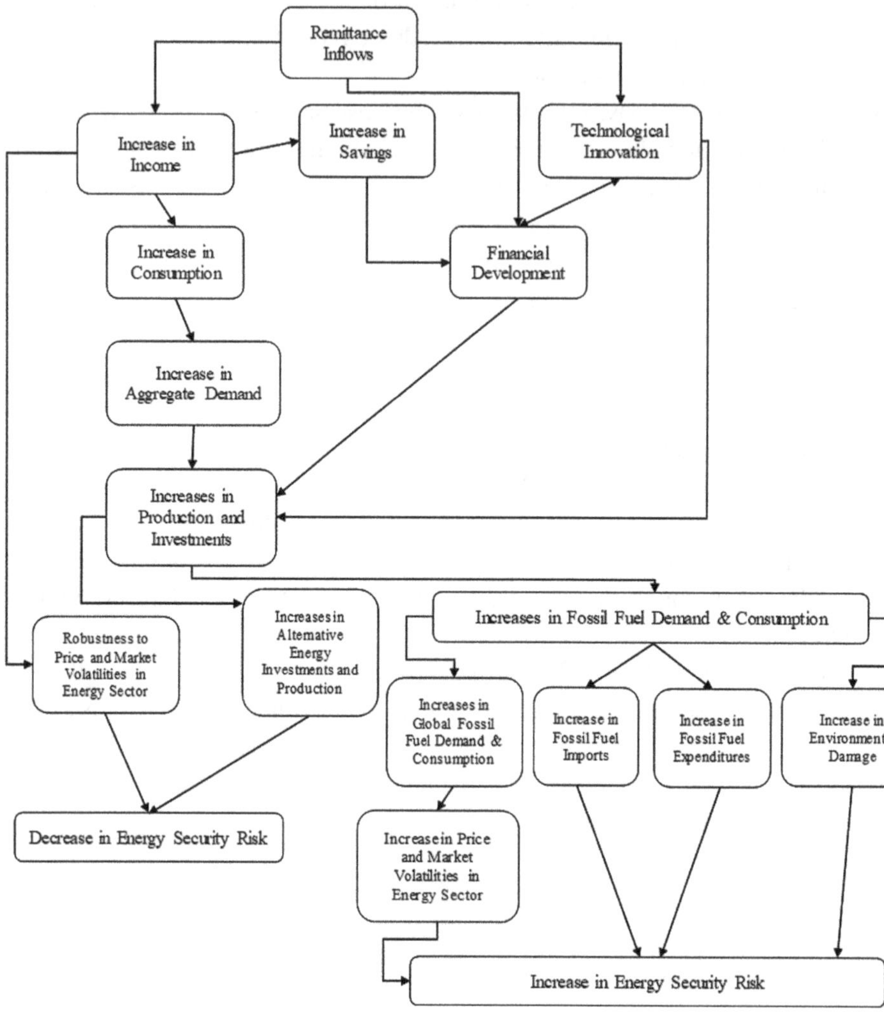

Literature Summary

In this sub-section, the findings of empirical research that studied either the remittances-energy consumption nexus or the remittances-environment nexus are summarized. In that respect, studies that investigated remittance inflows' impact on energy use are reviewed first, followed by the research that examined the remittances-environmental degradation link.

For Bangladesh, Das & McFarlane (2020) examined the remittances-disaggregated energy use nexus by employing the autoregressive distributed lag (ARDL) method and the Toda- Yamamoto (TY) causality test. Their results showed that natural gas usage and remittance inflows have a positive association, additionally, both electricity and coal consumption positively impact remittances. By employing the same causality test, Sahoo & Sethi (2020) investigated the impact of remittances sent to India on electricity consumption. Their findings indicated that remittances boost electricity use. Moreover, Rahman, Hosan, Karmaker, Chapman, & Saha (2021) studied how remittances impact energy use in South Asian countries via panel data methods. Their study's results illustrated that remittances cause energy consumption in the long run. In another study on Bangladesh, Das, McFarlane, & Carels (2021) analyzed remittances and renewable energy consumption link by using the TY causality test and the vector error correction model (VECM). They concluded that remittances cause renewable energy use, and there exists bi-directional causality between renewable energy and remittances in the short run. Lastly, Das & McFarlane (2021) examined remittance inflows' impact on Jamaica's electricity consumption and power losses using the VECM. Their findings mainly showed that remittances to Jamaica increase legal electricity consumption yet curb illegal electricity usage in the long run.

Regarding the studies that analyze the remittance-environmental damage link at a macro level, it is possible to make the following distinction: *(i)* single-country studies and *(ii)* multiple-country studies. The former studies are summarized in the next couple of paragraphs, and eventually, the latter studies are reviewed.

Ahmad et al. (2019) studied the effect of remittances to China on its carbon emissions in a bi-variate setting using the non-linear ARDL (NARDL) method. As expectedly, their results demonstrated that positive shocks to remittances boost emissions; nonetheless, negative shocks to remittances limit carbon emissions. Sharma, Bhattarai, & Ahmed (2019) examined the same phenomenon for Nepal via the ARDL technique. Their findings indicated that remittances to Nepal reduce carbon emissions in the long run. Using the same method, Villanthenkodath & Mahalik (2020) examined the technological innovations' role on environmental pollution by considering the effect of remittance inflows to India. According to their findings, remittances to India have a U-shape impact on carbon dioxide emissions. Brown, McFarlane, Campbell, & Das (2020) examined how remittances sent to Jamaica impact its carbon emissions via the ARDL and NARDL methods in a bi-variate setting. Their results revealed that remittances impose an inverted-U-shape impact on carbon emissions; moreover, positive developments in remittances increase emissions, yet negative developments curb emissions. Another study on India conducted by Neog & Yadava (2020) and employed the NARDL method showed that positive shocks to remittances contribute to environmental pollution; however, negative shocks curb pollution.

Li et al. (2021) analyzed how foreign financial inflows to China affected its pollution levels in a multivariate setting by utilizing the ARDL and NARDL techniques. Their findings on remittance inflows revealed that negative shocks to remittances significantly decrease the environmental pollution. Kibria (2021) investigated how foreign aid and remittances to Bangladesh impacted its environmental quality by taking asymmetries into account via the NARDL method. The findings in this study illustrated that positive and negative shocks to remittances have increasing and decreasing effects on carbon emissions, respectively. Mahalik, Villanthenkodath, Mallick, & Gupta (2021) studied the same subject for India by employing the ARDL approach. Their findings also confirmed that remittances to India contribute to its carbon emission level.

Further, Karasoy (2021) analyzed the environmental impact of remittances to the Philippines in a multivariate setting by employing the augmented ARDL and VECM methods. The findings showed that remittances sent to the Philippines intensify ecological footprint in the long run. Lastly, Karasoy (2022) examined remittances' impact on Egypt's environmental quality by using the ARDL method. The results showed that remittances increase environmental damage in the long run and in the short run.

To the best of the author's knowledge, the first multiple-country study on the remittances-environment nexus is conducted by Rahman et al. (2019). Rahman et al. (2019) examined the foreign investments-energy-remittances-environment nexus for 6 Asian countries via the ARDL method. Their individual long-run findings for each country revealed that remittances increase pollution in the Philippines, Bangladesh, and Sri Lanka. Khan, Ahmad, & Khan (2020) tested the remittances-led emissions hypothesis for the BRICS (Brazil, Russia, India, China, and South Africa) countries by using panel data methods. Their panel results confirmed that remittances increase environmental pollution. Moreover, the remittances-led emissions hypothesis is validated for Brazil, Russia, and China, whereas it is rejected for India. Qingquan et al. (2020) assessed the effect of monetary policy on pollution for 14 Asian economies via panel data methods. Their model also included remittances, and their panel data analysis showed that remittances increase carbon emissions. Moreover, they also found a positive association between remittances and pollution in China, India, Indonesia, Iran, Pakistan, and Sri Lanka; however, a negative link exists between the two variables in Korea, Malaysia, the Philippines, and Thailand. Usman & Jahanger (2021) examined how remittances and institutional quality impact ecological footprint in a sample of 93 countries via the panel quantile regression technique. Their findings regarding remittances revealed that from 5th

to 70th quantiles, remittances have an increasing impact on ecological footprint, yet this impact turns negative in higher quantiles.

Wawrzyniak & Doryń (2020) studied the same subject for 93 emerging and developing economies via the generalized method of moments (GMM) approach. Their results disclosed that remittances do not impose any significant effect on carbon emissions. For a sample of 97 countries, Yang, Jahanger, & Khan (2020) studied how remittances change carbon emissions by using the two-step GMM and the two-stage least squares (2SLS) techniques. Their findings showed that remittances increase emissions in both the whole panel and the sub-samples. Also, they claimed that remittances' intensifying impact on pollution is higher in developing and emerging countries compared to developed ones. Yang, Jahanger, & Ali (2021) provided further evidence regarding the remittances' impact on the ecological footprint of the BICS (Brazil, India, China, and South Africa) countries by utilizing panel data methods. Their panel data results supported the majority of the previous findings indicating that remittances increase ecological footprint. In addition, their country-specific findings revealed that remittances to Brazil, India, and South Africa intensify their ecological footprints, yet remittances to China do not have any significant impact on its ecological footprint. Finally, Elbatanony et al. (2021) analyzed the environmental effects of remittances for a sample of 51 countries via the method of moments quantile regression with fixed effects. Their results revealed that remittances have mixed effects on carbon emissions, depending on the income level of the sample countries.

To summarize, the reviewed studies above indicate that remittance inflows generally intensify energy consumption and the majority of these studies revealed that remittances result in greater environmental damage. Moreover, this literature review points out the following gaps that are aimed to be filled by this chapter. First, although there are some studies that analyzed the effects of remittances on energy use and environmental performance, there is not a single study that examined the remittances' effect on energy security. The energy security risk index employed in this chapter is a multifaceted variable covering numerous indicators, including various energy use and environmental metrics. The author believes that examining remittances' long-run influence on energy security may provide fresh insights. Second, no country-specific or regional studies on the MENA region investigated the remittances' impact on energy security. This chapter attempts to fill this gap by considering 3 MENA countries. Last, the majority of studies on this subject do not account for structural breaks, or structural breaks are considered to a limited extend. To alleviate this issue, unit root and cointegration tests that are employed in this chapter contain the Fourier-frequency components that allow for unknown numbers and forms of structural breaks (see the next section). Given that the sample countries analyzed in this chapter had experienced many different economic, political, and energy-related crises, the author believes that the methods selected are more appropriate.

DATA, MODEL, AND METHODOLOGY

In this section, data, model, and methods that are utilized in this chapter are explained. In the first sub-section, data and model are described. The second sub-section outlines the methods.

Data and Model

As a result of data availability, this chapter uses yearly time series spanning the 1980-2018 period for Egypt, Morocco, and Tunisia. In line with many studies conducted on this subject, the econometric model utilized for each country is a bi-variate one, and it is mathematically shown in the following equation:

$$LESRI_t = \alpha_0 + \beta_1 LREM_t + u_t \tag{1}$$

In Equation (1), the dependent variable ESRI and the independent variable REM show the energy security risk index and received personal remittances (% of GDP), respectively. Also, α_0 is the constant term, and β_1 shows the (long run) coefficient of remittances. Additionally, L indicates that the logarithms of the time series are used. The subscript t corresponds to time, and u_t is the error term. Time series for the energy security risk indices are obtained from the Global Energy Institute (Global Energy Institute - International Energy Security Risk Index, 2021), and time series for remittances are collected from the World Bank's databank (World Bank, 2020). Descriptive statistics are presented in the appendix.

Methodology

In this section, the methods employed in this chapter are summarized. In the first sub-section, the Fourier-Dickey-Fuller (*Fourier-DF*) stationarity test is explained. In the second sub-section, the Fourier-cointegration test is summarized. Lastly, in the third sub-section, the canonical cointegrating regression (CCR), the dynamic, and the fully modified ordinary least squares (DOLS and FMOLS) estimation methods are summed up.

The Fourier-DF Unit Root Test

In this study, besides the commonly used unit root tests of Phillips-Perron (PP) (Phillips & Perron, 1988) and Augmented Dickey-Fuller (ADF) (Dickey & Fuller, 1979), the Fourier-DF test of Enders & Lee (2012) is utilized. In lieu of selecting a specific number, date, and form for breaks, this unit root test incorporates a Fourier frequency component into its regression equation to account for (various) breaks (Enders & Lee, 2012: 196). Given that only one frequency (*k*) is utilized, the testing regression for the Fourier-DF test becomes:

$$\Delta z_t = \rho z_{t-1} + a_1 + a_2 t + a_3 \sin\left(2\pi kt/T\right) + a_4 \cos\left(2\pi kt/T\right) + e_t \quad (2)$$

Based on Equation (2), t-statistic for the null hypothesis that $\rho=0$ is denoted as τ_{DF} (Enders & Lee, 2012: 197). If the calculated test statistic is above the critical value proposed by the Enders & Lee (2012) study, this null hypothesis will be rejected. In other words, it will be concluded that the variable of interest is stationary.

The Fourier-cointegration Test

This study employs the Fourier-cointegration test proposed by Tsong et al. (2016) to examine whether a long run association exists between energy security risk (*LESRI*) and remittance inflows (*LREM*). The testing procedure for this method is based on the following regression:

$$z_t = d_t + x_t'\beta + \psi_t, t = 1,2,\ldots,T \quad (3)$$

In Equation (3), Where $\psi_t = \gamma_t + v_{1t}, \gamma_t = \gamma_{t-1} + \varepsilon_t$ with $\gamma_0=0$ and $x_t = x_{t-1} + v_{2t}$. Furthermore, ε_t is *iid* with variance σ_ε^2 and mean 0, indicating that γ_t is a stochastic process with 0 mean. The deterministic component, which is d_t, is shown as

$$d_t = \sum_{i=0}^{m} \delta_i t^i + f_t \; with \, m = 0 \, or \, 1 \quad (4)$$

Also,

$$f_t = \alpha_k \sin\left[2k\pi t / T\right] + \beta_k \cos\left[2k\pi t / T\right] \quad (5)$$

Based on this setting, in addition to υ_{1t}, υ_{2t} is also stationary, and y_t and x_t are $I(1)$. Definitely, given that $\sigma_\varepsilon^2 = 0$, $\psi_t = \upsilon_{1t}$ is a stationary process. This result means y_t (*LESRI*) and x_t (*LREM*) are cointegrated. Accordingly, the null hypothesis of cointegration and the alternative hypothesis of non-cointegration can be stated as

$$H_o : \sigma_\varepsilon^2 = 0 \, against \, H_1 : \sigma_\varepsilon^2 > 0 \quad (6)$$

To summarize, if the calculated test statistic is below the proposed critical value, then the null hypothesis of cointegration will be accepted. In addition, an F-test on the Fourier component is also conducted to determine whether the Fourier component should be included in this testing procedure. The significance of this F-test shows that the Fourier component should be kept in the testing procedure (Tsong et al., 2016: 1087-1088, 1092).

Estimating the Long Run Impacts: The FMOLS, DOLS, and CCR Procedures

If the long run association between remittance inflows and energy security risk is confirmed, the long run impact that remittance inflows impose on energy security will be estimated by three procedures: the FMOLS of Phillips & Hansen (1990), the CCR which was proposed by Park (1992), and lastly, the DOLS that was put forward by Saikkonen (1992) and Stock & Watson (1993) studies.

In order to summarize these procedures[1], the following cointegrating equation and system of equations should be proposed for the $\left(y_t, X_t^{'}\right)$ time series vector process:

$$y_t = X_t^{'}\beta + D_{1t}^{'}\gamma_1 + u_{1t} \quad (7)$$

$$X_t = \amalg_{21}^{'} D_{1t} + \amalg_{22}^{'} D_{2t} + \varepsilon_{2t} \, with \, \Delta\varepsilon_{2t} = u_{2t} \quad (8)$$

Where $D_t = \left(D_{1t}^{'}, D_{2t}^{'}\right)^{'}$ is the deterministic trend regressors, and Equation (8) shows the system of equations that govern the X_t regressors. Additionally, u_{1t} is the cointegrating equation error, and u_{2t} shows the regressor innovations.

It is assumed that strictly stationary and ergodic innovations, $u_t = \left(u_{1t}, u_{2t}^{'}\right)^{'}$, with 0 mean have the following (one sided) long-run covariance (**Λ**), (nonsingular long-run) covariance (**Ω**), and contemporaneous covariance (**Σ**) matrices:

$$\Lambda = \sum_{j=0}^{\infty} E\left(u_t u_{t-j}^{'}\right) = \begin{pmatrix} \lambda_{11} & \lambda_{12} \\ \lambda_{21} & \Lambda_{22} \end{pmatrix}$$

$$\Sigma = E\left(u_t u_t^{'}\right) = \begin{pmatrix} \sigma_{11} & \sigma_{12} \\ \sigma_{21} & \Sigma_{22} \end{pmatrix} \tag{9}$$

$$\Omega = \sum_{j=-\infty}^{\infty} E\left(u_t u_{t-j}^{'}\right) = \begin{pmatrix} \omega_{11} & \omega_{12} \\ \omega_{21} & \Omega_{22} \end{pmatrix} = \Lambda + \Lambda' - \Sigma$$

If the time series in consideration (*LREM* and *LESRI*) are *I(1)* and cointegrated, the static ordinary least squares (SOLS) estimates are consistent and converge faster than standard. However, the problem arises if there is a long-run correlation between the cointegrating equation errors (u_{1t}) and regressor innovations, u_{2t} (ω_{12}) or between the regressor innovations (λ_{12}) and u_{1t}. If this is the case, then the conventional testing processes are invalid because the SOLS estimates have non-Gaussian, asymmetric, and biased asymptotic distribution, in addition to including non-scalar nuisance parameters. To resolve this issue, the aforementioned three estimation methods (FMOLS, DOLS, and CCR) are proposed. The first method, the FMOLS, solves this issue by transforming only the dependent variable:

$$y_t^{+} = y_t - \hat{\omega}_{12}\hat{\Omega}_{22}^{-1}\hat{u}_{2t} \tag{10}$$

On the other hand, the second estimation method (the CCR) resolves this problem by modifying both the dependent and the independent variables:

$$y_t^{*} = y_t - \left(\hat{\Sigma}^{-1}\hat{\Lambda}_2\tilde{\beta} + \begin{pmatrix} 0 \\ \hat{\Omega}_{22}^{-1}\hat{\omega}_{21} \end{pmatrix}\right)^{'} \hat{u}_t \text{ and } X_t^{*} = X_t - \left(\hat{\Sigma}^{-1}\hat{\Lambda}_2\right)^{'}\hat{u}_t \tag{11}$$

where $\tilde{\beta}$ is a consistent estimator of cointegrating equation coefficient and $\hat{\Lambda}_2 = \left(\hat{\Lambda}_{12}, \hat{\Lambda}_{22}^{'}\right)'$.

Lastly, the third method, namely the DOLS estimators, use the lag and the lead of ΔX_t to augment the regression. By this augmentation, the long-run correlation between u_{1t} and u_{2t} is soaked up. This DOLS procedure is summarized in the following equation:

$$y_t = X_t^{'}\beta + D_{1t}^{'}\gamma_1 + \sum_{j=-l}^{p} \Delta X_{t+j}^{'}\delta + e_{1t} \tag{12}$$

In Equation (12), l and p show the lag(s) and the lead(s), respectively.

EMPIRICAL FINDINGS

In this section, empirical findings for each country are revealed. First, the unit root test results are shown in Table 2.

Table 2. Stationarity tests' results

Variables	ADF	PP	Fourier-ADF
	t-statistic (L)	t-statistic (B)	τ_{DF} (L)
Egypt			
LREM	-1.413 (0)	-1.394 (3)	-2.688 (3)
LESRI	-2.246 (1)	-2.014 (3)	-3.565 (3)
ΔLREM	-4.985*** (1)	-5.599*** (5)	-5.288*** (1)
ΔLESRI	-4.831*** (0)	-4.849*** (2)	-4.684** (0)
Morocco			
LREM	-2.843 (0)	-2.843 (0)	-3.224 (2)
LESRI	-1.443 (0)	-1.499 (2)	-3.542 (3)
ΔLREM	-5.534*** (1)	-7.265*** (8)	-5.868*** (1)
ΔLESRI	-5.762*** (0)	-5.777*** (5)	-5.942*** (3)
Tunisia			
LREM	-3.449* (1)	-2.663 (8)	-2.775 (3)
LESRI	-1.336 (1)	-0.762 (0)	-3.722 (3)
ΔLREM	-5.523*** (0)	-8.844*** (36)	-5.428*** (2)
ΔLESRI	-4.487*** (0)	-4.388*** (8)	-3.230*** (3)

Notes: *, **, and *** show 10%, 5% and 1% significance levels, respectively. The ADF and PP test equations include linear trend and constant in levels and contain only constant in the first differences. For the ADF test, the lag length selection is based on the Akaike information criterion (AIC). For the PP test, the Bartlett kernel spectral estimation method is chosen, and the bandwidth selection is based on the Newey-West approach. The Fourier frequency for the Fourier-ADF test is 1. L: lag length, B: bandwidth.

According to the results of the typical ADF and PP tests presented in Table 2, for each country, both LREM and LESRI variables are stationary in their first differences at 1% significance level. In other terms, the selected time series are *I(1)*. Additionally, the Fourier-ADF test results agree with the results of the ADF and PP tests. They also indicate that the time series are *I(1)*. Because the time series are stationary in their first differences, the existence of the long run association can be examined for the proposed model. In that respect, the results of the cointegration test are displayed in Table 3.

Table 3. Results of the Tsong et al. (2016) cointegration test

	Model	Model Type	Test Statistic	F-test
Egypt	*LESRI= f(LREM)*	*Level Shift*	0.082	3.814*
		Level Shift with Trend	0.026	18.408***
Morocco	*LESRI= f(LREM)*	*Level Shift*	0.050	74.323***
		Level Shift with Trend	0.029	49.160***
Tunisia	*LESRI= f(LREM)*	*Level Shift*	0.036	72.540***
		Level Shift with Trend	0.035	40.464***

Notes: The Fourier frequency for the co-integration test is 1. The dynamic ordinary least squares (DOLS) estimates are used to retrieve the co-integration test results. The critical values for the co-integration test are as follows. For the level shift model: 0.198 (1%), 0.124 (5%), and 0.095 (10%) and for the level shift with trend model: 0.063 (1%), 0.048 (5%), and 0.042 (10%). The critical values for the F-test are 3.352 (10%), 4.066 (5%), and 5.774 (1%) for the level shift model and 3.306 (10%), 4.019 (5%), and 5.860 (1%) for the level shift model with trend. The critical values are retrieved from the Tsong et al. (2016) study. * and *** show significance at 10% and 1% levels.

In the last column of Table 3, F-test results for the significance of the Fourier approximations are reported. All the F-statistics are significant, indicating that the Fourier components should be kept in the estimation procedure. The test statistics for the cointegration are reported for 2 different specifications. Moreover, none of these statistics is significant for all three sample countries; the null hypothesis of cointegration cannot be rejected. In other words, for each country, the ESRI and remittances have a long-run association. At this stage, the long run effect of remittances on energy security can be estimated for each country. In this regard, three different estimation methods (DOLS, FMOLS, and CCR) are used. The estimation results for each method and country are shown in Table 4.

Table 4. DOLS, FMOLS, and CCR estimation results

A) Constant				B) Constant and Trend		
	Method	Constant (p-value)	LREM's Coefficient (p-value)	Constant (p-value)	Trend (p-value)	LREM's Coefficient (p-value)
Egypt	*DOLS*	6.5805*** (0.0000)	0.2886*** (0.0005)	6.6329*** (0.0000)	-0.0027 (0.2437)	0.2876*** (0.0002)
	FMOLS	6.5655*** (0.0000)	0.2981*** (0.0000)	6.6207*** (0.0000)	-0.0037 (0.1506)	0.3073*** (0.0000)
	CCR	6.6766*** (0.0000)	0.2389*** (0.0000)	6.7369*** (0.0000)	-0.0036 (0.1672)	0.2438*** (0.0000)
Morocco	*DOLS*	6.8552*** (0.0000)	0.1070 (0.4989)	6.6649*** (0.0000)	0.0096*** (0.0000)	0.1087* (0.0825)
	FMOLS	6.3728*** (0.000)	0.3713 (0.1135)	6.5303*** (0.0000)	0.0102*** (0.0000)	0.1751** (0.0187)
	CCR	6.4334*** (0.0000)	0.3377 (0.1226)	6.5164*** (0.0000)	0.0101*** (0.0000)	0.1837** (0.0196)
Tunisia	*DOLS*	6.2089*** (0.0000)	0.5649*** (0.0036)	6.2484*** (0.0000)	0.0062*** (0.0006)	0.4529*** (0.0027)
	FMOLS	5.9586*** (0.0000)	0.7331*** (0.0001)	6.2846*** (0.0000)	0.0095*** (0.0000)	0.3781*** (0.002)
	CCR	6.0119*** (0.0000)	0.6976*** (0.0001)	6.3269*** (0.0000)	0.0092*** (0.0000)	0.3537*** (0.0023)

Notes: The dependent variable is LESRI. DOLS: Dynamic least squares, FMOLS: Fully modified least squares, and CCR: Canonical co-integrating regression. Maximum lag length and the Fourier frequency are set to 1. The optimum lag length selection process is based on the AIC for all models. ***, **, and * indicate significance at 1%, 5%, and 10% levels. The Fourier approximations (i.e., sine and cosine terms) are included in the equations as deterministic regressors. The estimation results of the Fourier terms for each model are presented in the appendix section.

Panel A of Table 4 presents the long run coefficients of LREM (remittance inflows) for the constant model specification. In Panel A, all the estimated coefficients are positive indicating that remittance inflows increase ESRIs of these countries. However, the coefficients are significant only for Egypt and Tunisia. In other words, in the constant model specification, remittance inflows significantly increase energy insecurities (i.e., energy security risks) in Egypt and Tunisia in the long run.

According to the DOLS findings in Panel A of Table 4, a 1% increase (decrease) in remittances leads to 0.29% and 0.56% increases (decreases) in the energy security risk indices of Egypt and Tunisia, respectively in the long run. Similarly, the FMOLS findings show that a positive (negative) 1% change in remittance inflows increases (decreases) Egypt's and Tunisia's long run energy insecurities approximately by 0.3% and 0.73%, respectively. Lastly, the CCR estimates indicate that a 1% change (increase) in remittances corresponds to 0.24% and 0.7% changes (increases) in energy security risks of Egypt and Tunisia, respectively.

In Table 4's Panel B, the long run impacts of remittances are reported with respect to the constant & trend model specification for selected MENA countries. Similar to the findings in Panel A, the long run results in Panel B reveal that remittances to selected MENA countries positively and significantly impact energy insecurity.

The DOLS estimates in Panel B of Table 4 show that a 1% increase (decrease) in remittances positively (negatively) changes the energy security risk indices of Egypt by 0.29% of Morocco by 0.11%, and of Tunisia by 0.45%. Likewise, the FMOLS results reveal that a 1% increase (decrease) in remittances leads to 0.31%, 0.18%, and 0.38% increases (decreases) in energy security risks of Egypt, Morocco, and Tunisia, respectively. Finally, in consonant to the previous results, the CCR estimates confirm that a 1% increase (decrease) in remittances sent to Egypt, Morocco, and Tunisia deepens (lessens) the long run energy insecurities of these countries by 0.24%, 0.18%, and 0.35%, respectively.

To sum up, the empirical findings in this chapter show that a long run association between remittance inflows and energy (in)security existed in Egypt, Morocco, and Tunisia for the 1980-2018 period. Moreover, in these countries, remittance inflows unanimously increased energy security risks (i.e., energy insecurities) in the long run. Last, the results revealed that the long run increasing impact of remittances on energy insecurity is greater in Tunisia than in Egypt and Morocco.

Given the fact that the ESRI used in this chapter also includes energy consumption and environmental variables, and all three sample countries are heavily dependent on fossil fuels, it can be stated that the findings in this chapter generally agree with the majority of the previous studies that showed remittance inflows increase energy consumption and/or environmental damage. However, the empirical findings in this chapter disagree with the results in Sharma et al. (2019) and Wawrzyniak & Doryń (2020) because the former found that remittance inflows decrease environmental pollution in the long run, and the latter did not find any significant association. Moreover, the findings in Das et al. (2021) do not confirm the empirical results of this chapter because their results showed that remittances to Bangladesh increase renewable energy use. Nonetheless, the differences in empirical findings are not unexpected, given the fact that the former studies considered different countries, periods, models and employed varying methods.

CONCLUSION

In order to develop agile energy security strategies, alternative perspectives are needed. Therefore, exploring new venues for solving the energy security issues, which have been a constant theme in the MENA countries, may provide new policy options and instruments to policymakers. Besides energy insecurity, another common feature of some MENA countries (such as Egypt, Morocco, and Tunisia) is that they receive significant amounts of remittances. In this respect, this chapter empirically studied how remittance inflows to Egypt, Morocco, and Tunisia impacted their energy insecurities for the 1980-2018 period from a macro perspective by utilizing time series analysis.

The empirical results of this chapter revealed that remittance inflows and energy security risk index are cointegrated. Additionally, inward remittances to Egypt, Morocco, and Tunisia increase the energy security risks of these countries in the long run. Also, compared to Egypt and Morocco, remittances' worsening impact on energy security is higher in Tunisia. These findings clearly indicate that remittances are linked to continuing energy insecurity issues in these countries. Further, these results also indicate that remittances can be instrumentalized to deal with energy security issues in the MENA region in the long run. Given that remittance inflows are viewed as a lifeline for these countries (UN, 2020: 22) and another SDGs -*Goal 10*- includes lowering remittance costs to 3% (UN, 2021), it may not be viable to offer that these countries should lower their dependency on remittance inflows. However, our findings show that some components of the 10th and 7th SDGs are not compatible with each other at least for Egypt, Morocco, and Tunisia. In that respect, some policy suggestions can be propounded to limit the remittances' increasing influence on energy insecurity in these countries.

First, remittance-receiving households should be encouraged to implement and utilize renewable energy sources (such as solar and wind) to increase the share of renewables in these countries. This encouragement can be achieved through various incentives. For instance, lower-cost rates can be applied to remittance-receiving households that implement such renewable technologies. Although the effects of these implementations may be limited in the short run, the effects would definitely be significant in the long run. In the long run, this approach would limit fossil fuel consumption in these households, and at a macro level, would eventually increase renewable energy use in these countries. As fossil fuel consumption falls and renewable energy use increases, this situation would significantly contribute to limiting energy insecurity in the long run.

Second, households that use remittances in innovative investments should be incentivized by easier access to credits, lower interest rate, credit, and transfer costs. This would lead to the development and usage of novel environmentally friendly technologies. Consequently, in the long run, this may positively affect the energy securities of these countries.

Last, a proper financial system that is compatible with the international financial system should be developed to encourage remittance-receiving households to save and discourage unofficial remittance inflows. Increasing savings mean more credits, and these potential credits can be granted to suitable investors to develop and produce renewable energy and environmentally harmless production methods.

Future research should investigate other countries and regions to analyze whether remittances have a similar impact on energy security. Additionally, besides country-specific examination, panel data methods can be utilized for regional analysis.

REFERENCES

Ahmad, M., Ul Haq, Z., Khan, Z., Khattak, S. I., Ur Rahman, Z., & Khan, S. (2019). Does the inflow of remittances cause environmental degradation? Empirical evidence from China. *Economic Research-Ekonomska Istraživanja, 32*(1), 2099–2121. doi: 10.1080/1331677X.2019.1642783

Amuedo-Dorantes, C. (2014). The good and the bad in remittance flows. *IZA World of Labor: Evidence-Based Policy Making*, (97). Advance online publication. doi:10.15185/izawol.97

Brown, L., McFarlane, A., Campbell, K., & Das, A. (2020). Remittances and CO2 emissions in Jamaica: An asymmetric modified environmental kuznets curve. *Journal of Economic Asymmetries*, *22*, e00166. doi:10.1016/j.jeca.2020.e00166

Das, A., & McFarlane, A. (2020). Remittances and disaggregated energy consumption in Bangladesh. *International Review of Economics*, *67*(2), 251–268. doi:10.100712232-019-00339-4

Das, A., & McFarlane, A. (2021). Remittances, electricity consumption and electric power losses in Jamaica. *Journal of Economic Studies*. doi:10.1108/JES-09-2020-0466

Das, A., McFarlane, A., & Carels, L. (2021). Empirical exploration of remittances and renewable energy consumption in Bangladesh. *Asia-Pacific Journal of Regional Science*, *5*(1), 65–89. doi:10.100741685-020-00180-6

Dickey, D. A., & Fuller, W. A. (1979). Distribution of the Estimators for Autoregressive Time Series With a Unit Root. *Journal of the American Statistical Association*, *74*(366), 427. doi:10.2307/2286348

Elbatanony, M., Attiaoui, I., Ali, I. M. A., Nasser, N., & Tarchoun, M. (2021). The environmental impact of remittance inflows in developing countries: Evidence from method of moments quantile regression. *Environmental Science and Pollution Research International*, *28*(35), 48222–48235. Advance online publication. doi:10.100711356-021-13733-9 PMID:33907951

Enders, W., & Lee, J. (2012). The flexible Fourier form and Dickey-Fuller type unit root tests. *Economics Letters, 117*(1), 196–199. doi:10.1016/j.econlet.2012.04.081

Global Energy Institute. (2020). *International Index of Energy Security Risk - Assessing Risk in a Global Energy Market.* International Index of Energy Security Risk 2018 Edition. Retrieved from https://www.globalenergyinstitute.org/sites/default/files/IESRI-Report_2020_4_20_20.pdf

Global Energy Institute - International Energy Security Risk Index. (2021). Retrieved July 21, 2021, from https://www.globalenergyinstitute.org/international-energy-security-risk-index

IEA. (2021). *Energy security*. Retrieved July 21, 2021, from https://www.iea.org/areas-of-work/ensuring-energy-security

Jushi, E., Hysa, E., Cela, A., Panait, M., & Voica, M. C. (2021). Financing Growth through Remittances and Foreign Direct Investment: Evidences from Balkan Countries. *Journal of Risk and Financial Management, 14*(3), 1–17. doi:10.3390/jrfm14030117

Karasoy, A. (2021). How do remittances to the Philippines affect its environmental sustainability? Evidence based on the augmented ARDL approach. *Natural Resources Forum, 45*(2), 120–137. doi:10.1111/1477-8947.12218

Karasoy, A. (2022). How Do Consuming Alternative Energy Sources and Remittance Inflows Impact Egypt's Ecological Footprint? *Beykent Üniversitesi Sosyal Bilimler Dergisi, 15*(1), 8–28. doi:10.18221/bujss.1060051

Khan, Z. U., Ahmad, M., & Khan, A. (2020). On the remittances-environment led hypothesis: Empirical evidence from BRICS economies. *Environmental Science and Pollution Research International, 27*(14), 16460–16471. doi:10.100711356-020-07999-8 PMID:32124280

Kibria, M. G. (2021). Environmental downfall in Bangladesh: Revealing the asymmetric effectiveness of remittance inflow in the presence of foreign aid. *Environmental Science and Pollution Research International.* Advance online publication. doi:10.100711356-021-15619-2 PMID:34338983

Le, T.-H., & Nguyen, C. P. (2019). Is energy security a driver for economic growth? Evidence from a global sample. *Energy Policy, 129*, 436–451. doi:10.1016/j.enpol.2019.02.038

Li, J., Jiang, T., Ullah, S., & Majeed, M. T. (2021). The dynamic linkage between financial inflow and environmental quality: Evidence from China and policy options. *Environmental Science and Pollution Research International*. Advance online publication. doi:10.100711356-021-15616-5 PMID:34341938

Mahalik, M. K., Villanthenkodath, M. A., Mallick, H., & Gupta, M. (2021). Assessing the effectiveness of total foreign aid and foreign energy aid inflows on environmental quality in India. *Energy Policy*, *149*, 112015. doi:10.1016/j.enpol.2020.112015

Naeem Nawaz, S. M., & Alvi, S. (2018). Energy security for socio-economic and environmental sustainability in Pakistan. *Heliyon*, *4*(10), 1–21. doi:10.1016/j.heliyon.2018.e00854 PMID:30364534

Neog, Y., & Yadava, A. K. (2020). Nexus among CO2 emissions, remittances, and financial development: A NARDL approach for India. *Environmental Science and Pollution Research International*, *27*(35), 44470–44481. doi:10.100711356-020-10198-0 PMID:32770469

Park, J. Y. (1992). Canonical Cointegrating Regressions. *Econometrica*, *60*(1), 119–143. doi:10.2307/2951679

Phillips, P. C. B., & Hansen, B. E. (1990). Statistical Inference in Instrumental Variables Regression with I(1) Processes. *The Review of Economic Studies*, *57*(1), 99–125. doi:10.2307/2297545

Phillips, P. C. B., & Perron, P. (1988). Testing for a Unit Root in Time Series Regression. *Biometrika*, *75*(2), 335–346. doi:10.1093/biomet/75.2.335

Qingquan, J., Khattak, S. I., Ahmad, M., & Ping, L. (2020). A new approach to environmental sustainability: Assessing the impact of monetary policy on CO2 emissions in Asian economies. *Sustainable Development*, *28*(5), 1331–1346. doi:10.1002d.2087

Rahman, M. M., Hosan, S., Karmaker, S. C., Chapman, A. J., & Saha, B. B. (2021). The effect of remittance on energy consumption: Panel cointegration and dynamic causality analysis for South Asian countries. *Energy*, *220*, 119684. doi:10.1016/j.energy.2020.119684

Rahman, Z. U., Cai, H., & Ahmad, M. (2019). A New Look at The Remittances-FDI-Energy-Environment Nexus in The Case of Selected Asian Nations. *The Singapore Economic Review*, 1–19. doi:10.1142/S0217590819500176

Sahoo, M., & Sethi, N. (2020). Does remittance inflow stimulate electricity consumption in India? An empirical insight. *South Asian Journal of Business Studies*. doi:10.1108/SAJBS-05-2020-0158

Saikkonen, P. (1992). Estimation and Testing of Cointegrated Systems by an Autoregressive Approximation. *Econometric Theory*, *8*(01), 1–27. doi:10.1017/S0266466600010720

Sayeh, A., & Chami, R. (2020). Lifeliness in danger: The COVID-19 pandemic threatens to dry up a vital source of income for poor and fragile countries. *Finance & Development*, *57*(2), 16–19. https://www.imf.org/external/pubs/ft/fandd/2020/06/COVID19-pandemic-impact-on-remittance-flows-sayeh.htm

Sharma, K., Bhattarai, B., & Ahmed, S. (2019). Aid, Growth, Remittances and Carbon Emissions in Nepal. *Energy Journal*, *40*(1), 129–141. doi:10.5547/01956574.40.1.ksha

Stock, J. H., & Watson, M. W. (1993). A Simple Estimator of Cointegrating Vectors in Higher Order Integrated Systems. *Econometrica*, *61*(4), 783–820. doi:10.2307/2951763

The World Bank. (2021). *World Bank Open Data*. Retrieved July 22, 2021, from https://data.worldbank.org/

Tsong, C.-C., Lee, C.-F., Tsai, L.-J., & Hu, T.-C. (2016). The Fourier approximation and testing for the null of cointegration. *Empirical Economics*, *51*(3), 1085–1113. doi:10.100700181-015-1028-6

UN. (2020). *The Sustainable Development Goals Report 2020*. United Nations. Retrieved from https://unstats.un.org/sdgs/report/2020/The-Sustainable-Development-Goals-Report-2020.pdf

UN. (2021). *SDG: Goal 10*. Retrieved September 4, 2021, from https://sdgs.un.org/goals/goal10

United States Chamber of Commerce. (2014). *Index of U.S. Energy Security Risk: Assessing America's Vulnerabilities in a Global Energy Market*. Retrieved from https://www.globalenergyinstitute.org/sites/default/files/2020-10/024036 Global Energy Institute US Index_Web.pdf

Usman, M., & Jahanger, A. (2021). Heterogeneous effects of remittances and institutional quality in reducing environmental deficit in the presence of EKC hypothesis: A global study with the application of panel quantile regression. *Environmental Science and Pollution Research International*, *28*(28), 37292–37310. Advance online publication. doi:10.100711356-021-13216-x PMID:33713269

Villanthenkodath, M. A., & Mahalik, M. K. (2020). Technological innovation and environmental quality nexus in India: Does inward remittance matter? *Journal of Public Affairs*. Advance online publication. doi:10.1002/pa.2291

von Hippel, D., Suzuki, T., Williams, J. H., Savage, T., & Hayes, P. (2011). Energy security and sustainability in Northeast Asia. *Energy Policy*, *39*(11), 6719–6730. doi:10.1016/j.enpol.2009.07.001

Wang, Q., & Wu, N. (2012). Long-Run Covariance and its Applications in Cointegration Regression. *The Stata Journal: Promoting Communications on Statistics and Stata*, *12*(3), 515–542. doi:10.1177/1536867X1201200312

Wawrzyniak, D., & Doryń, W. (2020). Does the quality of institutions modify the economic growth-carbon dioxide emissions nexus? Evidence from a group of emerging and developing countries. *Economic Research-Ekonomska Istraživanja*, *33*(1), 124–144. doi:10.1080/1331677X.2019.1708770

WDI. (2021). *The World Bank's World Development Indicators*. Retrieved August 29, 2021, from https://databank.worldbank.org/source/world-development-indicators

Woldemariam, K., & Yiheyis, Z. (2017). Remittances and Economic Development in Africa: A Review of Empirical Evidence. In Foreign Capital Flows and Economic Development in Africa (pp. 401–421). Palgrave Macmillan US. doi:10.1057/978-1-137-53496-5_18

World Bank. (2020). *World Development Indicators*. Retrieved August 9, 2020, from https://databank.worldbank.org/data/home.aspx

Xu, X., Yu, J., Zhang, D., & Ji, Q. (2021). Energy insecurity, economic growth, and the role of renewable energy: A cross-country panel analysis. *The Singapore Economic Review*, *66*(2), 323–343. doi:10.1142/S021759081943001X

Yang, B., Jahanger, A., & Ali, M. (2021). Remittance inflows affect the ecological footprint in BICS countries: Do technological innovation and financial development matter? *Environmental Science and Pollution Research International*, *28*(18), 23482–23500. doi:10.100711356-021-12400-3 PMID:33449318

Yang, B., Jahanger, A., & Khan, M. A. (2020). Does the inflow of remittances and energy consumption increase CO2 emissions in the era of globalization? A global perspective. *Air Quality, Atmosphere & Health*, *13*(11), 1313–1328. doi:10.100711869-020-00885-9

ENDNOTE

1 This summary is based on the Wang & Wu (2012: 532-534) study.

APPENDIX

Table 5. Estimation results for the Fourier terms

	A) Constant			B) Constant and Trend	
	Method	sin (p-value)	cos (p-value)	sin (p-value)	cos (p-value)
Egypt	*DOLS*	-0.0997*** (0.0029)	-0.0458 (0.2044)	-0.1312*** (0.0018)	-0.0488 (01354)
	FMOLS	-0.1111*** (0.0006)	-0.0387 (0.2477)	-0.1599*** (0.0011)	-0.0422 (0.2212)
	CCR	-0.0929*** (0.0008)	-0.0152 (0.5726)	-0.1384*** (0.002)	-0.0171 (0.5404)
Morocco	*DOLS*	-0.1667*** (0.0000)	0.1205*** (0.0001)	-0.0477** (0.0183)	0.1214*** (0.0000)
	FMOLS	-0.1345** (0.013)	0.1219*** (0.0096)	-0.0316 (0.2116)	0.1196*** (0.0000)
	CCR	-0.1385*** (0.0094)	0.1203** (0.0108)	-0.0317 (0.1976)	0.1197*** (0.0000)
Tunisia	*DOLS*	0.0339 (0.1206)	0.1104*** (0.0000)	0.1003*** (0.0002)	0.1198*** (0.0000)
	FMOLS	0.0443 (0.1096)	0.0749*** (0.0084)	0.1301*** (0.0001)	0.1093*** (0.0000)
	CCR	0.0411 (0.1323)	0.0827*** (0.0028)	0.1237*** (0.0001)	0.1129*** (0.0000)

Notes: The dependent variable is LESRI. DOLS: Dynamic least squares, FMOLS: Fully modified least squares, and CCR: Canonical co-integrating regression. ***, **, and * indicate significance at 1%, 5%, and 10% levels.

Table 6. Descriptive statistics

	Egypt		Morocco		Tunisia	
	LREM	**LESRI**	**LREM**	**LESRI**	**LREM**	**LESRI**
Mean	1.877	7.118	1.824	7.041	1.430	7.008
Median	1.804	7.129	1.862	6.974	1.476	6.982
Maximum	2.679	7.282	2.141	7.337	1.615	7.311
Minimum	1.049	6.911	1.538	6.827	1.116	6.792
Std. Dev.	0.440	0.096	0.165	0.174	0.135	0.133
Skewness	-0.036	-0.608	-0.151	0.602	-0.746	0.515
Kurtosis	1.981	2.893	2.019	1.795	2.483	2.601
J-B	1.695	2.421	1.712	4.715*	4.054	1.988
Probability	0.428	0.297	0.424	0.094	0.131	0.370
Correlation Matrices:	*LREM*	*LESRI*	*LREM*	*LESRI*	*LREM*	*LESRI*
LREM	*1*	*0.547*	*1*	*0.504*	*1*	*0.513*
LESRI	*0.547*	*1*	*0.504*	*1*	*0.513*	*1*

Notes: J-B is Jarque-Bera, and * shows significance at 10%.

Chapter 8
Renewable Energy in the Framework of a Sustainable Urbanism:
Solutions Based on a Spanish Legislation Model

María Jesús Garcia García García
University of Valencia, Spain

ABSTRACT

Urban environments, areas in which multiple activities developed by human beings are brought together, have a strong environmental impact. Therefore, in these areas, measures aimed at rational use of natural resources and efficient consumption of energy must be promoted. To this end, energy efficiency must also be accompanied by a policy of diversifying the sources used in energy production, which opens the door to renewable energy promotion and use. The much discussed energy efficiency cannot consist only of measures aimed at saving and containing demand, but also requires the regulation and management of supply that promotes the introduction of renewable energy sources, which is in turn clean energy given its low level of emissions into the atmosphere.

DOI: 10.4018/978-1-7998-9502-2.ch008

INTRODUCTION

The idea of sustainable development has to relate to other concepts and ideas within the framework of its justification and content. We cannot talk about sustainable development without first placing it in a broader context from which it owes. In this sense, a reference to the principle of sustainable development, energy efficiency and the idea of urban rehabilitation are necessary. These are generic concepts within the framework of which energy rehabilitation is framed and which also lend its justification and reason for being.

The first term to which it is necessary to allude to explain the idea of energy rehabilitation has to do with the concept of sustainable development, which for some time now has been used in our domestic legislation and whose legal implementation has been carried out mainly through Law 2/2011, of March 4, on Sustainable Economy.

Sustainable development is a concept that combines three interrelated elements and has traditionally been treated differently in Spanish legislation. Perhaps that is why it can be said that in our legal system, the concept is relatively new in that it interrelates different objectives that converge in the different sectoral areas, emphasizing the interdependence between them rather than in their singular treatment (Chu and Majumdar, 2012). These three elements or pillars that converge in the concept of sustainable development are: social, economic and environmental. "Sustainable development is a type of development that advocates first of all the harmonization between economic development and environmental protection, adding social progress; it would therefore be development in which high and stable growth in the production of goods and services is compatible with widespread social progress, environmental protection and prudent and efficient use of natural resources."; it is therefore a delicate building supported by three main pillars: social, economic and environmental in which none of them prevails over the others ".

Accordingly, sustainability has an integrated character, preceded as it is by the objectives of economic recovery, environmental sustainability and social cohesion. It is a general principle with transversal or horizontal projection, capable of crossing different and varied sectoral areas, as many as are relevant to achieve its integrating objectives (Curry and Pillay, 2012).

Among the different sectoral areas transferred by the idea of sustainable development is undoubtedly the field of urban planning and housing. The activity generated in cities has an important environmental impact, so it is necessary to orient urban structures, homes, and buildings under premises that are as respectful as possible with the environment, taking advantage of its economic potential and its effect on the social fabric inhabits it. It is about promoting integrated actions in the urban environment that are in tune with the objectives also integrating sustainable development .

Since sustainable development is an integrative concept, its influence on the urban and housing sector could not fail to have these same connotations, insofar as sustainable urbanism is nothing but a projection of sustained development in a specific area . Urban planning must respond to the requirements of sustainable development, minimizing its commitment to growth and betting on the regeneration of the existing city in order to achieve a sustainable and inclusive urban model, environmentally, socially and economically that improves the quality of life of citizens in urban spaces. In short, we are talking about the integrated objectives of sustainability applied to urban planning, in which the duty of conservation plays a fundamental role.

Spanish Constitution is devoid of explicit references to sustainable development, since it is a relatively new concept in our domestic law. However, the relationship between this principle and certain constitutional rights is evident, since the idea of sustainable urbanism advocates the preservation of urban spaces and buildings in which the constitutional right to decent housing and an adequate environment can be realized. From the perspective of the principle of sustainability, urban planning must be oriented towards rehabilitation in order to comply with article 45 of the Constitution, since rehabilitation does not consume land and makes it possible to exploit and use, rather to reuse, the existing heritage. In this sense, the implications between sustainable development and protection and the right to an adequate environment that guarantees the quality of life are fully established. Sustainable urban development would comply with article 45 of the Constitution in its two aspects: as a requirement imposed on the public authorities to ensure their protection and as the right of citizens to enjoy an adequate environment in order to ensure a certain quality of life. The observation that the urban environment is also the environment, or in other words, that the urban environment also integrates the environment is clearly deduced from this and other recently adopted laws in our legal system. Thus, article 1 of Royal Legislative Decree 7/2015, of October 30, approving the revised text of the Land Law, when defining its object, shows that the actions that are carried out try to ensure citizens an adequate quality of life, and the effectiveness of their right to enjoy decent and proper housing. Article 5 of the act also establishes the rights of the citizen to whose satisfaction the act is subject, including the right to enjoy a decent, adequate and accessible home that is free of noise or other polluting immissions, as well as the right to enjoy an environment and a suitable landscape (Droege, 2011).

On the other hand, Law 7/85 of April 2, regulating the bases of the local regime, in its most recent amendment, also highlighted this point when it speaks in its article 25 of the competence of the municipalities in matters relating to the urban environment.

The connections with the constitutional right to housing are also clear in the extent of maintaining the housing in proper condition for use, ensuring the maintenance of the existing population to promote the social and economic texture of this urban area.

In short, this is indicative that the actions of rehabilitation, renewal and regeneration referred to from the point of view of sustainable development has its legal basis in articles 45 and 46 of the Constitution, without denying the implications that such a treatment of environmental protection will have on the economic and social aspects, such as the revival of economic activity, among others.

And precisely with this integrative purpose, the law on rehabilitation, renovation and urban regeneration is enacted, which combines the treatment of urban development actions in the urban environment and actions on housing and the architectural park from the point of view of the umbrella of sustainability. The point of view from which this law rushing to the processing of such sectoral domains was certainly novel, as if, until then, we were accustomed to witnessing the treatment sector of the subjects based on its consideration of the substantive, that is, as a sectoral domain where the material determined the content of the law, in the case of the mentioned norm, the unifying element of the different precepts contained in it is not the substantive scope or material, but the principle under which aúnaba treatment, that acted as an element of cohesion of the different precepts. We refer to the principle of sustainability, which justified that we found regulated under the umbrella of the same law sectoral areas as varied as urban legislation, precepts on urban rehabilitation, urban leasing legislation or horizontal property. Nothing new if we take into account Law 2/2011 of March 4, Sustainable Economy. The Real Decreto Legislativo 7/2015, de October 30, approving the revised text of the Land Law and Urban Rehabilitation serves now to a more specific, as it regulates subjects with a greater connection to each other that are all grouped under the generic concept of "urban", or more accurately, "urban environment". The content of this law seems to consider the regulation of urban planning, as it affects consolidated urban land, susceptible to rehabilitation, regeneration or renovation actions or new urbanization actions. But the truth is that the law seems to be aimed at regulating urban development actions based on the principle of sustainable development.

BACKGROUND

Obviously, within this perspective of environmental sustainability present in the idea of sustainable urbanism one of the most relevant aspects from the point of view of its impact on the environment is that it has to do with the use and exploitation of energy and the use of certain forms of energy that maximize their consumption causing the minimum environmental impact. And focused as sustainable urbanism is on the idea of maintenance and conservation of the already made city and the existing buildings, this objective is to accommodate the buildings and the existing real estate park to these new energy demands. That is why we talk about energy rehabilitation, meaning a set of actions in the urban environment and in existing buildings and homes that try to adapt the pre-existing elements (housing, but not only this, but also endowments, services and equipment) to the principles of efficiency and energy saving in accordance with the legal framework provided by Law 38/1999, of November 5, on Building Planning and Royal Decree 314/2006, of March 17, Technical Building Code.

Directive 2012/27 / EU states that buildings represent 40% of energy consumption in the European Union, so it becomes clear the need to influence the building sector and improve its energy performance (Lund, 2012) .

These objectives, we find in article 3 of Royal Legislative Decree 7/2015, of October 30, consolidated text of the Law of Soil and Urban Rehabilitation, which speaks of minimizing polluting emissions and greenhouse gases, water and energy consumption and waste production. Also in paragraph i) of the same article these objectives are highlighted, when the law refers to the need to prioritize renewable energies over the use of fossil energy sources and combat energy poverty with measures in favor of efficiency and energy saving . And finally, subparagraph (h) refers to the need to promote the protection of the atmosphere and the use of clean materials, products and technologies that reduce pollutant emissions and greenhouse gases from the construction sector, as well as reused and recycled materials that contribute to improving resource efficiency.

Therefore, energy efficiency is one of the goals to be achieved in the urban environment to achieve the objectives of environmental sustainability that underlie the generic concept of sustainable urban development. Thus, when article 3 of Royal Legislative Decree 7/2015, of October 30, refers to the purposes to which public policies will tend in the urban environment, it bets on a concept of sustainability from the economic, social, environmental point of view that promotes the rational use of natural resources and energy efficiency. The law seems to understand that energy efficiency can promote sustainable environmental conditions by itself and contribute to economic sustainability by generating jobs and employment, which is why its treatment is singled out as an emerging sector with great potential, not only environmental but also economical.

To this end, energy efficiency must also be accompanied by a policy of diversifying the sources used in energy production, which opens the door to the use, promotion and use of renewable energies. The much discussed energy efficiency cannot consist only of measures aimed at saving and containing demand, but also requires the regulation and management of supply that promotes the introduction of renewable energy sources, which is in turn clean energy given its low level of emissions into the atmosphere (Purohit et al., 2013).

In accordance with the above, energy rehabilitation, based on the principle of energy efficiency, would be aimed at:

- Prioritize renewable energies.
- Promote energy savings.
- Reduce pollutant emissions and waste production.

ENERGY EFFICIENCY IN THE CONTEXT OF EUROPEAN UNION REGULATIONS

European legislation should not be forgotten to set specific targets for energy efficiency and emission reductions. In this sense it is necessary to make reference to the Directive 2010/31/EU of the European Parliament and of the Council of May 19 2010 on the energy performance of buildings (object transposition part in the Royal Decree 235/2013 of April 5) and Directive 2012/27/EU of the European Parliament and of the Council of October 25 2012 on energy efficiency, which aims to update the community legal order in the context of the general objective of the Europe 2020 Strategy, this implies the purposes of a 20% reduction in greenhouse gas emissions, an increase in the contribution of renewable energies by 20% and a 20% improvement in energy efficiency. The Directive also sets more ambitious targets for 2050 aimed at reducing the level of CO2 emissions by 80-90% compared with 1990 levels. The measures identified by the said Directive concern key areas and sectors for achieving energy efficiency targets, such as building renovation. In this context, the Directive requires not only a significant percentage of central Government buildings to be renovated annually in order to improve their energy efficiency but also that Member States also establish a strategy to mobilize investment in the renovation of residential buildings in order to improve the energy efficiency of the entire housing stock. The formulas to achieve these objectives are through the control of demand, promoting actions aimed at ensuring savings measures, but also through actions of management and regulation of supply, which includes its diversification, prioritizing the use of renewable energies for energy production over the use of fossil energies.

In this context, the Renewable Energy Plan (PER) 2011-2020 has been approved by resolution of the Council of Ministers of November 11, 2011, setting objectives that are in line with Directive 2009/28/EC of the European Parliament and of the Council of April 23 2009 on the promotion of the use of energy from renewable sources, and in response to the mandates of the Royal Decree 661/2007, which regulates the activity of electricity production under the special regime and the Law 2/2011 of March 4, Sustainable Economy. The PER has the objective of achieving, as indicated by the Community Directive, that in the year 2020 at least 20% of the gross final consumption of energy in Spain comes from the use of renewable sources.

Law 8/2013 of June 26 on urban rehabilitation, regeneration and renewal served these purposes by partially transposing Directive 2012/27/EU, on energy efficiency, so that the principle of energy rationality is one of the pillars of the standard. The aforementioned standard established among its objectives those of minimizing the consumption of energy in homes constituting habitual residence and prioritizing renewable energies with measures in favor of efficiency and energy saving. This applies not only to new buildings and installations, but also to existing ones that are the subject of intervention, as we shall see below. As we already know, the content of this rule has become part of the Royal Legislative Decree 7/2015, of October 30, consolidated text of the Land and Urban Rehabilitation Law. In this context, the urban duty of conservation is considered as an instrument at the service of energy rehabilitation, to the extent that its content is divided into a series of strata that support actions aimed at adapting the existing real estate to the demands of energy efficiency that the rules that regulate the building are demanding.

Some Clarifications on the Concept of Rehabilitation in the Current Regulations

Law 8/2013, of June 26, established a new nomenclature when formulating actions in the urban environment that differs from the nomenclature used in previous legislation, where the concept of urban rehabilitation was comprehensive of a set of isolated or integrated actions that could even involve the demolition of buildings and urban elements incompatible with actions of sanitation or urban regeneration. Thus, two types of rehabilitation were distinguished: isolated rehabilitation and integrated rehabilitation . The first considered the real estate in its own individuality, without implications or connections with the surrounding elements. On the contrary, integrated rehabilitation considered the affected heritage as part of a larger space in which it was integrated. Integrated rehabilitation consisted of the rehabilitation of urban spaces, which without dispensing with the rehabilitation of individual elements, transcended their effects to project on other elements of the environment, considering it as a whole.

According to the new nomenclature introduced by Law 8/2013, of June 26, rehabilitation is the term that refers to the actions singularizadas or individualizas in buildings, while those that affect the urban fabric is called regeneration or urban renewal, according to entail the demolition of any of the elements of urban pre-existing . Based on this distinction, article 2 of Royal Legislative Decree 7/2015 of

October 30, establishes possible actions to be carried out in the urban environment, differentiating between rehabilitation of buildings and regeneration and urban renewal. According to the cited article, performances on the urban environment are those that are intended to perform rehabilitation work edificicatoria, when there are situations of failure or degradation of the basic requirements of functionality, safety and habitability of buildings, and the regeneration and urban renewal will be those that affect, both buildings as urban fabric, ppudiendo come to include works of new building in the replacement of buildings previously demolished.

With reference to the basic requirements of functionality, safety, and livability, the idea of rehabilitation of building is spacious and accommodates works and activities related to energy efficiency of the buildings, which introduces the concept of energy rehabilitation, which involves performing actions on the real estate existing in a situation of failure or degradation in relation to the basic requirements of the building to adapt it to the demands of efficiency and energy savings .

EFFICIENCY AND ENERGY REHABILITATION WITHIN THE FRAMEWORK OF THE URBAN DUTY OF CONSERVATION AND THE USE OF RENEWABLE ENERGIES

Both the rehabilitation of buildings and the integrated actions of urban renewal and regeneration are actions of urban significance, so on these actions must be projected the urban instruments and techniques that affect the situation of buildings and homes. So what sets the Real Decreto legislativo 7/2015, de October 30, which in its article 2.1 states that all of these actions will apply to the statutory scheme basic duties and charges to which they are entitled in accordance with the performance of urban transformation or building that behave, in accordance with the provisions of article 7.

One of these duties is the duty of urban conservation, whose content is integrated by reference to Law 38/1999, of November 5, on Building Planning, which in turn refers to Royal Decree 314/2006, of March 17, which approves the Technical Building Code. The requirements of energy efficiency are incorporated into the duty of conservation through the regulation made in the Technical Building Code in its reference to living conditions, since they are applicable not only to new buildings, but also to existing ones when they are the subject of some intervention. Among these requirements are those related to the use of renewable energies.

But in addition, the Building Planning Law configures the duty of conservation in open terms by reference to the regulations that may be applicable in each case. The act points to the Technical Building Code may be supplemented with the requirements of other regulations issued by the competent Authorities and is regularly updated according to the evolution of technology and the demand of the society, which emphasizes the open-ended character of the setting of the duty of conservation.

The Content of the Duty of Conservation: The Three Levels

The duty of urban conservation is regulated in article 15 of the consolidated text of the Law on Land and Urban Rehabilitation, article that reformulates the wording of the duty of conservation established by Royal Legislative Decree 2/2008, of June 20 in the terms in which it was drafted after Law 8/2013, of June 26 of urban rehabilitation, regeneration and renewal.

The aforementioned precept establishes different duties and burdens that configure the urban content of the property right, among which is, as we say, the duty of conservation . According to the statement of reasons of the law on urban rehabilitation, regeneration and renovation, of which the aforementioned article 15 brings cause, the duty of conservation is articulated in three levels depending on the content that results from it for the owner.

At its basic level, the duty of conservation entails the duty to allocate the buildings to uses compatible with territorial and urban planning and the need to guarantee the safety, health, accessibility and decoration of the real estate . In this way, the duty of conservation includes the three basic requirements of the building, referring to functionality, safety and habitability . So far the duty of conservation coincides with its traditional formulation. However, the law of rehabilitation, regeneration and renewing urban added to article 9 of the Consolidated Text of the Land Act, Royal legislative Decree 2/2008 June 20, a new section under which the duty of conservation also includes the realization of the work and the necessary works in order to satisfy, in general, the basic requirements of the building set forth in article 3.1 of the Law 38/1999, de 5 de November de Ordenación de la Edificación. This explicit reference to the previous law is no longer included in article 15 of Royal Legislative Decree 7/2015 of October 30, but is implicit in said article insofar as the aforementioned precept establishes the duty to preserve buildings in the legal conditions of safety, health, universal accessibility, ornament and the others required by the laws to serve as support for the uses for which they are intended and are compatible with territorial and urban planning, from which an implicit reference to the Building Planning Law and the Technical Building Code is inferred.

A second level, in which the duty of conservation includes the works and works necessary to adapt and progressively update the buildings, in particular the facilities, to the legal norms that are explicitly required at all times. Law 8/2013, of June 26, started from the consideration of an open conservation duty, whose content was not delimited by article 9 of Royal Legislative Decree 2/2008, of June 20, consolidated text of the Land Law, but subject to variations or extensions from other subsequent laws. Thus, Article 9, paragraph 2, established the competent administration could impose at any time the realization of works to comply with the legal duty of conservation in accordance with the provisions of the applicable state and autonomous community legislation. This provision, due to its obviousness, is also absent in article 15 of Royal Legislative Decree 7/2015 of October 30, which does not prevent considering the duty of conservation as a duty of an evolutionary nature and subject to the variation of the technical requirements that are foreseen in the different applicable regulations. In this way, the duty of conservation will also include the necessary works to adapt the buildings and update their facilities to the legal norms that are required at any time, as the regulations of the sector introduce modifications in order to maintain the conditions of use of the buildings. The duty of conservation is thus in open terms, by reference to the regulations that may be applicable in each case.

Finally, a third level includes additional works, carried out for reasons of general interest, in respect of which the law distinguishes two cases:

- those carried out for tourist or cultural reasons that constitute an assumption already included in the previous legislation, although under the consideration of improvement and forced rehabilitation works.
- those carried out for the improvement of the quality and sustainability of the urban environment, assumption that now introduces law of rehabilitation, regeneration and urban renewal in the modification that makes of the Consolidated Text of the Law of the Land.

Energy Efficiency, Duty of Conservation, and Rehabilitation

The Initial Content of the Duty of Conservation: Energy Saving and the Use of Renewable Energy

The preservation of buildings in conditions of safety, health and public decoration constitutes a legal duty imposed on the owner on the basis of the social function of the property in accordance with article 33 of the Constitution. To the duty of the owner to maintain the buildings in the aforementioned conditions, it is now also added that of universal accessibility, and in addition, it is implicitly integrated into the duty of conservation the realization of the works and the works necessary to satisfy in general, the basic requirements of the building established in article 3.1 of Law 38/1999, of November 5, of Building Planning.

Article 3.1 of the Building Planning Act establishes three basic requirements for buildings aimed at ensuring the safety of people, the welfare of society and the protection of the environment. These are requirements relating to functionality, safety and habitability. And precisely in this last section the measures relating to energy saving and thermal insulation are established so that a rational use of the energy necessary for the proper use of the building is achieved. The efficiency and energy savings are part of the duty of conservation to the extent that this duty entails keeping the buildings in terms of serving to its use, which leads to the need to satisfy the basic requirements of the building, within which are located relative to energy savings, thermal insulation and rational use of energy, and that is conducive to the simultaneous execution of conservation work directed to maintain the buildings in terms of safety, sanitation and beautification works and aimed to enhance the energy efficiency of the same. Therefore, we wanted to take advantage of the realization of conservation works in the most traditional sense of the term to promote the realization of actions related to energy rehabilitation, which are now integrated into the duty of conservation.

The requirements of energy efficiency have been incorporated into the duty of conservation through the regulation carried out by the Technical Building Code, Royal Decree 314/2006, of March 17. and are applicable not only to newly constructed buildings, but also to those undergoing repair or rehabilitation . The aforementioned Royal Decree is issued in implementation of Law 38/1999 of November 5, on Building Planning, whose second provision empowers the Government to approve a regulatory norm that establishes the basic requirements that buildings must meet in relation to the requirements relating to safety and habitability listed in paragraphs b) and c) of article 3.1 of the Building Planning Law.

Among the requirements relating to habitability, Article 15 of the said Code regulates the basic requirements of energy saving . These basic requirements apply and must be met both by new buildings and by existing buildings that are subject to modification, reform, extension or rehabilitation and are basic, so that their requirements are mandatory throughout the national territory . The eleventh Final Provision of the Law 8/2013, of June 26, has widened the scope of application of Technical Building Code to modify its article 1 that the basic requirements must be met, as established by the regulations set in the project, the construction, the maintenance, preservation and use of buildings and facilities, as well as in interventions in existing buildings . Likewise, the aforementioned rule has modified Article 2 of the Technical Code, so that its basic requirements will be required not only to new buildings but also to all those existing buildings that are the subject of intervention .

The basic requirements for energy savings set out in the Technical Code are five, namely: limitation of energy demand, performance of thermal installations, energy efficiency of lighting installations, minimum solar contribution of domestic hot water and minimum photovoltaic contribution of electrical energy . The purpose for which it is intended with this energy saving enters into what was formerly the definíamos as energy efficiency, as it is to get a rational and sustainable use of the energy used in the building or home, and also enhance the consumption comes from renewable energy sources.

For this reason, among these basic requirements we must highlight those that refer to solar contributions, and above all, those that refer to the production of domestic hot water, which " have the character of minimums, and can be extended as a result of additional provisions issued by the competent administrations"

And this raises the question of which Administration is competent to regulate this question. Some autonomous communities have adopted their own regulations on renewable energy. This is the case of the Autonomous Community of Andalusia and its Law 2/2007, on the promotion of renewable energies and energy savings and efficiency in Andalusia. The aforementioned law is developed by Decree 169/2011, of May 31, whose Chapter II " Basic requirements for the use of renewable energies, savings and energy efficiency", complies with the provisions of the Technical Building Code by adapting these requirements to the energy needs and climatic characteristics of Andalusia . However, the Andalusian regulations do not establish additional requirements of contribution to the minimum contributions of thermal energy required in the state regulations through the Technical Building Code.

By applying the principle of linking negative in the interpretation of the principle of legality, the judgments of the Supreme Court of 22 may 2015 (RJ 2015/2620 and 2015/2016) consider such other competent authorities are, in addition to the autonomous communities, local authorities and that the criteria listed in the Technical Building Code constitute minimum values that the local authorities have to respect, but that can increase.

According to the jurisprudential criterion established after the judgments of the Supreme Court of May 22, 2015 (RJ 2015/2620 and 2015/2016), this reference to the competent administrations, also includes local entities that can thus adapt the state provisions to their own peculiarities and the needs that conform their specific local interests.

Thus, some municipalities, in the context of sustainable urbanism, have adopted ordinances aimed at promoting energy savings and promoting the use of renewable energies such as solar energy based on the competences that local legislation recognizes them in matters such as urbanism and the environment . Such is the case of the municipal Ordinance on the collection and use of solar thermal energy in buildings, of the City of Burgos . Also the Ordinance of the same name of the City of Pamplona as well as the municipal Ordinance of eco-energy efficiency and use of renewable energy in buildings and their facilities, of the City of Zaragoza among others.

The approval of the Technical Building Code, Royal Decree 314/2006, of March 17, has come to provide the legal authorization whose absence motivated the challenge of the local ordinances on energy use to understand that the Law of Bases of Local

Regime did not provide sufficient legal authorization for municipalities to approve ordinances in this sense. The judgments of the Supreme Court of May 22, 2015 (RJ 2015/2620 and 2015/2016) indicate that from the approval of the Technical Building Code the normative coverage of the ordinances relating to the use of solar energy is accommodated in the aforementioned legal text, which has the character of basic legislation

Energy Efficiency as an Additional Content to the Duty of Conservation: The Improvement of Energy Efficiency and the Use of Renewable Energy Sources

The duty of conservation will also include the adaptation of the building to the use of renewable energies indirectly when rehabilitation actions are carried out that update the duty of conservation.

The duty to initial conservation is added to the duty to perform additional work for tourism or cultural, or to improve the quality and sustainability of the urban environment, since the distinction between the duty of conservation based on the fulfillment of the social function of property and the duty of conservation based on reasons of general interest has been a constant in the planning legislation that the different laws have been addressed in a different way, but always on the basis of their recognition . The content of these additional works is defined by reference to the Technical Building Code, and may consist of partial or complete adaptation to all or some of the basic requirements established therein. In accordance with this article, the Administration may order the owners to carry out works aimed at improving the quality or sustainability of the urban environment, including works to improve energy efficiency and including those aimed at encouraging the installation of renewable energy sources in buildings.

The attention to renewable energies is framed by both in the works additional ordering the administration to improve the quality and sustainability of the urban environment imposed by reasons of general interest, and is justified in the community legislation and in particular Directive 2012/27/EU, noting that buildings represent 40% of energy consumption in the European Union compels the member states to develop strategies that include the realization of investment in the renovation of residential and commercial buildings . The Directive requires not only the annual renovation of a percentage of public buildings of the central State Administration (thus underlining the exemplary nature of public actions), but also the mobilisation of investments aimed at the renovation of buildings for commercial or residential uses with the aim of improving their energy efficiency.

The characteristic of these actions is that the law empowers the public administrations to impose them for reasons of general interest beyond the limits that govern the duty of conservation, in which case, the law indicates, the ordering Administration will be responsible for the works that exceed this limit to obtain improvements of general interest . We are talking about additional conservation, which involves carrying out additional works and financing measures to improve the energy efficiency of buildings. This implies the renovation of buildings through internal strategies of mobilization of real estate investments where the legal limit for the owner is that of the duty of conservation and the administration has to contribute to the financing of the rest, since they are improvements of general interest. In this case, the duty of conservation goes beyond the particular objective of adaptation of the building in question to identify with a broader objective of general interest focused on the consolidated urban environment where it is located and on the fulfillment of energy efficiency objectives that are deduced from European policies, for which the mobilization of investments aimed at the renovation of residential and commercial buildings is foreseen. The additional content of the duty of conservation must be framed within integral actions and policies of economic, social, environmental regeneration and cohesion of the city as a whole, perspective from which the duty of conservation transcends the individual level to achieve improvements of general interest. And from this perspective, not only building rehabilitation actions are imposed, but also actions that include actions to implement renewable energy not only at the building level, but also through urban development actions in urban fabrics that involve urban regeneration or renewal.

However, article 15 of Royal Legislative Decree 7/2015, of October 30, does not seem to require any additional requirements for the imposition of such additional conservation works. Since it is an additional duty that transcends the individual plan of the property to which it refers to have an impact on the urban environment, it could be considered consubstantial to the general interest alleged that the property was included in some type of legal instrument of rehabilitation. Well established, for example in the article 111 of the Law of Sustainable Economy, repealed by the Law 8/2013, of June 26, in which it is stated that the competent authority could command, in the form, terms and deadlines set by the applicable law, the execution of works of improvement up to the maximum amount of statutory duty, in addition to for reasons cultural and tourist collected by the applicable legislation, in the course of the construction or the building that were to become affected by a program, plan or any other legal instrument for the rehabilitation of housing approved and in force, and refers to works designed to guarantee the rights recognized by law to individuals, or to be imposed by legal norms supervened for reasons of safety, adequacy of facilities and minimum services, reduction of polluting emissions and immissions of any kind and those necessary to reduce water and energy consumption.

Energy Rehabilitation and the Limit of the Duty of Conservation

The Quantitative Limits of the Duty of Conservation

Law 8/2013, of June 26, introduced some modifications regarding the limits of the duty of conservation and that were fixed by reference to quantitative criteria. These amendments are now incorporated in article 15 of the consolidated text of the Law on Land and Urban Rehabilitation. In this sense, the limit of the duty of conservation is established at half of the current construction value of a new plant property, equivalent to the original in relation to the constructive characteristics and the useful surface, carried out in the necessary conditions so that its occupation is authorized, or where appropriate, is in a position to be legally destined for its own use. By establishing this limit, the state legislation configures the content of the duty of conservation by reference to an objective criterion, thus preventing the autonomous legislation from establishing its own limits with respect to that duty. Law 8/2013, of June 26, introduced an important novelty in the duty of urban conservation and in the regulation that until now had been carried out of it by the Consolidated Text of the Land Law of 2008, by quantifying the limit of the duty of conservation, which were doing the autonomous regulations in some cases . This modification should be subject to positive evaluation, as in this way, the limit is set up with basic character and binding on the legislator autonomous, so that in accordance with the provisions of article 149.1.1 of the Constitution, defines the legal positions of the owners in relation to their duty of conservation, and imposes rules equal and uniform to all property owners of buildings.

The duty of conservation is thus established by reference to an objective limit and individualized by reference to each particular property, which is fixed by reference to a percentage applied on the current value of a new construction of similar functional and structural characteristics and equal useful surface. In short, the law addresses the value or cost of replacement by taking into account not the value of the damaged building (current value of the building), but a new building whose valuation will serve to contrast the cost of repair works and set the limit of the duty of conservation. The valuation system deals exclusively with the replacement value, that is, the hypothetical value of a construction of the same structure and building typology that allows a use of similar characteristics to that of the construction with respect to which the declaration of ruin is intended. The evaluation of these repairs will have to be carried out based on current techniques and materials that allow to maintain or return to the construction the original functionality. The legislation takes into account the economic profitability of the extension of the useful life of the constructions through the conservation or rehabilitation of the same ones. In this sense, the limit of the profitability of the conservation of the buildings is fixed in the disbursement by the owner of expenses that remain below half of the cost of a new construction with characteristics similar to the existing one .

But the limit of the duty of conservation is solved based on economic criteria where the cost of replacement works is only one of the terms of the equation. The second is related to the content and extent of the works to be made to understand fulfilled the aforementioned duty. In this sense, it should be noted that the conservation works are not merely the works necessary to maintain the property in conditions of safety, health and ornament. We have already indicated in the previous section that the duty of conservation also involves the realization of works necessary to ensure universal accessibility, as well as those others that are necessary to meet the basic requirements of the building established in article 3.1 of Law 38/1999 of November 5 on Building Planning, where interventions related to energy efficiency are registered, including both those that are deduced from the current regulations and those others that are explicitly required to adapt and update their facilities to the standards. In short, this means going to the Technical Building Code, in which the basic requirements of the building are specified, which in terms of energy saving also implies the need to include among these works, those relating to the installation of collection systems, transformation, storage and use of solar energy. In addition, the values derived from this basic requirement will be considered minimum, without prejudice to stricter values that may be established by the competent administrations and that contribute to sustainability according to the characteristics of their location and territorial scope.

It could be thought that since one of the terms of the equation expands with respect to what was the traditional duty of conservation, the logical consequence will be that the limit of the duty of conservation will be easily exceeded by including more demanding actions in the maintenance of buildings that also meet criteria of saving and energy efficiency. However, this possible consequence is remedied if we take into account that the second comparative term focuses on the replacement cost of the property. This clarification is important, since urban planning legislation has not always made use of this criterion to delimit the cessation of the duty of conservation. Thus, for example, article 247 of the Consolidated Text of the Land Act of 1992, when referring to the declaration of ruin (which implied the cessation of the duty of conservation), referred to the present value of the building or plants concerned, excluding the value of the land. This meant taking as a reference the value of the building at the time when it was planned to carry out the repair works on it, which implied valuing the building applying criteria indicative of the depreciation suffered by the property depending on age and state of conservation among others. Obviously, this means that the building is quantified at a lower magnitude easily exceeded by the cost of maintenance and rehabilitation works carried out in it. As we said, the option to include the replacement cost of a similar building of new plant allows increasing the second term of the equation quantitatively.

On the other hand, it should also be noted that although Article 15 of the consolidated text of the Land and Urban Rehabilitation Law refers to conservation works, such conservation works also include rehabilitation works, which are now classified as such when there are situations of insufficiency or degradation of the basic requirements of functionality, safety, and habitability of buildings as established in article 2 of the same legal text. In these cases, and as we have noted above, rehabilitation updates the duty of conservation and is embedded in it.

The Regime of Additional Works to Improve Quality and Environmental Sustainability

The Law 8/2013 introduced within the duty of urban conservation the realization of additional works that had as a common denominator the fact that its imposition is based on reasons of general interest and that the Consolidated Text of the Land Law of 2008 considered as works of improvement. Royal Legislative Decree 7/2015, of October 30, refers to them in article 15.1 c). These additional works include those carried out for tourism and cultural reasons, to which are added those focused on improving the quality and sustainability of the urban environment, introduced by Law 2/2011 of March 4 on Sustainable Economy and where those actions related to energy efficiency are framed. The content of these additional works for the improvement of quality and sustainability is determined as we have indicated, by reference to the Technical Building Code

Well, these works additional to the duty of conservation which now also include those related to the improvement of the quality and sustainability of the urban environment, has a particular legal regime, since the law establishes the possibility for the administration to force the landlord to the realization of the same, even once exceeded the limit of the duty of conservation. In this case, the administration that orders or imposes the realization of such works must pay the economic excess that entails the realization of the same. And this economic excess is fixed by reference to the limit of the duty of conservation referred to above. In this way, the limit of the duty of conservation (half of the current value of construction of a building of a new plant, equivalent to the original in connection with the construction features and the useful surface, taken with the conditions necessary for their occupation is approvable or, in your case is in a position to be legally intended for use by own) it is also the limit of the works to be executed at the expense of the owners when the Administration of the order to the improvement of the quality or sustainability of the urban environment. The law assumes the criterion of the joint participation of the owner and the Administration in the maintenance of the property, as deduced from article 9.1 of the Land Law after the wording given to it by the law on urban rehabilitation, regeneration and renovation .

Royal Legislative Decree 7/2015, of October 30, considers the importance of these rehabilitation actions in the urban environment and the need to adapt them to the limits of the legal duty of conservation through the economic viability report, regulated in article 22. This memory, it is expected not only in cases in which they carry out activities of regeneration and renewal, but also in the case of actions of rehabilitation of building, isolated or included in a scope of work (by areas or spaces), establishing as one of its aims to ensure the least possible impact on the personal wealth of the individuals, adjusted in any case the limits of the duty to bequeath conservation.

REFERENCES

Chu, S., & Majumdar, A. (2012). Opportunities and challenges for a sustainable energy future. *Nature, 488*(7411), 294-303.

Curry, N., & Pillay, P. (2012). Biogas prediction and design of a food waste to energy system for the urban environment. *Renewable Energy*, *41*, 200–209. doi:10.1016/j.renene.2011.10.019

Droege, P. (2011). *Urban energy transition: From fossil fuels to renewable power.* Elsevier.

Lund, P. (2012). Large-scale urban renewable electricity schemes–Integration and interfacing aspects. *Energy Conversion and Management*, *63*, 162–172. doi:10.1016/j.enconman.2012.01.037

Purohit, I., Purohit, P., & Shekhar, S. (2013). Evaluating the potential of concentrating solar power generation in Northwestern India. *Energy Policy*, *62*, 157–175. doi:10.1016/j.enpol.2013.06.069

ADDITIONAL READING

UN-HABITAT, Local Governments for Sustainability, and the United Nations Environment Programme. (2009). *Sustainable Urban Energy Planning: A handbook for cities and towns in developing countries*. UNEP.

Chapter 9
Progress and Determinants of Renewable Energy Development in India

Sebak Kumar Jana
https://orcid.org/0000-0002-3532-4350
Vidyasagar University, India

Khushboo Singh
Vidyasagar University, India

ABSTRACT

For sustainable development of energy, renewable energy can play a pivotal role. In order to meet the huge requirement of electricity in India, massive addition to the installed electricity-generating capacity is required. The major objectives of the present study are (1) to find the progress of renewable energy development in different states of India, (2) to find the determinants of renewable energy development in India, and (3) to assess the scope of renewable energy development in the states of India. The study finds that there is state-wise disparity in the development of renewable energy in India and also reveals that renewable energy potential (MW) and real gross state domestic product of the states are significant factors in state-wise renewable energy development in India.

DOI: 10.4018/978-1-7998-9502-2.ch009

INTRODUCTION

To meet the huge requirement of electricity, massive addition to the installed generating capacity in India is required. In addition to the capacity installed, renewable energy could play a major role in India. For sustainable energy development, renewable energy can play a pivotal role (Bhaskar 2013, Bhattacharya et al 2016, Bhattacharya and Jana 2009, IBEF, 2018, Jana 2022, Roy and Jana 1998). India is richly bestowed with renewable energy resources like sunlight, wind and biomass. These renewable sources are increasingly contributing to the national energy mix in India. The renewable share in India's total electricity generation has increased from 0.4% in 2001-02 to 5.4% in 2017. Electricity production from renewable sources as percentage of total production for some countries, including India, is presented in Table 1.

Solar power, one of the potential energy sources, is a fast-developing industry in India. The country's solar installed capacity has reached 28.18 GW as on 31.03.2019 as compared to 21.65 GW on 31.03.2018. Over the years, India has expanded its renewable source of electricity generation capacity, leading to a downward trend in cost and increased usage (MOSPI 2020). However, India has still untapped potential to provide clean energy sources. Renewable energy can supply clean energy that is accessible to all (Bhaskar 2013). There is high potential for renewable energy generation from various sources- wind, solar, biomass, small hydro and cogeneration based on bagasse. The total potential for renewable power generation in India as on March, 2019 was estimated at 1097,465 MW. This includes solar power potential of 748,990 MW (68.25%), wind power potential of 302,251 MW (27.54%), SHP (small-hydro power) potential of 21,134 MW (1.93%), Biomass power of 17,536 MW(1.60%), 5,000 MW (0.46%) from bagasse-based cogeneration in sugar mills and 2554 MW (0.23%)from waste to energy (MOSPI 2020). Installed Electricity Generation Capacity in Utilities and Non-utilities (MW) in India is presented in Table 2. The present chapter studies renewable energy development in the states of India.

Table 1. Electricity production from renewable sources, excluding hydroelectric (% of total)

Sl. No.	Country/Region	Share of Renewable	Sl. No.	Country/Region	Share of Renewable
1	Denmark	65.4	11	United Kingdom	*23.0*
2	Kenya	48.3	12	Italy	*22.5*
3	Nicaragua	43.6	13	Belgium	*20.3*
4	El Salvador	35.2	14	Finland	*20.1*
5	Lithuania	31.2	15	Euro area	*18.7*
6	Portugal	30.6	16	European Union	18.6
7	Uruguay	28.4	17	United States	7.4
8	Iceland	26.7	18	World	6.8
9	Germany	26.3	19	India	5.4
10	Guatemala	25.3	20	South Asia	4.8

Source: World Bank

Table 2. Installed electricity generation capacity in utilities and non-utilities (MW)

	Renewable Energy (Utility)	Renewable (Non Utility)	Total Renewable Energy RES	Total Electricity	% RES
31.03.2010	15,521	454	15,975	1,90,915	8.4
31.03.2011	18,455	567	19,022	2,08,071	9.1
31.03.2012	24,503	872	25,375	2,39,252	10.6
31.03.2013	27,542	1,124	28,666	2,64,070	10.9
31.03.2014	34,988	1,259	36,247	2,90,812	12.5
31.03.2015	38,959	1,301	40,260	3,19,561	12.6
31.03.2016	45,924	1,368	47,292	3,53,442	13.4
31.03.2017	57,244	1,433	58,677	3,78,362	15.5
31.03.2018	69,022	1,726	70,748	3,98,935	17.7
31.03.2019	77,642	1,881	79,523	4,14,100	19.2

Source: MOSPI, 2020

Figure 1. Growth of Renewable Energy in India
Source: MOSPI, GOI

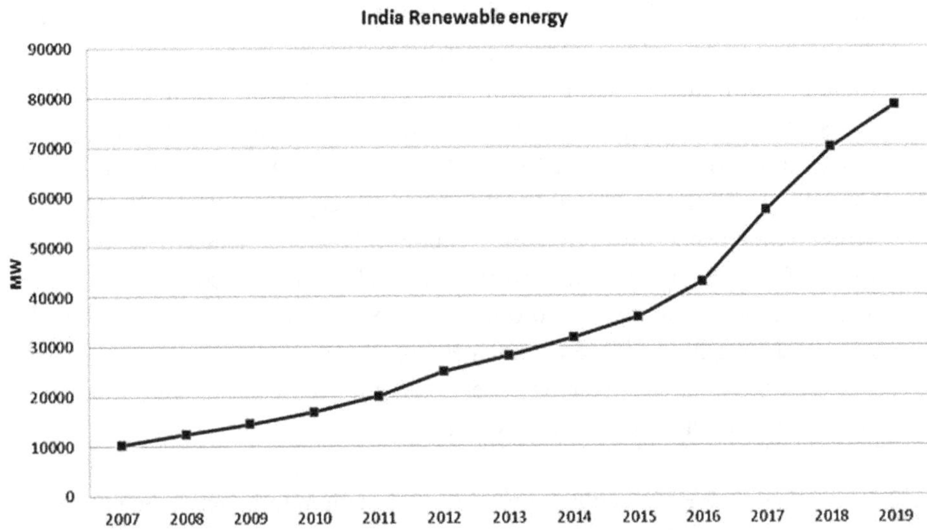

India's renewable energy sector is currently a desirable renewable energy market globally. The rank of India is fourth in renewable power installed capacity with fifth in wind power and fifth in solar power. The government plans to establish renewable energy capacity of 523 GW (including 73 GW of Hydropower) by 2030. As of May 2021, India had 95.7 GW of renewable energy capacity, representing about 25% of the overall installed power capacity. The installed generation capacity of renewable power in India has increased at a CAGR of 15.51% between 2016 and 2021. India had 94.4 GW of renewable energy capacity in FY21. About USD 42 billion has been invested in India's renewable energy sector from 2014 to 2021. Some of the private power producers involved in renewable power generation in India are Suzlon, Tata Power Adani, Green Energy Ltd. (AGEL), Virescent Infrastructure, JSW Energy Karcham Wangtoo hydropower plant, GE Power, Edelweiss Infrastructure Yield Plus (EIYP), Vikram Solar etc. IBEF (2021).

Some initiatives by the Government of India to boost India's renewable energy sector are as follows IBEF (2021):

1. Indian Renewable Energy Development Agency Ltd. (IREDA) from solar module manufacturers set up solar manufacturing units Production Linked Incentive (PLI) scheme.
2. The Central Electricity Authority (CEA) and CEEW's Centre for Energy Finance (CEEW-CEF) jointly launched the India Renewables Dashboard that

provides detailed operational information on renewable energy (RE) projects in India.

3. Ministry of Power (MoP) released the draft National Electricity Policy (NEP) 2021 and has invited suggestions from all stakeholders.
4. India has introduced Gram Ujala program to include cheap LED bulbs in rural areas climate change policy and bolstered its self-reliance credentials.
5. The Union Budget 2021-22, allocated Ministry for New and Renewable Energy (MNRE) US$ 41.12 million) for the 'Green Energy Corridor' scheme.
6. To encourage domestic production, customs duty on solar lanterns has increased from 5% to 15% and on solar inverters, it has been increased from 5% to 20%
7. The GOI has announced a production-linked incentive (PLI) scheme for manufacturing high-efficiency solar PV modules.
8. The GOI has announced a plan to set up an inter-ministerial committee under NITI Aayog to enhance research on energy modeling.

In Table 3, state-wise Renewable Energy Development in India has been presented.

Table 3. State-wise renewable energy development in India

		Small Hydro	Wind Power	Bio Power Total	Solar Power Total
		(MW)	(MW)	(MW)	(MW)
1	Andhra Pradesh	162.11	4096.65	536.04	4347.61
2	Arunachal Pradesh	131.105		0.00	11.17
3	Assam	34.11		2.00	68.38
4	Bihar	70.70		125.70	190.46
5	Chhatisgarh	76.00		270.31	500.77
6	Gujarat	82.69	8782.12	99.87	5219.66
7	Haryana	73.50		216.75	658.35
8	Himachal Pradesh	936.11		10.20	56.56
9	Jammu & Kashmir	146.34		0.00	42.30
10	Jharkhand	4.05		4.30	74.27
11	Karnataka	1280.73	4938.60	1901.92	7452.01
12	Kerala	236.02	62.50	2.50	293.07
13	Madhya Pradesh	99.71	2519.89	157.91	2627.70
14	Maharashtra	380.58	5012.83	2596.99	2428.77
15	Manipur	5.45		0.00	11.97
16	Meghalaya	32.53		13.80	4.01

Continued on following page

Table 3. Continued

		Small Hydro	Wind Power	Bio Power Total	Solar Power Total
		(MW)	(MW)	(MW)	(MW)
17	Mizoram	36.47		0.00	7.79
18	Nagaland	30.67		0.00	3.01
19	Odisha	88.63		66.67	426.75
20	Punjab	174.35		488.03	1002.33
21	Rajasthan	23.85	4326.82	121.25	6920.40
22	Sikkim	52.11		20.86	1.94
23	Tamil Nadu	123.05	9717.04	1023.63	4594.33
24	Telangana	90.87	128.10	205.90	3973.14
25	Tripura	16.01		59.47	14.62
26	Uttar Pradesh	49.10		2126.48	2019.20
27	Uttarakhand	214.32		131.39	433.25
28	West Bengal	98.50		319.92	162.66

Source: Government of India

LITERATURE REVIEW

Some studies relevant to the present chapter are mentioned here. Khare et al. (2013) presented the major constraints that hamper renewable energy development in India despite India having adequate sunshine and balanced wind speed. Mishra and Behera (2015) found that renewable energy leads to improvement in people's overall standard of living in rural Odisha. Hrishikesh and Debnath (2017) made a systematic cost-benefit analysis with data from selected 12 institutions in the Kamrup district of Assam to show the economic feasibility of such a project in replacing the existing state-supplied electricity source a payback period of 5 years. Kanase-Patil et al. (2010) analyzed the off-grid electrification by utilizing Integrated Renewable Energy System (IRES) to satisfy the electrical and cooking needs of the seven unelectrified villages in the Almora district of Uttarakand state, India. Sangroya and Nayak (2014) found that wind energy has been India's fastest-growing renewable energy sector. Nathan (2014) suggested the appropriate land procurement and rooftop policies and manufacturing strategies and technology. Jain and Patwardhan (2013) attempted to characterize jobs created for renewable energy technologies as a function of multiple attributes to observe the employment effects. Nayar (2016) discussed the renewable purchase obligation mechanism.

Kar and Gopakumar (2015) identify the opportunities and challenges of increasing the share of renewable energy to address energy security in India. Kocsis & Kiss (2014) examine the relationship between the ratio of renewable energy consumption in gross energy consumption, per capita GDP, and R&D expenditures for the countries in the European Union. According to Kumar et al. (2010) India will need an assured supply of energy of 3–4 times more than the present total energy consumption to meet the growing energy requirement. Osmani (2014) finds a wide gap between India's energy production and energy consumption. According to Rathore et al. (2019) the development of solar rooftop PV currently is much below the desired level despite the fact that the Government of India has set a target of installing 40 GW of solar rooftop PV within 2022. According to Sharma (2012), India should prepare a carefully thought-out strategy for sustainable energy supply options. Yadav et al. (2017) present modeling of a system with a wind-biogas-based hybrid system operating in parallel with a limited grid supply system to meet the rural demand.

OBJECTIVES OF THE STUDY

The major objectives of the study are to assess the following:

1. Progress of renewable energy development in different states of India
2. Finding the determinants of renewable energy development in India
3. The scope of renewable energy development

METHODOLOGY

The panel data analysis method has been applied to get the relationship between the growths of Renewable energy resources, Renewable energy potential, and GSDP (gross state domestic product) of the different states in India. The data have been collected from Government Sources such as MOSPI (Government of India), MNRE (Ministry of New and Renewable Energy), etc.

Panel Data Analysis

Social scientists widely use panel data analysis because it permits the inclusion of data for N cross-sections (for example, households, countries, firms, individuals and so on) and T time periods (for example, years, quarters, months and so on). If the panel has the same number of time observations for every variable and every individual, it is known as a balanced panel. In general, simple linear panel data models can be estimated using three different methods: (b) allowing for fixed effects and (c) allowing for random effects.

The Fixed Effects Method

The fixed effects method treats the constant as group (section)-specific. This implies that the model permits for different constants for each group (section). So the model is similar to that in Equation. The fixed effects estimator is also known as the least squares dummy variable (LSDV) estimator because, to allow for different constants for each group, it includes a dummy variable for each group. It essentially captures all effects that are specific to a particular individual and do not vary over time. To understand this better consider the following model:

$$Y_{it} = a_i + \beta_1 X_{1it} + \beta 2X_{2it} + \cdots + \beta kX_{kit} + u_{it}$$

Here, we have taken the following variables:

Y= Total of renewable resources of the state
X_1 (renpotential) = Renewable energy potential (MW) of the state
X_2 (gsdp) = Gross state domestic product at constant price (In Cr. Rupees)

The Random Effects Method

An alternative method of the fixed effects method for estimating a model is the random effects model. The difference between the fixed effects and the random effects model is that the random effects model handles the constants for each section not as fixed, but as random parameters. Hence the variability of the constant for each section comes from:

$$a_i = a + v_i$$

Where v_i is a zero mean standard random variable. Therefore, the random effects model therefore takes the following form:

$$Y_{it} = (a + v_i) + \beta_1 X_{1it} + \beta_2 X_{2it} + \cdots + \beta kXkit + u_{it}$$

$$Yit = a + \beta_1 X_1 it + \beta 2X2it + \cdots + \beta kXkit + (v_i + u_{it})$$

Y= Total of renewable resources of the state

X_1 *(renpotential)* = Renewable energy potential (MW) of the state
X_2 *(gsdp)* = Gross state domestic product at constant price (In Cr. Rupees)

The Hausman Test

The Hausman test is formulated to assist in making a choice between the fixed effects and random effects models. Hausman (1978) adapted a test based on the idea that under the hypothesis of no correlation, both OLS and GLS are consistent, but OLS is inefficient, while under the alternative OLS is consistent but GLS is not. The Hausman test is a test of H_0: That random effects would be consistent and efficient, versus H_1: That random effects would be inconsistent. So, if the Hausman test statistic is significant (that is, low p-value), one must reject REM and use Fixed Effects. If the statistic is not significant (that is, high p-value), the null hypothesis will be accepted and the one may use the Random Effects (REM)

RESULTS OF PANEL DATA ESTIMATION

In Table 4 summary statistics of the variables taken have been presented.

Table 4. Summary statistics of the variables

Variables	Obs	Mean	Std. Dev.	Min	Max	Skew.	Kurt.
Ren	290	1187.317	2282.766	0	12933.23	2.512	9.241
Renpot	290	16314.155	31387.553	0	167276	2.945	11.541
Gsdp	290	320000	364000	4249	2039074	1.845	7.078

Source: Own Estimation from MOSPI

Estimation Results of Fixed Effect Model

Table 5 presents the regression results of the fixed effect model. The overall F-test for 28 individual differences shows that there are significant differences between at least some individuals. The results of estimation of FEM show that the estimated coefficients have expected signs and are statistically significant. The estimation results show that both GDP and renewable energy potential have significant and positive impact on renewable energy development of the states.

Table 5. Regression results of the fixed effects model

ren	Coef.	St.Err.	t-Value	p-Value	[95% Conf	Interval]	Sig
renpotential	.015	.002	7.39	0	.011	.018	***
gsdp	.004	0	12.71	0	.004	.005	***
Constant	-432.682	102.52	-4.22	0	-634.561	-230.802	***
Mean dependent var	1187.317			SD dependent var	2282.766		
R-squared	0.652			Number of obs	290.000		
F-test	242.877			Prob > F	0.000		
Akaike crit. (AIC)	4592.414			Bayesian crit. (BIC)	4603.423		

*** p<.01, ** p<.05, * p<.1
Source: Own Estimation from MOSPI

Estimation Results of Random Effect Model

Table 6 presents the regression results of the rndom effect model. The results of estimation of REM show that the estimated coefficients have expected signs and statistically significant. The estimation results show that both GSDP and renewable energy potential have significant and positive impact on renewable energy development of the states. It may be noted the values of the estimated coefficients are different in FEM and REM, though these are significant in both the models.

Table 6. Regression results of the random effects model

ren	Coef.	St.Err.	t-Value	p-Value	[95% Conf	Interval]	Sig
renpotential	.015	.002	7.85	0	.011	.019	***
gsdp	.004	0	13.84	0	.004	.005	***
Constant	-421.049	232.292	-1.81	.07	-876.332	34.235	*
Mean dependent var	1187.317			SD dependent var	2282.766		
Overall r-squared	0.657			Number of obs	290.000		
Chi-square	539.585			Prob > chi2	0.000		
R-squared within	0.652			R-squared between	0.659		

*** p<.01, ** p<.05, * p<.1
Source: Own Estimation from MOSPI

The output of the Hausman test shows the coefficients common to both models and their estimated difference. The chi-square statistic comparing has small p value, leading us to reject the hypothesis that the coefficient estimates are equal to one another. This difference suggests that random estimator is inconsistent and choosing the fixed effect model. However, the results reveal that there not much difference between these two models.

CONCLUSION

The government has taken various programs to promote green energy development in India. In addition, renewable energy has the potential to create many employment opportunities at all levels, especially in rural areas. It is expected that by 2040, around 49% of the total electricity will be generated by renewable energy as more efficient batteries will be used to store electricity, which will further cut the solar energy cost by 66% as compared to the current price. The Government of India wants to develop a 'green city' in every state of the country, powered by renewable energy. Here we go through the different aspects of Renewable energy in India. From the above analysis, we know that India has a great potential of renewable energy that is growing year by year. India's geographical location makes it rich in two of the major renewable energy source namely solar energy and wind energy. The government is also trying to improve Renewable energy production to achieve the sustainable development goal and India's energy need. Almost all forms of renewable energy are present in India. We show that India's renewable energy sector grows day by day by time series analysis. Presently share of renewable power in total power capacity of India is 18.91%. That is making a significant difference from the energy security point of view and also from the point of carbon reduction.

We have applied panel data analysis to get the relationship between the growths of renewable energy resources, renewable energy potential and GSDP (gross state domestic product) of the different states in India. We got the result that both the renewable energy potential and GSDP of a state positively impact the overall renewable energy of that state. That means higher the GSDP and potential of the state and higher the renewable energy generation can be achieved and vice-versa.

REFERENCES

Bhaskar, B. (2013). *Energy Security and Economic Development in India: a holistic approach.* The Energy and Resources Institute (TERI).

Bhattacharya, M., Paramati, S. R., Ozturk, I., & Bhattacharya, S. (2016). The effect of renewable energy consumption on economic growth: Evidence from top 38 countries. *Applied Energy*, *162*, 733–741. doi:10.1016/j.apenergy.2015.10.104

Bhattacharya, S. C., & Jana, C. (2009). Renewable energy in India: Historical developments and prospects. *Energy*, *34*(8), 981–991. doi:10.1016/j.energy.2008.10.017

IBEF. (2018, September). *Renewable Energy*. IBEF.

IBEF. (2021). *Renewable Energy Industry in India*. https://www.ibef.org/industry/renewable-energy.aspx

Jain, M., & Patwardhan, A. (2013). Employment Outcomes of Renewable Energy Technologies: Implications of Policies in India. *Economic and Political Weekly*, 84–89.

Jana, S. K. (2022). Sustainable Energy Development in Emerging Economies: A Study on BRICS. In Environmental Sustainability, Growth Trajectory and Gender: Contemporary Issues of Developing Economies (pp. 23-35). Emerald Publishing Limited.

Kanase-Patil, A. B., Saini, R. P., & Sharma, M. P. (2010). Integrated renewable energy systems for off grid rural electrification of remote area. *Renewable Energy*, *35*(6), 1342–1349. doi:10.1016/j.renene.2009.10.005

Kapoor, K. K., & Dwivedi, Y. K. (2017). A take on solar power in India. *Economic and Political Weekly*, *52*(7), 21–24.

Kar, S. K., & Gopakumar, K. (2015). Progress of renewable energy in India. *Adv Energy Res*, *3*(2), 97–115. doi:10.12989/eri.2015.3.2.097

Khare, V., Nema, S., & Baredar, P. (2013). Status of solar wind renewable energy in India. *Renewable & Sustainable Energy Reviews*, *27*, 1–10. doi:10.1016/j.rser.2013.06.018

Kocsis, I., & Kiss, J. T. (2014). Renewable energy consumption, R&D and GDP in European Union countries. *Environmental Engineering and Management Journal*, *13*(11), 2825–2830. doi:10.30638/eemj.2014.316

Kumar, A., Kumar, K., Kaushik, N., Sharma, S., & Mishra, S. (2010). Renewable energy in India: Current status and future potentials. *Renewable & Sustainable Energy Reviews*, *14*(8), 2434–2442. doi:10.1016/j.rser.2010.04.003

Mishra, P., & Behera, B. (2016). Socio-economic and environmental implications of solar electrification: Experience of rural Odisha. *Renewable & Sustainable Energy Reviews*, *56*, 953–964. doi:10.1016/j.rser.2015.11.075

MNRE. (2018). *Annual Report: 2017-18*. Government of India. https://powermin.nic.in/en/content/annual-reports-year-wise-ministry

MOSPI. (2020). *Energy Statistics, 2020*. Government of India.

Nathan, H. S. K. (2014). Solar energy for rural electricity in India: A misplaced emphasis. *Economic and Political Weekly*, *49*(50), 60–67.

Nayar, R. (2016). Enforcing Renewable Purchase Obligations. *Economic and Political Weekly*, *51*(40), 21–23.

Osmani, A. R. (2014, July –December). Conventional energy to renewable energy: Perspectives for India. *The NEHU Journal*, *12*(2), 41–60.

Rathore, P. K. S., Chauhan, D. S., & Singh, R. P. (2019). Decentralized solar rooftop photovoltaic in India: On the path of sustainable energy security. *Renewable Energy*, *131*, 297–307. doi:10.1016/j.renene.2018.07.049

Roy, J., & Jana, S. (1998). Solar lanterns for rural households. *Energy*, *23*(1), 67–68. doi:10.1016/S0360-5442(97)00067-4

Sangroya, D., & Nayak, J. K. (2015). Development of wind energy in India. *International Journal of Renewable Energy Research*, *5*(1), 1–13.

Sharma, S. (2012). True electricity demand and the critical role of renewable energy sources. *Economic and Political Weekly*, *47*(24), 24–27.

Yadav, D. K., Bhatti, T. S., & Verma, A. (2017). Study of integrated rural electrification system using wind-biogas based hybrid system and limited grid supply system. *International Journal of Renewable Energy Research*, *7*(1), 1–11.

Glossary

Circularity: A circular economy (also referred to as "circularity") is an economic system that tackles global challenges like climate change, biodiversity loss, waste, and pollution. Most linear economy businesses take a natural resource and turn it into a product that is ultimately destined to become waste because it has been designed and made. This process is often summarised by "take, make, waste." By contrast, a circular economy uses reuse, sharing, repair, refurbishment, remanufacturing, and recycling to create a closed-loop system, minimize resource inputs, and create waste, pollution, and carbon emissions. The circular economy aims to keep products, materials, equipment, and infrastructure in use for longer, thus improving the productivity of these resources. Waste materials and energy should become input for other processes through waste valorization: either as a component or recovered resource for another industrial process or as regenerative resources for nature (e.g., compost). This regenerative approach contrasts with the traditional linear economy, which has a "take, make, dispose of" production model.

Dashboard of Sustainability (DS): Is software developed by the European Union's Joint Research Centre at Ispra and represents the complicated relationship between environmental, economic, and social issues. Moreover, it provides information in a way palatable to decision-makers and the general public.

Direct Material Consumption (DMC): Measures the total sum of the domestic extraction flows, including imported but excludes exported.

Disaster: Is an unexpected event whose occurrence results in extensive damage, including properties, the environment, animals, and humans inhabiting the affected region. Since disasters occur unexpectedly, there is a need to understand and determine strategies to minimize the effects caused by such disasters.

Eco Commerce: Eco commerce is a business, investment, and technology-development model that employs market-based solutions to balancing the world's energy needs and environmental integrity. Through green trading and green finance, eco-commerce promotes the further development of "clean technologies" such as wind power, solar power, biomass, and hydropower.

Ecological Footprint (EF): Is the area of productive land needed to maintain its current consumption ratio while using the prevailing technology to absorb wastes and calculated for a specific population.

Eco-Tariffs: An eco-tariff, also known as an environmental tariff, is a trade barrier erected to reduce pollution and improve the environment. These trade barriers may take the form of import or export taxes on products with a large carbon footprint or imported from countries with lax environmental regulations.

Emissions Trading: Emissions trading (also known as cap and trade, emissions trading scheme, or ETS) is a market-based approach to controlling pollution by providing economic incentives for reducing the emissions of pollutants.

Environmental Enterprise: An environmental enterprise is an environmentally friendly/compatible business. Specifically, an environmental enterprise is a business that produces value in the same manner which an ecosystem does, neither producing waste nor consuming unsustainable resources. In addition, an environmental enterprise rather finds alternative ways to produce one's products instead of taking advantage of animals for the sake of human profits. To be closer to being an environmentally friendly company, some environmental enterprises invest their money to develop or improve their technologies which are also environmentally friendly. In addition, environmental enterprises usually try to reduce global warming, so some companies use environmentally friendly materials to build their stores. They also set in environmentally friendly place regulations. All these efforts of the environmental enterprises can bring positive effects both for nature and people. The concept is rooted in the well-enumerated theories of natural capital, the eco-economy, and cradle-to-cradle design. Examples of environmental enterprises would be Seventh Generation, Inc., and Whole Foods.

Environmental Sustainability Index (ESI): Measures overall progress towards environmental sustainability for 142 countries measured by the World Economic Forum. It provides a more analytical approach to environmental decision-making and allows comparison of the progress among the nations.

Green Economy: A green economy is an economy that aims at reducing environmental risks and ecological scarcities and that aims for sustainable development without degrading the environment. It is closely related to ecological economics but has a more politically applied focus. The 2011 UNEP Green Economy Report argues "that to be green, and an economy must be not only efficient but also fair. Fairness implies recognizing global and country-level equity dimensions, particularly in assuring a Just Transition to an economy that is low-carbon, resource-efficient, and socially inclusive."

Green Politics: Green politics, or ecopolitics, is a political ideology that aims to foster an ecologically sustainable society often, but not always, rooted in environmentalism, nonviolence, social justice, and grassroots democracy. It began taking shape in the western world in the 1970s; since then, Green parties have developed and established themselves in many countries around the globe and have achieved some electoral success.

Gross Domestic Product (GDP): Gives the total value of goods and services produced in a specific time frame, usually a year. It is a popular indicator in the economic sphere but does not apply in the social sphere.

Hazard: Is the potential for a sudden event to occur and whose occurrence can lead to unforeseen consequences and emergencies.

Human Development Index (HDI): Measures the average achievement in the three primary aspects of human development; knowledge, longevity, and improved living standards. Life expectancy defines longevity, enrollment in schools, and adult literacy provides awareness, and the GDP per capita defines the living standard. The UN development program published it.

Living Planet Index (LPI): Assesses the overall global state of the ecosystem using national and international data on the impact of human activities on the environment.

Low-Carbon Economy: A low-carbon economy (LCE) or decarbonized economy is based on low-carbon power sources with minimal greenhouse gas (GHG) emissions into the atmosphere, specifically carbon dioxide. GHG emissions due to anthropogenic (human) activity are the dominant cause of observed climate change since the mid-20th century. Continued emission of greenhouse gases may cause long-lasting changes worldwide, increasing the likelihood of severe, pervasive, and irreversible effects for people and ecosystems.

Mitigation: Is measures taken before a disaster happens to neutralize or decrease its impact on society or the environment.

Natural Resource Economics: Natural resource economics deals with the supply, demand, and allocation of the Earth's natural resources. One main objective of natural resource economics is to understand better the role of natural resources in the economy to develop more sustainable methods of managing those resources to ensure their future generations. Resource economists study interactions between economic and natural systems intending to develop a sustainable and efficient economy.

Pandemic: Is a disease outbreak that spreads across countries or continents that affects more people and takes more lives than an epidemic (Robinson, 2020).

Risk: Is the probability, depending on how high or low a hazard will cause harm. Risk is determined by vulnerable conditions associated with physical surroundings, social setup, environmental factors, or economic factors.

Salience, Credibility, and Legitimacy of GDP and HDI: Salience means that the indicators are useful, applicable, and attractive to the user. Credibility means that the pointers are valid and make scientific sense. Finally, legitimacy touches pointers' perception from the perspectives of users, stakeholders, businesses, trade unions, and environmental non-governmental organizations.

Sustainability in Terms of Energy and Environment: Defines developing affordable and sustainable protocols for various applications of clean and green energy aligned with environmental requirements with minimum greenhouse gas emission over time with contemplating some indicators such as deployment diversity, policy developments, technology costs, and investment in renewable energy (Danish et al., 2016).

Sustainability Pillars: Have counted the parameters to be analyzed to balance the proposed system or solution in accordance with resiliency and sustainability pillars within these criteria: technical sustainability, economic sustainability, institutional sustainability, environmental sustainability, social sustainability, etc. (Danish et al., 2019a).

Sustainable Development: Sustainable development is an organizing principle for meeting human development goals while simultaneously sustaining the ability of natural systems to provide the natural resources and ecosystem services on which the economy and society depend. The desired result is a state of society where living conditions and resources are used to continue to meet human needs without undermining the integrity and stability of the natural system. Sustainable development can be defined as development that meets the needs of the present without compromising the ability of future generations to meet their own needs. Sustainability goals, such as the current UN-level Sustainable Development Goals, address the global challenges, including poverty, inequality, climate change, environmental degradation, peace, and justice.

Well-Being Index (WI): Is called stress index, combines two types of indicators then combines them. The first type has thirty-six pointers for health, population, wealth, freedom, peace, crime, equity, communication, and education. The other type has fifty-one land, water, air, and energy.

Compilation of References

Abdelzaher, H., Saleh, B. M., Ismail, H. A., Hafiz, M., Gabal, M. A., Mahmoud, M., Hashish, S., Gawad, R. M. A., Gharieb, R. Y., & Abdelnaser, A. (2020). COVID-19 genetic and environmental risk factors: A look at the evidence. *Frontiers in Pharmacology*, *11*, 579415. doi:10.3389/fphar.2020.579415 PMID:33117174

Ahmadi, M., Lotfy, M. E., Danish, M. S. S., Ryuto, S., Yona, A., & Senjyu, T. (2019). Optimal multi-configuration and allocation of SVR, capacitor, centralised wind farm, and energy storage system: A multi-objective approach in a real distribution network. *IET Renewable Power Generation*, *13*(5), 762–773. Advance online publication. doi:10.1049/iet-rpg.2018.5057

Ahmadi, M., Lotfy, M. E., Shigenobu, R., Howlader, A. M., & Senjyu, T. (2019). Optimal Sizing of Multiple Renewable Energy Resources and PV Inverter Reactive Power Control Encompassing Environmental, Technical, and Economic Issues. *IEEE Systems Journal*, *13*(3), 3026–3037. doi:10.1109/JSYST.2019.2918185

Ahmad, M., Ul Haq, Z., Khan, Z., Khattak, S. I., Ur Rahman, Z., & Khan, S. (2019). Does the inflow of remittances cause environmental degradation? Empirical evidence from China. *Economic Research-Ekonomska Istraživanja*, *32*(1), 2099–2121. doi:10.1080/1331677X.2019.1642783

Akin, C. S. (2014). The impact of foreign trade, energy consumption and income on CO2 emissions. *International Journal of Energy Economics and Policy*, *4*(3), 465.

Alami, H., Lehoux, P., Fleet, R., Fortin, J.-P., Liu, J., Attieh, R., Cadeddu, S. B. M., Abdoulaye Samri, M., Savoldelli, M., & Ag Ahmed, M. A. (2021). How can health systems better prepare for the next pandemic? Lessons learned from the management of COVID-19 in Quebec (Canada). *Frontiers in Public Health*, *9*, 1–15. doi:10.3389/fpubh.2021.671833 PMID:34222176

Aldieri, L., & Vinci, C. P. (2018). Green economy and sustainable development: The economic impact of innovation on employment. *Sustainability*, *10*(10), 3541. doi:10.3390u10103541

Al-Mulali, U., Ozturk, I., & Solarin, S. A. (2016). Investigating the environmental Kuznets curve hypothesis in seven regions: The role of renewable energy. *Ecological Indicators*, *67*, 267–282. doi:10.1016/j.ecolind.2016.02.059

Al-Soud, M. S., & Hrayshat, E. S. (2009). A 50 MW concentrating solar power plant for Jordan. *Journal of Cleaner Production*, *17*(6), 625–635. doi:10.1016/j.jclepro.2008.11.002

Ameri, T., Dennler, G., Lungenschmied, C., & Brabec, C. J. (2009). Organic tandem solar cells: A review. *Energy & Environmental Science*, *2*(4), 347–363. doi:10.1039/b817952b

Amuedo-Dorantes, C. (2014). The good and the bad in remittance flows. *IZA World of Labor: Evidence-Based Policy Making*, (97). Advance online publication. doi:10.15185/izawol.97

Apergis, N., Payne, J. E., Menyah, K., & Wolde-Rufael, Y. (2010). On the causal dynamics between emissions, nuclear energy, renewable energy, and economic growth. *Ecological Economics*, *69*(11), 2255–2260. doi:10.1016/j.ecolecon.2010.06.014

Ari, A., & Cergibozan, R. (2018). Sustainable growth in Turkey: The role of trade openness, financial development, and renewable energy use. *Industrial Policy and Sustainable Growth*, 1-21.

Aristizábal, A. J., & Páez, C. A. (2018). Power quality analysis of 6 kW building-integrated photovoltaic system and economic evaluation using RETScreen™. *Intl J Ambient Energy*, *39*(8), 813–822. doi:10.1080/01430750.2017.1354319

Asian Development Bank. (2021). *Accelerating sustainable development after COVID-19: The role of SDG bonds*. Asian Development Bank. https://www.adb.org/publications/sustainable-development-after-covid-19-sdg-bonds

Asumadu-Sarkodie, S., & Owusu, P. A. (2016). The potential and economic viability of solar photovoltaic power in Ghana. *Energ Source Part A*, *38*(5), 709–716. doi:10.1080/15567036.2015.1122682

Awerbuch, S. (2006). Portfolio-based electricity generation planning: Policy implications for renewables and energy security. *Mitigation and Adaptation Strategies for Global Change*, *11*(3), 693–710. doi:10.100711027-006-4754-4

Bailey, I., & Caprotti, F. (2014). The green economy: Functional domains and theoretical directions of enquiry. *Environment & Planning A*, *46*(8), 1797–1813. doi:10.1068/a130102p

Bakhshi, R., & Sadeh, J. (2018). Economic evaluation of grid–connected photovoltaic systems viability under a new dynamic feed–in tariff scheme: A case study in Iran. *Renewable Energy*, *119*, 354–364. doi:10.1016/j.renene.2017.11.093

Barbier, E. (2011). The policy challenges for green economy and sustainable economic development. *Natural Resources Forum*, *35*(3), 233–245. doi:10.1111/j.1477-8947.2011.01397.x

Bashshur, R., Doarn, C. R., Frenk, J. M., Kvedar, J. C., & Woolliscroft, J. O. (2020). Telemedicine and the COVID-19 pandemic, lessons for the future. *Telemedicine Journal and e-Health*, *26*(5), 571–573. doi:10.1089/tmj.2020.29040.rb PMID:32275485

Bessette, A., & Hanan, G. S. (2014). Design, synthesis and photophysical studies of dipyrromethene-based materials: Insights into their applications in organic photovoltaic devices. *Chemical Society Reviews*, *43*(10), 3342–3405. doi:10.1039/C3CS60411J PMID:24577078

Bhaskar, B. (2013). *Energy Security and Economic Development in India: a holistic approach*. The Energy and Resources Institute (TERI).

Bhattacharya, M., Paramati, S. R., Ozturk, I., & Bhattacharya, S. (2016). The effect of renewable energy consumption on economic growth: Evidence from top 38 countries. *Applied Energy*, *162*, 733–741. doi:10.1016/j.apenergy.2015.10.104

Bhattacharya, S. C., & Jana, C. (2009). Renewable energy in India: Historical developments and prospects. *Energy*, *34*(8), 981–991. doi:10.1016/j.energy.2008.10.017

Bi, D., Yi, C., Luo, J., Décoppet, J. D., Zhang, F., Zakeeruddin, S. M., Li, X., Hagfeldt, A., & Grätzel, M. (2016). Polymer-templated nucleation and crystal growth of perovskite films for solar cells with efficiency greater than 21%. *Nature Energy*, *1*(10), 16142. doi:10.1038/nenergy.2016.142

Blinken, A. J., & Becerra, X. (2021). Strengthening global health security and reforming the international health regulations: Making the world safer from future pandemics. *Journal of the American Medical Association*, *326*(13), 1255–1256. doi:10.1001/jama.2021.15611 PMID:34464446

Bordeianu, S. (1995). Political risk services and political risk yearbook: Edited William D. Coplin and Michael K. O'Leary. Syracuse, NY: Political Risk Services, 1994. 7 vols. $5,570.00 commercial subscribers (to the services), $750.00 (to the Yearbook). Journal of Government Information, 22(2), 186-188.

Borges-Martínez, M., Alvarez, D., Montenegro-Pohlhammer, N., Menendez, M. I., Lopez, R., & Cardenas-Jiron, G. (2019). Assessment of bodipy–oxasmaragdyrin dyads for dye-sensitized solar cells: Aromaticity, photosensitization capability, and charge transport. *The Journal of Physical Chemistry C*, *123*(32), 19362–19375. doi:10.1021/acs.jpcc.9b05136

Bovenberg, A. L., & van der Ploeg, F. (1994). Environmental policy, public finance and the labour market in a second-best world. *Journal of Public Economics*, *55*(3), 349–390. doi:10.1016/0047-2727(93)01398-T

Browne, J., Coffey, B., Cook, K., Meiklejohn, S., & Palermo, C. (2018). A guide to policy analysis as a research method. *Health Promotion International*. Advance online publication. doi:10.1093/heapro/day052 PMID:30101276

Brown, L., McFarlane, A., Campbell, K., & Das, A. (2020). Remittances and CO2 emissions in Jamaica: An asymmetric modified environmental kuznets curve. *Journal of Economic Asymmetries*, *22*, e00166. doi:10.1016/j.jeca.2020.e00166

Brunn, M. (2019, July 3). Waste management and the elderly: A neglected problem? *Rycycling Magazine*. https://www.recycling-magazine.com/2019/07/03/waste-management-and-the-elderly-a-neglected-problem/

Bucher, L., Desbois, N., Koukaras, E. N., Devillers, C. H., Biswas, S., Sharma, G. D., & Gros, C. P. (2018). BODIPY–diketopyrrolopyrrole–porphyrin conjugate small molecules for use in bulk heterojunction solar cells. *Journal of Materials Chemistry. A, Materials for Energy and Sustainability*, *6*(18), 8449–8461. doi:10.1039/C8TA01291A

Büyükzeren, R., Altintaş, H. B., Martin, K., & Kahraman, A. (2015). Binalardaki Fotovoltaik Uygulamasının Teknik, Çevresel ve Ekonomik İncelenmesi: Meram Tıp Fakültesi Hastanesi Örneği. *EMO Bilimsel Dergi, 5*(10), 9–14.

Cakmak, Y., Kolemen, S., Buyuktemiz, M., Dede, Y., & Erten-Ela, S. (2015). Synthesis and dye sensitized solar cell applications of bodipy derivatives with bisdimethylfluorenyl amine donor groups. *New Journal of Chemistry, 39*(5), 4086–4092. doi:10.1039/C4NJ02393E

Cao, Y., Bai, Y., Yu, Q., Cheng, Y., Liu, S., Shi, D., Gao, F., & Wang, P. (2009). Dye-sensitized solar cells with a high absorptivity ruthenium sensitizer featuring a 2-(hexylthio) thiophene conjugated bipyridine. *The Journal of Physical Chemistry C, 113*(15), 6290–6297. doi:10.1021/jp9006872

Carroll, D. (2021, May 17). Opinion: We need a new global surveillance system to detect and prevent the next pandemic. *Devex.* https://www.devex.com/news/sponsored/opinion-we-need-a-new-global-surveillance-system-to-detect-and-prevent-the-next-pandemic-99900

Cerna, L. (2013). The nature of policy change and implementation: A review of different theoretical approaches. *Organisation for Economic Cooperation and Development (OECD) Report*, 492–502.

Cheema, H., Younts, R., Gautam, B., Gundogdu, K., & El-Shafei, A. (2016). Design and synthesis of BODIY sensitizers with long alkyl chains tethered to N-carbazole and their application for dye sensitized solar cells. *Materials Chemistry and Physics, 184*, 57–63. doi:10.1016/j.matchemphys.2016.09.024

Chen, C. Y., Wu, S. J., Wu, C. G., Chen, J. G., & Ho, K. C. (2006). A ruthenium complex with superhigh light-harvesting capacity for dye-sensitized solar cells. *Angewandte Chemie International Edition, 45*(35), 5822–5825. doi:10.1002/anie.200601463 PMID:16865765

Cheng, Y. J., Yang, S. H., & Hsu, C. S. (2009). Synthesis of conjugated polymers for organic solar cell applications. *Chemical Reviews, 109*(11), 5868–5923. doi:10.1021/cr900182s PMID:19785455

Chen, J. D., Cui, C., Li, Y. Q., Zhou, L., Ou, Q. D., Li, C., Li, Y., & Tang, J. X. (2015). Single-junction polymer solar cells exceeding 10% power conversion efficiency. *Advanced Materials, 27*(6), 1035–1041. doi:10.1002/adma.201404535 PMID:25408480

Chen, J., & Cao, Y. (2009). Development of novel conjugated donor polymers for high-efficiency bulk-heterojunction photovoltaic devices. *Accounts of Chemical Research, 42*(11), 1709–1718. doi:10.1021/ar900061z PMID:19572607

Chen, W.-M., Kim, H., & Yamaguchi, H. (2014). Renewable energy in eastern Asia: Renewable energy policy review and comparative SWOT analysis for promoting renewable energy in Japan, South Korea, and Taiwan. *Energy Policy, 74*, 319–329. doi:10.1016/j.enpol.2014.08.019

Choi, W., Pate, M. B., Warren, R. D., & Nelson, R. M. (2018). An economic analysis comparison of stationary and dual-axis tracking grid-connected photovoltaic systems in the US Upper Midwest. *International Journal of Sustainable Energy, 37*(5), 455–478. doi:10.1080/14786451.2017.1294178

Chu, S., & Majumdar, A. (2012). Opportunities and challenges for a sustainable energy future. *Nature, 488*(7411), 294-303.

Claessens, S., & Yurtoglu, B. B. (2013). Corporate governance in emerging markets: A survey. *Emerging Markets Review, 15*, 1–33. doi:10.1016/j.ememar.2012.03.002

Coffin, R. C., Peet, J., Rogers, J., & Bazan, G. C. (2009). Streamlined microwave-assisted preparation of narrow-bandgap conjugated polymers for high-performance bulk heterojunction solar cells. *Nature Chemistry, 1*(8), 657–661. doi:10.1038/nchem.403 PMID:21378958

Cole, M. A. (2004). Trade, the pollution haven hypothesis and the environmental Kuznets curve: Examining the linkages. *Ecological Economics, 48*(1), 71–81. doi:10.1016/j.ecolecon.2003.09.007

Cook, T. (2005). Macroappraisal in Theory and Practice: Origins, Characteristics, and Implementation in Canada. *Archival Science, 5*(2), 101–161. doi:10.100710502-005-9010-2

Cooper, R. (2020). *Water security beyond Covid-19*. Institute of Development Studies.

Cortizo-Lacalle, D., Howells, C. T., Pandey, U. K., Cameron, J., Findlay, N. J., Inigo, A. R., Tuttle, T., Skabara, P. J., & Samuel, I. D. W. (2014). Solution processable diketopyrrolopyrrole (DPP) cored small molecules with BODIPY end groups as novel donors for organic solar cells. *Beilstein Journal of Organic Chemistry, 10*, 2683–2695. doi:10.3762/bjoc.10.283 PMID:25550732

Cox, E., Royston, S., & Selby, J. (2019). From exports to exercise: How non-energy policies affect energy systems. *Energy Research & Social Science, 55*, 179–188. doi:10.1016/j.erss.2019.05.016

Cristea, A., Hummels, D., Puzzello, L., & Avetisyan, M. (2013). Trade and the greenhouse gas emissions from international freight transport. *Journal of Environmental Economics and Management, 65*(1), 153–173. doi:10.1016/j.jeem.2012.06.002

Cuomo, F., Mallin, C., & Zattoni, A. (2016). Corporate governance codes: A review and research agenda. *Corporate Governance, 24*(3), 222–241. doi:10.1111/corg.12148

Curry, N., & Pillay, P. (2012). Biogas prediction and design of a food waste to energy system for the urban environment. *Renewable Energy, 41*, 200–209. doi:10.1016/j.renene.2011.10.019

Dağtekin, M., Kaya, D., Öztürk, H. H., & Kiliç, F. Ç. (2014). A study of techno-economic feasibility analysis of solar photovoltaic (PV) power generation in the province of Adana in Turkey. *Energ Explor Exploit, 32*(4), 719–735. doi:10.1260/0144-5987.32.4.719

Dai, S. Y., & Song, Q. H. (2012). New 2,6-modified BODIPY sensitizers for dye-sensitized solar cells. *Chemistry, an Asian Journal, 7*(4), 696–700. doi:10.1002/asia.201100779 PMID:22250129

Danish, M. S. S., & Senjyu, T. S. (Eds.). (2021). System of green resilience eco-oriented land uses in urban socio-ecosystems. In Y. Wang (Ed.), Eco-Friendly Energy Processes and Technologies for Achieving Sustainable Development: (pp. 1–23). IGI Global. doi:10.4018/978-1-7998-4915-5

Danish, M. S. S., Bhattacharya, A., Stepanova, D., Mikhaylov, A., Grilli, M. L., Khosravy, M., & Senjyu, T. (2020). A systematic review of metal oxide applications for energy and environmental sustainability. *Metals*, *10*(12), 1604. doi:10.3390/met10121604

Danish, M. S. S., Elsayed, M. E. L., Ahmadi, M., Senjyu, T., Karimy, H., & Zaheb, H. (2020). A strategic-integrated approach for sustainable energy deployment. *Energy Reports*, *6*, 40–44. doi:10.1016/j.egyr.2019.11.039

Danish, M. S. S., Estrella, L. L., Alemaida, I. M. A., Lisin, A., Moiseev, N., Ahmadi, M., Nazari, M., Wali, M., Zaheb, H., & Senjyu, T. (2021a). Photocatalytic applications of metal oxides for sustainable environmental remediation. *Metals*, *11*(1), 80. doi:10.3390/met11010080

Danish, M. S. S., Ibrahimi, A. M., Yaqobi, M. A., Udagawa, S., Mikhaylov, A., Faisal, N., & Senjyu, T. (2021b). Afghanistan aquaculture and fishery sectors—A foresight outlooks. *Journal of Ecoscience and Plant Revolution*, *2*(1), 17–37. doi:10.37357/1068/jepr.2.1.03

Danish, M. S. S., Matayoshi, H., Howlader, H. O. R., Chakraborty, S., Mandal, P., & Senjyu, T. (2019a). Microgrid Planning and Design: Resilience to Sustainability. *2019 IEEE PES GTD Grand International Conference and Exposition Asia (GTD Asia)*, 253–258. 10.1109/GTDAsia.2019.8716010

Danish, M. S. S., Sabory, N. R., Danish, S. M. S., Ludin, G. A., Yona, A., & Senjyu, T. (2016a). An Open-door Immature Policy for Rural Electrification: A Case Study of Afghanistan. *International Journal of Sustainable and Green Energy*, *6*(3), 8–13.

Danish, M. S. S., Sabory, N. R., Danish, S. M. S., Ludin, G. A., Yona, A., & Senjyu, T. (2016a). An open-door immature policy for rural electrification: A case study of Afghanistan. *International Journal of Sustainable and Green Energy*, *6*(3), 8–13. doi:10.11648/j.ijrse.s.2017060301.12

Danish, M. S. S., Sabory, N. R., Danish, S. M. S., Senjyu, T., Ludin, G. A., Noorzad, A. S., & Yona, A. (2017c). Electricity Sector Development Trends in an After-war Country: Afghanistan Aspiration for an Independent Energy Country. *Journal of Energy and Power Engineering*, *11*, 553–557.

Danish, M. S. S., Sabory, N. R., Ershad, A. M., Danish, S. M. S., Ohta, R., Sediqi, M., Ahmadi, M., & Senjyu, T. (2016b). The Least Developed Countries Needs for Changing the Passive Trend of Renewable Energy Exploitation to a Proactive Trend. *International Journal of Energy and Power Engineering*, *5*(6), 215–221. doi:10.11648/j.ijepe.20160506.17

Danish, M. S. S., Sabory, N. R., Ershad, A. M., Danish, S. M. S., Yona, A., & Senjyu, T. (2016). Sustainable Architecture and Urban Planning trough Exploitation of Renewable Energy. *International Journal of Sustainable and Green Energy*, *6*(3), 1. doi:10.11648/j.ijrse.s.2017060301.11

Danish, M. S. S., & Senjyu, T. (2020). Green Building Efficiency and Sustainability Indicators. In *Green Building Management and Smart Automation* (1st ed., pp. 128–145). IGI Global. doi:10.4018/978-1-5225-9754-4.ch006

Danish, M. S. S., Senjyu, T., Ahmadi, M., Ludin, G. A., Ahadi, M. H., Karimy, H., & Khosravy, M. (2021c). A review on energy efficiency for pathetic environmental trends mitigation. *Journal of Sustainability Outreach*, *2*(1), 1–8. doi:10.37357/1068/jso.2.1.01

Danish, M. S. S., Senjyu, T., Faisal, N., Stannikzai, M. Z., Nazari, A. M., & Vargas-Hernández, J. G. (2021d). A review on environmental-friendly energy multidisciplinary exposition from goals to action. *Journal of Environmental Science Revolution*, *2*(1), 1–9. doi:10.37357/1068/jesr.2.1.01

Danish, M. S. S., Senjyu, T., Funabashi, T., Ahmadi, M., Ibrahimi, A. M., Ohta, R., Howlader, H. O. R., Zaheb, H., Sabory, N. R., & Sediqi, M. M. (2018). A sustainable microgrid: A sustainability and management-oriented approach. *Applied Energy Symposium and Forum, Renewable Energy Integration with Mini/Microgrids*, *159*, 160–167. 10.1016/j.egypro.2018.12.045

Danish, M. S. S., Senjyu, T., Ibrahimi, A. M., Ahmadi, M., & Howlader, A. M. (2019b). A managed framework for energy-efficient building. *Journal of Building Engineering*, *21*, 120–128. doi:10.1016/j.jobe.2018.10.013

Danish, M. S. S., Senjyu, T., Ibrahimi, A. M., Bhattacharya, A., Nazari, Z., Danish, S. M. S., & Ahmadi, M. (2021e). Sustaining energy systems using metal oxide composites as photocatalysts. *Journal of Sustainable Energy Revolution*, *2*(1), 6–15. doi:10.37357/1068/jser.2.1.02

Danish, M. S. S., Senjyu, T., Nazari, M., Zaheb, H., Nassor, T. S., Danish, S. M. S., & Karimy, H. (2021f). Smart and sustainable building appraisal. *Journal of Sustainable Energy Revolution*, *2*(1), 1–5. doi:10.37357/1068/jser.2.1.01

Danish, M. S. S., Senjyu, T., Sabory, N. R., Danish, S. M. S., Ludin, G. A., Noorzad, A. S., & Yona, A. (2017a). Afghanistan's aspirations for energy independence: Water resources and hydropower energy. *Renewable Energy*, *113*, 1276–1287. doi:10.1016/j.renene.2017.06.090

Danish, M. S. S., Senjyu, T., Sabory, N. R., Khosravy, M., Grilli, M. L., Mikhaylov, A., & Majidi, H. (2021g). A forefront framework for sustainable aquaponics modeling and design. *Sustainability*, *13*(16), 9313. doi:10.3390u13169313

Danish, M. S. S., Senjyu, T., Urasaki, N., Rahmany, N. A., Ershad, A. M., Sabory, N. R., Zarabi, K., Anwarzai, M. A., Karimy, H., & Zaheb, H. (2021h). Developing nations as a foremost allotment for achieving 2030 SDGs—A case study. *Journal of Engineering and Technology Revolution*, *2*(1), 1–10. doi:10.37357/1068/jetr.2.1.01

Danish, M. S. S., Senjyu, T., Zaheb, H., Sabory, N. R., Ibrahimi, A. M., & Matayoshi, H. (2019a). A Novel Transdisciplinary Paradigm for Solid Waste to Sustainable Energy. *Journal of Cleaner Production*. Advance online publication. doi:10.1016/j.jclepro.2019.05.402

Danish, M. S. S., Yona, A., & Senjyu, T. (2014). Pre-design and life cycle cost analysis of a hybrid power system for rural and remote communities in Afghanistan. *The Journal of Engineering-IET*, *2014*(8), 438–444. doi:10.1049/joe.2014.0172

Danish, M. S. S., Zaheb, H., Sabory, N. R., Karimy, H., Faiq, A. B., Fedayi, H., & Senjyu, T. (2019b, June). The road ahead for municipal solid waste management in the 21st century: A novel-standardized simulated paradigm. *IOP Conference Series. Earth and Environmental Science*, *291*(1), 012009. doi:10.1088/1755-1315/291/1/012009

Das, A., & McFarlane, A. (2021). Remittances, electricity consumption and electric power losses in Jamaica. *Journal of Economic Studies*. doi:10.1108/JES-09-2020-0466

Das, A., & McFarlane, A. (2020). Remittances and disaggregated energy consumption in Bangladesh. *International Review of Economics*, *67*(2), 251–268. doi:10.100712232-019-00339-4

Das, A., McFarlane, A., & Carels, L. (2021). Empirical exploration of remittances and renewable energy consumption in Bangladesh. *Asia-Pacific Journal of Regional Science*, *5*(1), 65–89. doi:10.100741685-020-00180-6

Deshpande, R., Kibe, J., & Kaaria, L. (2020, April 15). *COVID-19 exposes risks and opportunities in Kenya's gig economy*. CGAP. https://www.cgap.org/blog/covid-19-exposes-risks-and-opportunities-kenyas-gig-economy

Dickey, D. A., & Fuller, W. A. (1979). Distribution of the Estimators for Autoregressive Time Series With a Unit Root. *Journal of the American Statistical Association*, *74*(366), 427. doi:10.2307/2286348

Dinçer, F. (2011). Overview of the photovoltaic technology status and perspective in Turkey. *Renewable & Sustainable Energy Reviews*, *15*(8), 3768–3779. doi:10.1016/j.rser.2011.06.005

Djankov, S., McLiesh, C., & Klein, M. U. (Eds.). (2004). *Doing business in 2004: understanding regulation* (Vol. 1). World Bank Publications.

Dobson, A. P., Pimm, S. L., Hannah, L., Kaufman, L., Ahumada, J. A., Ando, A. W., Bernstein, A., Busch, J., Daszak, P., Engelmann, J., Kinnaird, M. F., Li, B. V., Loch-Temzelides, T., Lovejoy, T., Nowak, K., Roehrdanz, P. R., & Vale, M. M. (2020). Ecology and economics for pandemic prevention. *Science*, *369*(6502), 379–381. Advance online publication. doi:10.1126cience.abc3189 PMID:32703868

Droege, P. (2011). *Urban energy transition: From fossil fuels to renewable power*. Elsevier.

Duman, A. C., & Güler, Ö. (2020). Economic analysis of grid-connected residential rooftop PV systems in Turkey. *Renewable Energy*, *148*, 697–711. doi:10.1016/j.renene.2019.10.157

Edalati, S., Ameri, M., Iranmanesh, M., Tarmahi, H., & Gholampour, M. (2016). Technical and economic assessments of grid-connected photovoltaic power plants: Iran case study. *Energy*, *114*, 923–934. doi:10.1016/j.energy.2016.08.041

Ejaz, N., Akhtar, H., Nisar, U., & Ali, N. (2010). *Environmental impacts of improper solid waste management in developing countries: A case study of Rawalpindi city*. doi:10.2495/SW100351

El Keshky, M. E. S., Basyouni, S. S., & Al Sabban, A. M. (2020). Getting through covid-19: The pandemic's impact on the psychology of sustainability, quality of life, and the global economy – A systematic review. *Frontiers in Psychology, 11*, 1–12. doi:10.3389/fpsyg.2020.585897 PMID:33281683

Elbatanony, M., Attiaoui, I., Ali, I. M. A., Nasser, N., & Tarchoun, M. (2021). The environmental impact of remittance inflows in developing countries: Evidence from method of moments quantile regression. *Environmental Science and Pollution Research International, 28*(35), 48222–48235. Advance online publication. doi:10.100711356-021-13733-9 PMID:33907951

Enders, W., & Lee, J. (2012). The flexible Fourier form and Dickey-Fuller type unit root tests. *Economics Letters, 117*(1), 196–199. doi:10.1016/j.econlet.2012.04.081

Energy and power baseline assessment working paper. (2010). *International Centre for Environmental Management (ICEM)*. https://www.mrcmekong.org/assets/Publications/Consultations/SEA-Hydropower/2-Energy-and-Power-baseline-assessment.pdf

Erdiwansyah, M., Mamat, R., Sani, M. S. M., & Sudhakar, K. (2019). Renewable energy in Southeast Asia: Policies and recommendations. *The Science of the Total Environment, 670*, 1095–1102. doi:10.1016/j.scitotenv.2019.03.273 PMID:31018425

Erten-Ela, S., Yilmaz, M. D., Icli, B., Dede, Y., Icli, S., & Akkaya, E. U. (2008). A panchromatic boradiazaindacene (bodipy) sensitizer for dye-sensitized solar cells. *Organic Letters, 10*(15), 3299–3302. doi:10.1021/ol8010612 PMID:18588306

European Commission. (2022, January 20). *Strict health regulations to ensure workers' protection in Azraq camp*. https://ec.europa.eu/trustfund-syria-region/strict-health-regulations-ensure-workers-protection-azraq-camp_en

Evrendilek, F., & Ertekin, C. (2003). Assessing the potential of renewable energy sources in Turkey. *Renewable Energy, 28*(15), 2303–2315. doi:10.1016/S0960-1481(03)00138-1

Fani, M., & Norouzi, N. (2020). Using Social and Economic Indicators for Modeling, Sensitivity Analysis and Forecasting the Gasoline Demand in the Transportation Sector: An ANN Approach in case study for Tehran metropolis. *Iranian Journal of Energy, 23*(2), 71–91.

Feleke, A., Intrator, K., Palmer, C., & Giles, M. (2021, June 10). *Five lessons for participatory planning during a pandemic*. ReliefWeb. https://reliefweb.int/report/world/five-lessons-participatory-planning-during-pandemic

Gakungu, N. K., Gitau, A., Njoroge, B. K., & Kimani, M. (2012). Solid waste management in Kenya: Acase study of public technical training institutions. *ICASTOR Journal of Engineering, 5*(3), 127–138.

Gao, F., Wang, Y., Shi, D., Zhang, J., Wang, M., Jing, X., Humphry-Baker, R., Wang, P., Zakeeruddin, S. M., & Grtzel, M. (2008). Enhance the optical absorptivity of nanocrystalline TiO2 film with high molar extinction coefficient ruthenium sensitizers for high performance dye-sensitized solar cells. *Journal of the American Chemical Society*, *130*(32), 10720–10728. doi:10.1021/ja801942j PMID:18642907

Geiger, T., Kuster, S., Yum, J. H., Moon, S. J., Nazeeruddin, M. K., Grätzel, M., & Nuesch, F. (2009). Molecular Design of Unsymmetrical Squaraine Dyes for High Efficiency Conversion of Low Energy Photons into Electrons Using TiO2 Nanocrystalline Films. *Advanced Functional Materials*, *21*(17), 2720–2727. doi:10.1002/adfm.200900231

Geneva Environment Network. (2022, January 20). *COVID-19 and the environment*. https://www.genevaenvironmentnetwork.org/resources/updates/updates-on-covid-19-and-the-environment/

GGKP News. (2020, April 29). *What does COVID-19 mean for sustainable development?* Green Growth Knowledge Platform. https://www.greengrowthknowledge.org/news/what-does-covid-19-mean-sustainable-development

Global Energy Institute - International Energy Security Risk Index. (2021). Retrieved July 21, 2021, from https://www.globalenergyinstitute.org/international-energy-security-risk-index

Global Energy Institute. (2020). *International Index of Energy Security Risk - Assessing Risk in a Global Energy Market*. International Index of Energy Security Risk 2018 Edition. Retrieved from https://www.globalenergyinstitute.org/sites/default/files/IESRI-Report_2020_4_20_20.pdf

Graf, K., Körzdörfer, T., Kümmel, S., & Thelakkat, M. (2013). Synthesis of donor-substituted meso-phenyl and meso-ethynylphenyl BODIPYs with broad absorption. *New Journal of Chemistry*, *37*(5), 1417–1426. doi:10.1039/c3nj00157a

Guía Práctica Sobre, P. E. (2016). *Energy policy: A practical Guidebook* (1st ed.). Latin American Energy Organization (OLADE).

Gungah, A., Emodi, N. V., & Dioha, M. O. (2019). Improving Nigeria's renewable energy policy design: A case study approach. *Energy Policy*, *130*, 89–100. doi:10.1016/j.enpol.2019.03.059

Guo, M., Nowakowska-Grunt, J., Gorbanyov, V., & Egorova, M. (2020). Green technology and sustainable development: Assessment and green growth frameworks. *Sustainability*, *12*(16), 6571. doi:10.3390u12166571

Gürtürk, M. (2019). Economic feasibility of solar power plants based on PV module with levelized cost analysis. *Energy*, *171*, 866–878. doi:10.1016/j.energy.2019.01.090

Hagberg, D. P., Edvinson, T., Marinado, T., Boschloo, G., Hagfeldt, A., & Sun, L. (2006). A novel organic chromophore for dye-sensitized nanostructured solar cells. *Chemical Communications*, (21), 2245–2247. doi:10.1039/b603002e PMID:16718317

Hagfeldt, A., Boschloo, G., Sun, L., Kloo, L., & Pettersson, H. (2010). Dye-sensitized solar cells. *Chemical Reviews*, *110*(11), 6595–6663. doi:10.1021/cr900356p PMID:20831177

Hagfeldt, A., & Graetzel, M. (1995). Light-Induced Redox Reactions in Nanocrystalline Systems. *Chemical Reviews*, *95*(1), 49–68. doi:10.1021/cr00033a003

Hagfeldt, A., & Gratzel, M. (2000). Molecular Photovoltaics. *Accounts of Chemical Research*, *33*(5), 269–277. doi:10.1021/ar980112j PMID:10813871

Hamdouch, A., & Depret, M. H. (2010). Policy integration strategy and the development of the 'green economy': Foundations and implementation patterns. *Journal of Environmental Planning and Management*, *53*(4), 473–490. doi:10.1080/09640561003703889

Hardee, K., Feranil, I., Boezwinkle, J., & Clark, B. (2004). *The policy circle: A framework for analyzing the components of family planning reproductive health maternal health and HIV/AIDS policies.* Academic Press.

Hart, S. L. (1992). An Integrative Framework for Strategy-Making Processes. *Academy of Management Review*, *17*(2), 327–351. doi:10.5465/amr.1992.4279547

Hattori, S., Ohkubo, K., Urano, Y., Sunahara, H., Nagano, T., Wada, Y., Tkachenko, N. V., Lemmetyinen, H., & Fukuzumi, S. (2005). Charge separation in a nonfluorescent donor–acceptor dyad derived from boron dipyrromethene dye, leading to photocurrent generation. *The Journal of Physical Chemistry B*, *109*(32), 15638–15375. doi:10.1021/jp050952x PMID:16852949

Hayek, F. A. (1945). The use of knowledge in society. *The American Economic Review*, *35*(4), 519–530.

Heindl, A.-B., & Liefner, I. (2019). The Analytic Hierarchy Process as a methodological contribution to improve regional innovation system research: Explored through comparative research in China. *Technology in Society*, *59*, 101197. doi:10.1016/j.techsoc.2019.101197

Hertel, S., & Keil, R. (2021, January 1). *After isolation: Urban planning and the Covid-19 pandemic.* Ontario Professional Planners Institute. https://ontarioplanners.ca/blog/planning-exchange/january-2021/after-isolation-urban-planning-and-the-covid-19-pandemic

He, Z., Xiao, B., Liu, F., Wu, F., We, H., Yang, Y., Xiao, S., Wang, C., Russell, T. P., & Cao, Y. (2015). Single-junction polymer solar cells with high efficiency and photovoltage. *Nature Photonics*, *9*(3), 174–179. doi:10.1038/nphoton.2015.6

Hou, G., Sun, H., Jiang, Z., Pan, Z., Wang, Y., Zhang, X., Zhao, Y., & Yao, Q. (2016). Life cycle assessment of grid-connected photovoltaic power generation from crystalline silicon solar modules in China. *Applied Energy*, *164*, 882–890. doi:10.1016/j.apenergy.2015.11.023

Hub, I. S. K. (2021, July 6). *Guest article: Preventing future pandemics starts with protecting our forests.* SDG Knowledge Hub. https://sdg.iisd.org:443/commentary/guest-articles/preventing-future-pandemics-starts-with-protecting-our-forests/

Hussein, B. A., Huynh, J. T., Prieto, P. L., Barran, C. P., Arnold, A. E., Sarycheva, O. V., Lough, A. J., & Koivisto, B. D. (2018). Molecular lemmings: Strategies to avoid when designing BODIPY ferrocene dyads for dye-sensitized solar cell applications. *Dalton Transactions (Cambridge, England)*, *47*(14), 4916–4920. doi:10.1039/C8DT00174J PMID:29546908

Hu, W., & Wang, D. (2020). How does environmental regulation influence China's carbon productivity? An empirical analysis based on the spatial spillover effect. *Journal of Cleaner Production, 257*, 120484. doi:10.1016/j.jclepro.2020.120484

Ian Guenn, T. (2020, July 6). Waste management in aged care facilities—Taking care of Australia's elderly. *Waster*. https://waster.com.au/waste-management-aged-care-facilities/

IBEF. (2018, September). *Renewable Energy*. IBEF.

IBEF. (2021). *Renewable Energy Industry in India*. https://www.ibef.org/industry/renewable-energy.aspx

Ibrahimi, A. M., Howlader, H. O. R., Danish, M. S. S., Shigenobu, R., Sediqi, M. M., & Senjyu, T. (n.d.). Optimal Unit Commitment with Concentrated Solar Power and Thermal Energy Storage in Afghanistan Electrical System. *International Journal of Emerging Electric Power Systems*.

IEA. (2021). *Energy security*. Retrieved July 21, 2021, from https://www.iea.org/areas-of-work/ensuring-energy-security

Ilyas, S., Srivastava, R. R., & Kim, H. (2020). Disinfection technology and strategies for COVID-19 hospital and bio-medical waste management. *The Science of the Total Environment, 749*, 141652. doi:10.1016/j.scitotenv.2020.141652 PMID:32822917

Imam, A. A., Al-Turki, Y. A., & R, S. K. (2019). Techno-Economic Feasibility Assessment of Grid-Connected PV Systems for Residential Buildings in Saudi Arabia—A Case Study. *Sustainability, 12*(1), 1–25. doi:10.3390u12010262

Iqbal, M. Z., Ali, S. R., & Khan, S. (2019). Progress in dye sensitized solar cell by incorporating natural photosensitizers. *Solar Energy, 181*, 490–509. doi:10.1016/j.solener.2019.02.023

Jadhav, T., Misra, R., Biswas, S., & Sharma, G. D. (2015). Bulk heterojunction organic solar cells based on carbazole–BODIPY conjugate small molecules as donors with high open circuit voltage. *Physical Chemistry Chemical Physics, 17*(40), 26580–26588. doi:10.1039/C5CP04807A PMID:26394373

Jain, M., & Patwardhan, A. (2013). Employment Outcomes of Renewable Energy Technologies: Implications of Policies in India. *Economic and Political Weekly*, 84–89.

Jalilian, H., Kirkpatrick, C., & Parker, D. (2007). The impact of regulation on economic growth in developing countries: A cross-country analysis. *World Development, 35*(1), 87–103. doi:10.1016/j.worlddev.2006.09.005

Jana, S. K. (2022). Sustainable Energy Development in Emerging Economies: A Study on BRICS. In Environmental Sustainability, Growth Trajectory and Gender: Contemporary Issues of Developing Economies (pp. 23-35). Emerald Publishing Limited.

Jansen, S. (2020). A focus area maturity model for software ecosystem governance. *Information and Software Technology, 118*, 106219. doi:10.1016/j.infsof.2019.106219 PMID:32490064

Jushi, E., Hysa, E., Cela, A., Panait, M., & Voica, M. C. (2021). Financing Growth through Remittances and Foreign Direct Investment: Evidences from Balkan Countries. *Journal of Risk and Financial Management, 14*(3), 1–17. doi:10.3390/jrfm14030117

Kalyanasundaram, K., & Gratzel, M. (1998). Applications of functionalized transition metal complexes in photonic and optoelectronic devices. *Coordination Chemistry Reviews, 177*(1), 347–414. doi:10.1016/S0010-8545(98)00189-1

Kanase-Patil, A. B., Saini, R. P., & Sharma, M. P. (2010). Integrated renewable energy systems for off grid rural electrification of remote area. *Renewable Energy, 35*(6), 1342–1349. doi:10.1016/j.renene.2009.10.005

Kan, B., Li, M., Zhang, Q., Liu, F., Wan, X., Wang, Y., Ni, W., Long, G., Yang, X., Feng, H., Zuo, Y., Zhang, M., Huang, F., Cao, Y., Russell, T. P., & Chen, Y. (2015). A Series of Simple Oligomer-like Small Molecules Based on Oligothiophenes for Solution-Processed Solar Cells with High Efficiency. *Journal of the American Chemical Society, 137*(11), 3886–3893. doi:10.1021/jacs.5b00305 PMID:25736989

Kaneza, N., Zhang, J., Liu, H., Archana, S., Shan, Z., Vasiliu, M., Polansky, S. H., Dixon, D. A., Adams, R. E., Schmehl, R. H., Gupta, A., & Pan, S. (2016). Electrochemical and spectroscopic properties of boron dipyrromethene-thiophenetriphenylamine-based dyes for dye-sensitized solar cells. *The Journal of Physical Chemistry C, 120*(17), 9068–9080. doi:10.1021/acs.jpcc.6b01611

Kapoor, K. K., & Dwivedi, Y. K. (2017). A take on solar power in India. *Economic and Political Weekly, 52*(7), 21–24.

Karasoy, A. (2021). How do remittances to the Philippines affect its environmental sustainability? Evidence based on the augmented ARDL approach. *Natural Resources Forum, 45*(2), 120–137. doi:10.1111/1477-8947.12218

Karasoy, A. (2022). How Do Consuming Alternative Energy Sources and Remittance Inflows Impact Egypt's Ecological Footprint? *Beykent Üniversitesi Sosyal Bilimler Dergisi, 15*(1), 8–28. doi:10.18221/bujss.1060051

Kar, S. K., & Gopakumar, K. (2015). Progress of renewable energy in India. *Adv Energy Res, 3*(2), 97–115. doi:10.12989/eri.2015.3.2.097

Kaufmann, D., Kraay, A., & Mastruzzi, M. (2005). Governance matters IV: governance indicators for 1996-2004. *World bank policy research working paper series*, (3630).

Kaufmann, D., Kraay, A., & Mastruzzi, M. (2004). Governance matters III: Governance indicators for 1996, 1998, 2000, and 2002. *The World Bank Economic Review, 18*(2), 253–287. doi:10.1093/wber/lhh041

Kay, A., & Gratzel, M. (1996). Low cost photovoltaic modules based on dye sensitized nanocrystalline titanium dioxide and carbon powder. *Solar Energy Materials and Solar Cells, 44*(1), 99–117. doi:10.1016/0927-0248(96)00063-3

Khan, A. R. (2016). Policy implementation: Some aspects and issues. *Journal of Community Positive Practices*, *13*(3), 3–12.

Khanna, R. C., Cicinelli, M. V., Gilbert, S. S., Honavar, S. G., & Murthy, G. S. V. (2020). COVID-19 pandemic: Lessons learned and future directions. *Indian Journal of Ophthalmology*, *68*(5), 703–710. doi:10.4103/ijo.IJO_843_20 PMID:32317432

Khan, R. A., & Khandaker, S. (2016). A critical insight into policy implementation and implementation performance. *Public Policy and Administration*, *15*(4), 538–548. doi:10.13165/VPA-16-15-4-02

Khan, Z. U., Ahmad, M., & Khan, A. (2020). On the remittances-environment led hypothesis: Empirical evidence from BRICS economies. *Environmental Science and Pollution Research International*, *27*(14), 16460–16471. doi:10.100711356-020-07999-8 PMID:32124280

Khare, V., Nema, S., & Baredar, P. (2013). Status of solar wind renewable energy in India. *Renewable & Sustainable Energy Reviews*, *27*, 1–10. doi:10.1016/j.rser.2013.06.018

Khoshnava, S. M., Rostami, R., Zin, R. M., Štreimikienė, D., Yousefpour, A., Strielkowski, W., & Mardani, A. (2019). Aligning the criteria of green economy (GE) and sustainable development goals (SDGs) to implement sustainable development. *Sustainability*, *11*(17), 4615. doi:10.3390u11174615

Kibria, M. G. (2021). Environmental downfall in Bangladesh: Revealing the asymmetric effectiveness of remittance inflow in the presence of foreign aid. *Environmental Science and Pollution Research International*. Advance online publication. doi:10.100711356-021-15619-2 PMID:34338983

Kim, H. S., Lee, C. R., Im, J. H., Lee, K. B., Moehl, T., Marchioro, A., Moon, S. J., Humphry-Baker, R., Yum, J. H., Moser, J. E., Grätzel, M., & Park, N. G. (2012). Lead iodide perovskite sensitized all-solidstate submicron thin film mesoscopic solar cell with efficiency exceeding 9%. *Scientific Reports*, *2*(1), 591. doi:10.1038rep00591 PMID:22912919

Kim, J., & Park, S. (2018). A contingent approach to energy mix policy. *Energy Policy*, *123*, 749–758. doi:10.1016/j.enpol.2018.08.028

Kiogora, M. (2020, December 9). *Why universal health care is crucial in managing future pandemics*. United Nations Kenya. https://kenya.un.org/en/109033-why-universal-health-care-crucial-managing-future-pandemics

Kirkendoll, S. M. (2021, July 17). *New universal coronavirus vaccine could prevent future pandemics*. The University of North Carolina at Chapel Hill. https://www.unc.edu/posts/2021/07/07/new-universal-coronavirus-vaccine-could-prevent-future-pandemics/

Kiviaho, J., Nikkinen, J., Piljak, V., & Rothovius, T. (2014). The co-movement dynamics of European frontier stock markets. *European Financial Management*, *20*(3), 574–595. doi:10.1111/j.1468-036X.2012.00646.x

Klfout, H., Stewart, A., Elkhalifa, M., & He, H. (2017). BODIPYs for dye-sensitized solar cells. *ACS Applied Materials & Interfaces*, *9*(46), 39873–39889. doi:10.1021/acsami.7b07688 PMID:29072443

Knag, H., Kee, S., Yu, K., Lee, J., Kim, G., Kim, J., Kim, J. R., Kong, J., & Lee, K. (2015). Simplified tandem polymer solar cells with an ıdeal self-organized recombination layer. *Advanced Materials*, *27*(8), 1408–1413. doi:10.1002/adma.201404765 PMID:25449142

Kocsis, I., & Kiss, J. T. (2014). Renewable energy consumption, R&D and GDP in European Union countries. *Environmental Engineering and Management Journal*, *13*(11), 2825–2830. doi:10.30638/eemj.2014.316

Kolemen, S., Bozdemir, O. A., Cakmak, Y., Barin, G., Erten-Ela, S., Marszalek, M., Yum, J. H., Zakeeruddin, S. M., Nazeeruddin, M. K., Grätzel, M., & Akkaya, E. U. (2011). Optimization of distyryl-Bodipy chromophores for efficient panchromatic sensitization in dye sensitized solar cells. *Chemical Science (Cambridge)*, *2*(5), 949–954. doi:10.1039/c0sc00649a

Kolemen, S., Cakmak, Y., Erten-Ela, S., Altay, Y., Brendel, J., Thelakkat, M., & Akkaya, E. U. (2010). Solid state dye-sensitized solar cells using red and near-IR absorbing Bodipy sensitizers. *Organic Letters*, *12*(17), 3812–3815. doi:10.1021/ol1014762 PMID:20704314

Kolemen, S., Cakmak, Y., Ozdemir, T., Erten-Ela, S., Buyuktemiz, M., Dede, Y., & Akkaya, E. U. (2014). Design and characterization of Bodipy derivatives for bulk heterojunction solar cells. *Tetrahedron*, *70*(36), 6229–6234. doi:10.1016/j.tet.2014.03.049

Kraner, S., Widmer, J., Benduhn, J., Hieckmann, E., Jageler-Hoheisel, T., Ullbrich, S., Schütze, D., Radke, K. S., Cuniberti, G., Ortmann, F., Lorenz-Rothe, M., Meerheim, R., Spoltore, D., Vandewal, K., Koerner, C., & Leo, K. (2015). Influence of side groups on the performance of infrared absorbing aza-BODIPY organic solar cells. *Physica Status Solidi. A, Applications and Materials Science*, *212*(12), 2747–2753. doi:10.1002/pssa.201532385

Krupa, J. (2012). Identifying barriers to aboriginal renewable energy deployment in Canada. *Energy Policy*, *42*, 710–714. doi:10.1016/j.enpol.2011.12.051

Kubo, Y., Eguchi, D., Matsumoto, A., Nishiyabu, R., Yakushiji, H., Shigaki, K., & Kaneko, M. (2014). Boron-dibenzopyrromethene-based organic dyes for application in dyesensitized solar cells. *Journal of Materials Chemistry. A, Materials for Energy and Sustainability*, *2*(15), 5204–5211. doi:10.1039/C3TA15340A

Kumar, A., Kumar, K., Kaushik, N., Sharma, S., & Mishra, S. (2010). Renewable energy in India: Current status and future potentials. *Renewable & Sustainable Energy Reviews*, *14*(8), 2434–2442. doi:10.1016/j.rser.2010.04.003

Kumaresan, D., Thummel, R.P., Bura, T., Ulrich, G., & Ziessel, R. (2009). Color tuning in new journal. metal-free organic sensitizers (bodipys) for dye-sensitized solar cells. *Chemistry A Europian Journal, 15*, 6335–6339.

Kuzman, M. K., Grošelj, P., Ayrilmis, N., & Zbašnik-Senegačnik, M. (2013). Comparison of passive house construction types using analytic hierarchy process. *Energy and Building*, *64*, 258–263. doi:10.1016/j.enbuild.2013.05.020

Kyeong, M., Lee, J., Lee, K., & Hong, S. (2018). BODIPY-based conjugated polymers for use as dopant-free hole transporting materials for durable perovskite solar cells: Selective tuning of HOMO/LUMO levels. *ACS Applied Materials & Interfaces*, *10*(27), 23254–23262. doi:10.1021/acsami.8b05956 PMID:29916246

La Porta, R., Lopez-de-Silanes, F., Shleifer, A., & Vishny, R. (2002). Investor protection and corporate valuation. *The Journal of Finance*, *57*(3), 1147–1170. doi:10.1111/1540-6261.00457

La Porta, R., Lopez-de-Silanes, F., Shleifer, A., & Vishny, R. W. (1997). Legal determinants of external finance. *The Journal of Finance*, *52*(3), 1131–1150. doi:10.1111/j.1540-6261.1997.tb02727.x

Lee, J., Chang, B., Aktas, C., & Gorthala, R. (2016). Economic feasibility of campus-wide photovoltaic systems in New England. *Renewable Energy*, *99*, 452–464. doi:10.1016/j.renene.2016.07.009

Le, T.-H., & Nguyen, C. P. (2019). Is energy security a driver for economic growth? Evidence from a global sample. *Energy Policy*, *129*, 436–451. doi:10.1016/j.enpol.2019.02.038

Lewis, N. S. (2007). Toward cost-effective solar energy use. *Science*, *315*(5813), 798–801. doi:10.1126cience.1137014 PMID:17289986

Lewis, N. S., & Nocera, D. G. (2006). Powering the planet: Chemical challenges in solar energy utilization. *Proceedings of the National Academy of Sciences of the United States of America*, *103*(43), 15729–15735. doi:10.1073/pnas.0603395103 PMID:17043226

Liaoa, J., Zhao, H., Cai, Z., Xu, Y., Qin, F. G. F., Zong, Q., Peng, F., & Fang, Y. (2018). BODIPY-based panchromatic chromophore for efficient organic solar cell. *Organic Electronics*, *61*, 215–222. doi:10.1016/j.orgel.2018.05.049

Liao, J., Zhao, H., Xu, Y., Cai, Z., Peng, Z., Zhang, W., Zhou, W., Li, B., Zong, Q., & Yang, X. (2016). Novel D–A–D type dyes based on BODIPY platform for solution processed organic solar cells. *Dyes and Pigments*, *128*, 131–140. doi:10.1016/j.dyepig.2016.01.020

Li, G., Zhu, R., & Yang, Y. (2012). Polymer solar cells. *Nature Photonics*, *6*(3), 153–161. doi:10.1038/nphoton.2012.11

Li, H., Fu, K., Boix, P. P., Wong, L. H., Hagfeldt, A., Grätzel, M., Mhaisalkar, S. G., & Grimsdale, A. C. (2014). Hole-transporting small molecules based on thiophene cores for high efficiency perovskite solar cells. *Chemistry-Sustainability-Energy-Materials*, *7*(12), 3420–3425. doi:10.1002/cssc.201402587 PMID:25233841

Li, J., Jiang, T., Ullah, S., & Majeed, M. T. (2021). The dynamic linkage between financial inflow and environmental quality: Evidence from China and policy options. *Environmental Science and Pollution Research International*. Advance online publication. doi:10.100711356-021-15616-5 PMID:34341938

Li, J., & Lin, B. (2016). Green economy performance and green productivity growth in China's cities: Measures and policy implication. *Sustainability*, *8*(9), 947. doi:10.3390u8090947

Lin, Y. D., Ke, B. Y., Chang, Y. J., Chou, P. T., Liau, K. L., Liu, C. Y., & Chow, T. J. (2015). Pyridomethene-BF2 complex/phenothiazine hybrid sensitizer with high molar extinction coefficient for efficient sensitized solar cells. *Journal of Materials Chemistry. A, Materials for Energy and Sustainability*, *3*(32), 16831–16842. doi:10.1039/C5TA03807C

Li, Q. S., Zhang, X. L., Wu, G. H., Dai, C. G., Ding, Y., Dai, S. Y., & Song, Q. H. (2017). Effects of donors of bodipy dyes on the performance of dye-sensitized solar cells. *Dyes and Pigments*, *141*, 148–160. doi:10.1016/j.dyepig.2017.02.017

Li, T. Y., Benduhn, J., Li, Y., Jaiser, F., Spoltore, D., Zeika, O., Ma, Z., Neher, D., Vandewal, K., & Leo, K. (2018). Boron dipyrromethene (BODIPY) with meso-perfluorinated alkyl substituents as near infrared donors in organic solar cells. *Journal of Materials Chemistry. A, Materials for Energy and Sustainability*, *6*(38), 18583–18591. doi:10.1039/C8TA06261G

Liu, C., Wang, K., Gong, X., & Heeger, A. J. (2016). Low bandgap semiconducting polymers for polymeric photovoltaics. *Chemical Society Reviews*, *45*(17), 4825–4846. doi:10.1039/C5CS00650C PMID:26548402

Liu, J. Y., Hou, X. N., Tian, Y., Jiang, L., Deng, S., Röder, B., & Ermilov, E. A. (2016). Photoinduced energy and charge transfer in a bis(triphenylamine)-BODIPY-C_{60} artificial photosynthetic system. *RSC Advances*, *6*(62), 57293–57305. doi:10.1039/C6RA06841C

Liu, W., Tang, A., Chen, J., Wu, Y., Zhan, C., & Yao, J. (2014). Photocurrent enhancement of bodipy-based solution-processed small-molecule solar cells by dimerization via the meso position. *ACS Applied Materials & Interfaces*, *6*(24), 22496–22505. doi:10.1021/am506585u PMID:25427221

Liu, W., Wang, C., & Mol, A. P. J. (2013). Rural public acceptance of renewable energy deployment: The case of Shandong in China. *Applied Energy*, *102*, 1187–1196. doi:10.1016/j.apenergy.2012.06.057

Liu, W., Yao, J., & Zhan, C. (2017). A novel bodipy-based low-band-gap small-molecule acceptor for efficient non-fullerene polymer solar cells. *Chinese Journal of Chemistry*, *35*(12), 1813–1823. doi:10.1002/cjoc.201700542

Liu, X., Wang, H. Q., Li, Y., Gui, Z., Ming, S., Usman, K., Zhang, W., & Fang, J. (2017). Regular organic solar cells with efficiency over 10% and promoted stability by ligand- and thermal annealing-free al-doped ZnO cathode interlayer. *Advancement of Science*, *4*(8), 1700053. doi:10.1002/advs.201700053 PMID:28852624

Liu, Y., Chen, C. C., Hong, Z., Gao, J., Yang, Y., Zhou, H., Dou, L., Li, G., & Yang, Y. (2013). Solution-processed small-molecule solar cells: Breaking the 10% power conversion efficiency. *Scientific Reports, 3*(1), 3356. doi:10.1038rep03356 PMID:24285006

Li, Y. (2012). Molecular design of photovoltaic materials for polymer solar cells: Toward suitable electronic energy levels and broad absorption. *Accounts of Chemical Research, 45*(5), 723–733. doi:10.1021/ar2002446 PMID:22288572

Loiseau, E., Saikku, L., Antikainen, R., Droste, N., Hansjürgens, B., Pitkänen, K., Leskinen, P., Kuikman, P., & Thomsen, M. (2016). Green economy and related concepts: An overview. *Journal of Cleaner Production, 139*, 361–371. doi:10.1016/j.jclepro.2016.08.024

Loudet, A., & Burgess, K. (2007). BODIPY dyes and their derivatives: Syntheses and spectroscopic properties. *Chemical Reviews, 107*(11), 4891–4932. doi:10.1021/cr078381n PMID:17924696

Lund, P. (2012). Large-scale urban renewable electricity schemes–Integration and interfacing aspects. *Energy Conversion and Management, 63*, 162–172. doi:10.1016/j.enconman.2012.01.037

Lu, Z., Liang, M., Dai, P., Miao, K., Zhang, C., Sun, Z., & Xue, S. (2016). A strategy for enhancing the performance of Borondipyrromethene dye-sensitized solar cells. *The Journal of Physical Chemistry C, 120*(45), 25657–25667. doi:10.1021/acs.jpcc.6b07356

Magnusson, T., & Berggren, C. (2018). Competing innovation systems and the need for redeployment in sustainability transitions. *Technological Forecasting and Social Change, 126*, 217–230. doi:10.1016/j.techfore.2017.08.014

Mahalik, M. K., Villanthenkodath, M. A., Mallick, H., & Gupta, M. (2021). Assessing the effectiveness of total foreign aid and foreign energy aid inflows on environmental quality in India. *Energy Policy, 149*, 112015. doi:10.1016/j.enpol.2020.112015

Mao, M., & Song, Q. H. (2016). The structure-property relationships of D-π-A BODIPY dyes for dye-sensitized solar cells. *Chemical Record (New York, N.Y.), 16*(2), 719–733. doi:10.1002/tcr.201500251 PMID:26846846

Mao, M., Wang, J., Xiao, Z., Dai, S., & Song, Q. (2012). New 2,6-modified BODIPY sensitizers for dye-sensitized solar cells. *Dyes and Pigments, 94*(2), 224–232. doi:10.1016/j.dyepig.2012.01.011

Mao, M., Zhang, X. L., Cao, L., Tong, Y., & Wu, G. H. (2015). Design of bodipy based organic dyes for high-efficient dye-sensitized solar cells employing double electron acceptors. *Dyes and Pigments, 117*, 28–36. doi:10.1016/j.dyepig.2015.02.001

Mao, M., Zhang, X. L., Fang, X. Q., Wu, G. H., Dai, S. Y., Song, Q. H., & Zhang, X. X. (2014). Highly efficient light-harvesting boradiazaindacene sensitizers for dye-sensitized solar cells featuring phenothiazine donor antenna. *Journal of Power Sources, 268*, 965–976. doi:10.1016/j.jpowsour.2014.05.079

Mao, M., Zhang, X. L., Fang, X. Q., Wu, G. H., Ding, Y., Liu, X. L., Dai, S. Y., & Song, Q. H. (2014). 2,6-Conjugated Bodipy sensitizers for high-performance dye-sensitized solar cells. *Organic Electronics, 15*(9), 2079–2090. doi:10.1016/j.orgel.2014.05.024

Marcus, A. A. (1992). *Controversial issues in energy policy* (1st ed., Vol. 2). SAGE Publications. doi:10.4135/9781483325729

Matin Ibrahimi, A., Masih Sediqi, M., Or Rashid Howlader, H., Sayed Shah Danish, M., Chakraborty, S., & Senjyu, T. (2019). Generation expansion planning considering renewable energy integration and optimal unit commitment: A case study of Afghanistan. AIMS Energy, 7(4), 441–464. doi:10.3934/energy.2019.4.441

Maynard-Moody, S., Musheno, M., & Palumbo, D. (1990). Street-wise social policy: Resolving the dilemma of street-level influence and successful implementation. *The Western Political Quarterly*, *43*(4), 833–848. doi:10.1177/106591299004300409

Melikoglu, M. (2013). Vision 2023: Feasibility analysis of Turkey's renewable energy projection. *Renewable Energy*, *50*, 570–575. doi:10.1016/j.renene.2012.07.032

Milani, B. (2000). *Designing the green economy: The postindustrial alternative to corporate globalization*. Rowman & Littlefield.

Min ReadTime. (2021, July 21). *Learning from Covid-19 and preparing for the future: Reflections from a collaborative partnership*. The Rockefeller Foundation. https://www.rockefellerfoundation.org/case-study/learning-from-covid-19-and-preparing-for-the-future-reflections-from-a-collaborative-partnership/

Ministry of Energy and Natural Resources. (2014). National Renewable Energy action plan for Turkey. Author.

Mirzahosseini, A. H., & Taheri, T. (2012). Environmental, technical and financial feasibility study of solar power plants by RETScreen, according to the targeting of energy subsidies in Iran. *Renewable & Sustainable Energy Reviews*, *16*(5), 2806–2811. doi:10.1016/j.rser.2012.01.066

Mishra, A. R., Mishra, S. A., & Tiwari, A. V. (2014). Solid waste management—Case study. *International Journal of Research in Advent Technology*, *2*(1), 1–4.

Mishra, P., & Behera, B. (2016). Socio-economic and environmental implications of solar electrification: Experience of rural Odisha. *Renewable & Sustainable Energy Reviews*, *56*, 953–964. doi:10.1016/j.rser.2015.11.075

Mishra, R., Basumatary, B., Singhal, R., Sharma, G. D., & Sankar, J. (2018). Corrole-BODIPY Dyad as Small-Molecule Donor for Bulk Heterojunction Solar Cells. *ACS Applied Materials & Interfaces*, *10*(37), 31462–31471. doi:10.1021/acsami.8b08519 PMID:30136584

Miyata, A., Mitioglu, A., Plochocka, P., Portugall, O., Wang, J. T. W., Stranks, S. D., Snaith, H. J., & Nicholas, R. J. (2015). Direct measurement of the exciton binding energy and effective masses for charge carriers in organic–inorganic tri-halide perovskites. *Nature Physics*, *11*(7), 582–587. doi:10.1038/nphys3357

MNRE. (2018). *Annual Report: 2017-18*. Government of India. https://powermin.nic.in/en/content/annual-reports-year-wise-ministry

Mohammadi, K., Naderi, M., & Saghafifar, M. (2018). Economic feasibility of developing grid-connected photovoltaic plants in the southern coast of Iran. *Energy, 156*, 17–31. doi:10.1016/j.energy.2018.05.065

MOSPI. (2020). *Energy Statistics, 2020*. Government of India.

Mthethwa, R. M. (2012). Critical dimensions for policy implementation. *African Journal of Public Affairs*, 1–12.

Muh, E., Amara, S., & Tabet, F. (2018). Sustainable energy policies in Cameroon: A holistic overview. *Renewable & Sustainable Energy Reviews*, *82*, 3420–3429. doi:10.1016/j.rser.2017.10.049

Naeem Nawaz, S. M., & Alvi, S. (2018). Energy security for socio-economic and environmental sustainability in Pakistan. *Heliyon*, *4*(10), 1–21. doi:10.1016/j.heliyon.2018.e00854 PMID:30364534

Nathan, H. S. K. (2014). Solar energy for rural electricity in India: A misplaced emphasis. *Economic and Political Weekly*, *49*(50), 60–67.

Nathanson, J. A. (2020, November 10). *Solid waste management* [Britannica]. https://www.britannica.com/technology/solid-waste-management

Nayar, R. (2016). Enforcing Renewable Purchase Obligations. *Economic and Political Weekly*, *51*(40), 21–23.

Nazeeruddin, M. K., Pchy, P., Renouard, T., Zakeeruddin, S. M., Humphry-Baker, R., Comte, P., Liska, P., Cevey, L., Costa, E., Shklover, V., Spiccia, L., Deacon, G. B., Bignozzi, C. A., & Grtzel, M. (2001). Engineering of efficient panchromatic sensitizers for nanocrystalline TiO2-based solar cells. *Journal of the American Chemical Society*, *123*(8), 1613–1624. doi:10.1021/ja003299u PMID:11456760

Neog, Y., & Yadava, A. K. (2020). Nexus among CO2 emissions, remittances, and financial development: A NARDL approach for India. *Environmental Science and Pollution Research International*, *27*(35), 44470–44481. doi:10.100711356-020-10198-0 PMID:32770469

Nikolaou, V., Charalambidis, G., Charisiadis, A., & Coutsolelos, A. G. (2017). Porphyrin–BODIPY-based hybrid model compounds for artificial photosynthetic reaction centers. *Comptes Rendus. Chimie*, *20*(3), 314–322. doi:10.1016/j.crci.2016.07.009

Norouzi, N. (2021a). The Pahlev Reliability Index: A measurement for the resilience of power generation technologies versus climate change. *Nuclear Engineering and Technology*, *53*(5), 1658–1663. doi:10.1016/j.net.2020.10.013

Norouzi, N. (2021b). Post-COVID-19 and globalization of oil and natural gas trade: Challenges, opportunities, lessons, regulations, and strategies. *International Journal of Energy Research*, *45*(10), 14338–14356. doi:10.1002/er.6762 PMID:34219899

Norouzi, N., & Ataei, E. (2021). Covid-19 Crisis and Environmental law: Opportunities and challenges. *Hasanuddin Law Review*, *7*(1), 46–60. doi:10.20956/halrev.v7i1.2772

Norouzi, N., de Rubens, G. Z., Choupanpiesheh, S., & Enevoldsen, P. (2020a). When pandemics impact economies and climate change: Exploring the impacts of COVID-19 on oil and electricity demand in China. *Energy Research & Social Science*, *68*, 101654. doi:10.1016/j.erss.2020.101654 PMID:32839693

Norouzi, N., & Fani, M. (2020a). The impacts of the novel corona virus on the oil and electricity demand in Iran and China. *Journal of Energy Management and Technology*, *4*(4), 36–48.

Norouzi, N., & Fani, M. (2020b). Black gold falls, black plague arise-An Opec crude oil price forecast using a gray prediction model. *Upstream Oil and Gas Technology*, *5*, 100015. doi:10.1016/j.upstre.2020.100015

Norouzi, N., & Fani, M. (2021). The prioritization and feasibility study over renewable technologies using fuzzy logic: A case study for Takestan plains. *Journal of Energy Management and Technology*, *5*(2), 12–22.

Norouzi, N., Fani, M., & Ziarani, Z. K. (2020b). The fall of oil Age: A scenario planning approach over the last peak oil of human history by 2040. *Journal of Petroleum Science Engineering*, *188*, 106827. doi:10.1016/j.petrol.2019.106827

Norouzi, N., & Kalantari, G. (2020). The sun food-water-energy nexus governance model a case study for Iran. *Water-Energy Nexus*, *3*, 72–80. doi:10.1016/j.wen.2020.05.005

Norouzi, N., Khanmohammadi, H. U., & Ataei, E. (2021b). The Law in the Face of the COVID-19 Pandemic: Early Lessons from Uruguay. *Hasanuddin Law Review*, *7*(2), 75–88. doi:10.20956/halrev.v7i2.2827

Norouzi, N., Zarazua de Rubens, G. Z., Enevoldsen, P., & Behzadi Forough, A. (2021a). The impact of COVID-19 on the electricity sector in Spain: An econometric approach based on prices. *International Journal of Energy Research*, *45*(4), 6320–6332. doi:10.1002/er.6259

Nozik, A. J. (1978). Photoelectrochemistry:Applications to solar energy conversion. *Annual Review of Physical Chemistry*, *29*(1), 189–222. doi:10.1146/annurev.pc.29.100178.001201

Numbi, B. P., & Malinga, S. J. (2017). Optimal energy cost and economic analysis of a residential grid-interactive solar PV system-case of eThekwini municipality in South Africa. *Applied Energy*, *186*, 28–45. doi:10.1016/j.apenergy.2016.10.048

O'Neill-Carrillo, E., Irizarry-Rivera, A. A., Colucci-Rios, J. A., Perez-Lugo, M., & Ortiz-Garcia, C. (2008). Sustainable Energy: Balancing the Economic, Environmental and Social Dimensions of Energy. *2008 IEEE Energy 2030 Conference*, 1–7. doi:10.1109/ENERGY.2008.4781010

O'Regan, B., & Gratzel, M. (1991). A low-cost, high-efficiency solar cell based on dye-sensitized colloidal TiO2 films. *Nature*, *353*(6346), 737–740. doi:10.1038/353737a0

Obeng, M., Gyamfi, S., Derkyi, N. S., Kabo-bah, A. T., & Peprah, F. (2020). Technical and economic feasibility of a 50 MW grid-connected solar PV at UENR Nsoatre Campus. *Journal of Cleaner Production, 247*, 119159. doi:10.1016/j.jclepro.2019.119159

OECD Policy Responses to Coronavirus (COVID-19). (2021). How will COVID-19 reshape science, technology and innovation? *OECD Science, Technology and Industry Policy Papers, 107*, 1–10. doi:10.1787/23074957

Okoye, C. O., & Oranekwu-Okoye, B. C. (2018). Economic feasibility of solar PV system for rural electrification in Sub-Sahara Africa. *Renewable & Sustainable Energy Reviews, 82*, 2537–2547. doi:10.1016/j.rser.2017.09.054

Ollo-López, A., & Aramendía-Muneta, M. E. (2012). ICT impact on competitiveness, innovation and environment. *Telematics and Informatics, 29*(2), 204–210. doi:10.1016/j.tele.2011.08.002

Omar, M. A., & Mahmoud, M. M. (2018). Grid connected PV-home systems in Palestine: A review on technical performance, effects and economic feasibility. *Renewable & Sustainable Energy Reviews, 82*, 2490–2497. doi:10.1016/j.rser.2017.09.008

Ooyama, Y., Hagiwara, Y., Mizumo, T., Harima, Y., & Oshita, J. (2013). Photovoltaic performance of dye-sensitized solar cells based on D–π–A type BODIPY dye with two pyridyl groups. *New Journal of Chemistry, 37*(8), 2479–2485. doi:10.1039/c3nj00456b

Ortiz, A. (2019). Triarylamine-BODIPY derivatives: A promising building block as hole transporting materials for efficient perovskite solar cells. *Dyes and Pigments, 171*, 107690. doi:10.1016/j.dyepig.2019.107690

Osmani, A. R. (2014, July –December). Conventional energy to renewable energy: Perspectives for India. *The NEHU Journal, 12*(2), 41–60.

Özcan, O., & Ersöz, F. (2019). Project and cost-based evaluation of solar energy performance in three different geographical regions of Turkey: Investment analysis application. *Eng Sci Technol Int J, 22*(4), 1098–1106. doi:10.1016/j.jestch.2019.04.001

Pahle, M., Pachauri, S., & Steinbacher, K. (2016). Can the Green Economy deliver it all? Experiences of renewable energy policies with socio-economic objectives. *Applied Energy, 179*, 1331–1341. doi:10.1016/j.apenergy.2016.06.073

Panwar, N., Kaushik, S., & Kothari, S. (2011). Role of renewable energy sources in environmental protection: A review. *Renewable & Sustainable Energy Reviews, 15*(3), 1513–1524. doi:10.1016/j.rser.2010.11.037

Park, J. Y. (1992). Canonical Cointegrating Regressions. *Econometrica, 60*(1), 119–143. doi:10.2307/2951679

Phillips, P. C. B., & Hansen, B. E. (1990). Statistical Inference in Instrumental Variables Regression with I(1) Processes. *The Review of Economic Studies, 57*(1), 99–125. doi:10.2307/2297545

Phillips, P. C. B., & Perron, P. (1988). Testing for a Unit Root in Time Series Regression. *Biometrika*, *75*(2), 335–346. doi:10.1093/biomet/75.2.335

Pillot, B., Muselli, M., Poggi, P., & Dias, J. B. (2017). *On the impact of the global energy policy framework on the development and sustainability of renewable power systems in Sub-Saharan Africa: The case of solar PV*. Academic Press.

Piret, J., & Boivin, G. (2021). Pandemics throughout history. *Frontiers in Microbiology*, *11*, 1–16. doi:10.3389/fmicb.2020.631736 PMID:33584597

Policy Constraints. (2019). *Online Learning for Sports Management*. http://www.leoisaac.com/policy/top124.htm

Porter, M. E., & Van der Linde, C. (1995). Toward a new conception of the environment-competitiveness relationship. *The Journal of Economic Perspectives*, *9*(4), 97–118. doi:10.1257/jep.9.4.97

Prakash Narain, J., Sodani, P. R., & Kant, L. (2021). COVID-19 pandemic: Lessons for the health systems. *Journal of Health Management*, *23*(1), 74–84. doi:10.1177/0972063421994982

Preparedness, prevention and control of coronavirus disease (COVID-19) for refugees and migrants in non-camp settings. (2020). *World Health Organization (WHO)*. https://apps.who.int/iris/bitstream/handle/10665/331777/WHO-2019-nCoV-Refugees_Migrants-2020.1-eng.pdf?sequence=1&isAllowed=y

Pricewaterhouse Coopers. (n.d.). *Post-Covid business recovery: Realising new opportunities*. PwC. Retrieved January 21, 2022, from https://www.pwc.com/gx/en/issues/crisis-solutions/covid-19.html

Purohit, I., Purohit, P., & Shekhar, S. (2013). Evaluating the potential of concentrating solar power generation in Northwestern India. *Energy Policy*, *62*, 157–175. doi:10.1016/j.enpol.2013.06.069

Qin, C., Mirloup, A., Leclerc, N., Islam, A., El-Shafei, A., Han, L., & Ziessel, R. (2014). Molecular engineering of new thienyl-bodipy dyes for highly efficient panchromatic sensitized solar cells. *Advanced Energy Materials*, *4*(11), 1400085. doi:10.1002/aenm.201400085

Qingquan, J., Khattak, S. I., Ahmad, M., & Ping, L. (2020). A new approach to environmental sustainability: Assessing the impact of monetary policy on CO2 emissions in Asian economies. *Sustainable Development*, *28*(5), 1331–1346. doi:10.1002d.2087

Rafique, M. M., & Bahaidarah, H. M. (2019). Thermo-economic and environmental feasibility of a solar power plant as a renewable and green source of electrification. *International Journal of Green Energy*, *16*(15), 1–14. doi:10.1080/15435075.2019.1677237

Rahaman, M. M., & Varis, O. (2008). *Central Asian waters: Social, economic, environmental and governance puzzle* (1st ed.). Helsinki University of Technology.

Rahman, M. M., Hosan, S., Karmaker, S. C., Chapman, A. J., & Saha, B. B. (2021). The effect of remittance on energy consumption: Panel cointegration and dynamic causality analysis for South Asian countries. *Energy, 220*, 119684. doi:10.1016/j.energy.2020.119684

Rahman, Z. U., Cai, H., & Ahmad, M. (2019). A New Look at The Remittances-FDI- Energy-Environment Nexus in The Case of Selected Asian Nations. *The Singapore Economic Review*, 1–19. doi:10.1142/S0217590819500176

Rajapakshe, W. (2017). Factors Affecting Human Resources Management Policy Implementation in Small and Medium Enterprises (SMEs) in Sri Lanka. *International Journal of Academic Research in Business & Social Sciences*, *7*(12), 1129–1144.

Ramalingam, B., & Prabh, J. (2020). Innovation, development and COVID-19: Challenges, opportunities and ways forward. *OECD Policy Responses to Coronavirus (COVID-19)*, 1–14.

Ramos, C. A. F., Alcaso, A. N., & Cardoso, A. J. M. (2019). Photovoltaic-thermal (PVT) technology: Review and case study. *IOP Conference Series. Earth and Environmental Science*, *354*(1), 012048. doi:10.1088/1755-1315/354/1/012048

Rathore, P. K. S., Chauhan, D. S., & Singh, R. P. (2019). Decentralized solar rooftop photovoltaic in India: On the path of sustainable energy security. *Renewable Energy*, *131*, 297–307. doi:10.1016/j.renene.2018.07.049

Regions for Sustainable Change: Baseline assessment report the regional climate confidence index. (2010). *INTERREG IVC*. http://www.rscproject.org/docs/RSC_Baseline_Assessment_Report.pdf

Rehman, S., Bader, M. A., & Al-Moallem, S. A. (2007). Cost of solar energy generated using PV panels. *Renewable & Sustainable Energy Reviews*, *11*(8), 1843–1857. doi:10.1016/j.rser.2006.03.005

RobinsonJ. (2020, April 2). *Pandemics*. WebMD. https://www.webmd.com/cold-and-flu/what-are-epidemics-pandemics-outbreaks

Rose, K. H. (2013). A Guide to the Project Management Body of Knowledge (PMBOK® Guide)—Fifth Edition. Project Management Journal, 44(3). doi:10.1002/pmj.21345

Rousseau, T., Cravino, A., Bura, T., Ulrich, G., Ziessel, R., & Roncali, J. (2009). BODIPY derivatives as donor materials for bulk heterojunction solar cells. *Chemical Communications*, *1*(13), 673–675. doi:10.1039/b822770e PMID:19294258

Rousseau, T., Cravino, A., Bura, T., Ulrich, G., Ziessel, R., & Roncali, J. (2009). Multi-donor molecular bulk heterojunction solar cells: Improving conversion efficiency by synergistic dye combinations. *Journal of Materials Chemistry*, *19*(16), 2298–2300. doi:10.1039/b903189h

Roy, J., & Jana, S. (1998). Solar lanterns for rural households. *Energy*, *23*(1), 67–68. doi:10.1016/S0360-5442(97)00067-4

Roy, P., Sinha, N. K., Tiwari, S., & Khare, A. (2020). A review on perovskite solar cells: Evolution of architecture, fabrication techniques, commercialization issues and status. *Solar Energy, 198*, 665–688. doi:10.1016/j.solener.2020.01.080

Rume, T., & Islam, S. M. D.-U. (2020). Environmental effects of COVID-19 pandemic and potential strategies of sustainability. *Heliyon, 6*(9), e04965. doi:10.1016/j.heliyon.2020.e04965 PMID:32964165

Saaty, T. L. (1997). That is not the analytic hierarchy process: What the AHP is and what it is not. *Journal of Multi-Criteria Decision Analysis, 6*(6), 324–335. doi:10.1002/(SICI)1099-1360(199711)6:6<324::AID-MCDA167>3.0.CO;2-Q

Sabory, N. R., Senjyu, T., Danish, M. S. S., Hosham, A., Noorzada, A., Amiri, A. S., & Muhammdi, Z. (2021). Applicable Smart City Strategies to Ensure Energy Efficiency and Renewable Energy Integration in Poor Cities: Kabul Case Study. *Sustainability, 13*(21), 11984. doi:10.3390u132111984

Sachs, J., Schmidt-Traub, G., Kroll, C., Lafortune, G., & Gray. (2021). *The sustainable development goals and Covid-19: Includes the SDG index and dashboards*. Cambridge University Press.

Sadiq, M. A. F., Sabory, N. R., Danish, M. S. S., & Senjyu, T. (2020). Role of micro-hydropower plants in socio-economic development of rural Afghanistan. *Journal of Sustainable Energy Revolution, 1*(1), 1–7. doi:10.37357/1068/jser.1.1.01

Sagani, A., Mihelis, J., & Dedoussis, V. (2017). Techno-economic analysis and life-cycle environmental impacts of small-scale building-integrated PV systems in Greece. *Energy and Building, 139*, 277–290. doi:10.1016/j.enbuild.2017.01.022

Sahoo, M., & Sethi, N. (2020). Does remittance inflow stimulate electricity consumption in India? An empirical insight. *South Asian Journal of Business Studies*. doi:10.1108/SAJBS-05-2020-0158

Sahoo, S. K., Manoharan, B., & Sivakumar, N. (2018). *Introduction: Why Perovskite and Perovskite Solar Cells? Perovskite Photovoltaics Basic to Advanced Concepts and Implementation.* Academic Press.

Saikkonen, P. (1992). Estimation and Testing of Cointegrated Systems by an Autoregressive Approximation. *Econometric Theory, 8*(01), 1–27. doi:10.1017/S0266466600010720

Samu, R., & Fahrioglu, M. (2017). An analysis on the potential of solar photovoltaic power. *Energy Source Part B, 12*(10), 883–889. doi:10.1080/15567249.2017.1319437

Sangroya, D., & Nayak, J. K. (2015). Development of wind energy in India. *International Journal of Renewable Energy Research, 5*(1), 1–13.

Sassoon, J. (2007). Sharing our story. An archaeology of archival thought. *ECU University Archives and Manuscripts, 2*(35), 1–40.

Savvidis, G., Siala, K., Weissbart, C., Schmidt, L., Borggrefe, F., Kumar, S., Pittel, K., Madlener, R., & Hufendiek, K. (2019). The gap between energy policy challenges and model capabilities. *Energy Policy, 125*, 503–520. doi:10.1016/j.enpol.2018.10.033

Sayeh, A., & Chami, R. (2020). Lifeliness in danger: The COVID-19 pandemic threatens to dry up a vital source of income for poor and fragile countries. *Finance & Development*, *57*(2), 16–19. https://www.imf.org/external/pubs/ft/fandd/2020/06/COVID19-pandemic-impact-on-remittance-flows-sayeh.htm

Sharma, G. D., Siddiqui, S. A., Nikiforou, A., Zervaki, G. E., Georgakaki, I., Ladomenou, K., & Coutsolelos, A. G. (2015). A mono(carboxy)porphyrin-triazine-(bodipy)$_2$ triad as a donor for bulk heterojunction organic solar cells†. *Journal of Materials Chemistry. C, Materials for Optical and Electronic Devices*, *3*(24), 6209–6217. doi:10.1039/C5TC01076D

Sharma, K., Bhattarai, B., & Ahmed, S. (2019). Aid, Growth, Remittances and Carbon Emissions in Nepal. *Energy Journal*, *40*(1), 129–141. doi:10.5547/01956574.40.1.ksha

Sharma, S. (2012). True electricity demand and the critical role of renewable energy sources. *Economic and Political Weekly*, *47*(24), 24–27.

Shi, Z., & Ahalapitiya, J. H. (2018). Perovskites-based solar cells: A review of recent progress, materials and processing methods. *Materials (Basel)*, *11*(5), 729. doi:10.3390/ma11050729 PMID:29734667

Shubbak, M. H. (2019). Advances in solar photovoltaics: Technology review and patent trends. *Renewable & Sustainable Energy Reviews*, *115*, 109383. doi:10.1016/j.rser.2019.109383

Shum, K. L. (2017). Renewable energy deployment policy: A transition management perspective. *Renewable & Sustainable Energy Reviews*, *73*, 1380–1388. doi:10.1016/j.rser.2017.01.005

Signé, L. (2017). Policy implementation: A synthesis of the study of policy implementation and the causes of policy failure. OCP Policy Center.

Si, L., Chang, C. W., He, H., Sykes, A., Lin, C. Y., & Diau, E. W. G. (2012). Dual functionality of bodipy chromophore in porphyrin-sensitized nanocrystalline solar cells. *The Journal of Physical Chemistry C*, *116*(19), 10451–10460. doi:10.1021/jp210458j

Simsek, Y., Lorca, Á., Urmee, T., Bahri, P. A., & Escobar, R. (2019). Review and assessment of energy policy developments in Chile. *Energy Policy*, *127*, 87–101. doi:10.1016/j.enpol.2018.11.058

Singh, N., Ogunseitan, O. A., & Tang, Y. (2021). Medical waste: Current challenges and future opportunities for sustainable management. *Critical Reviews in Environmental Science and Technology*, 1–23. doi:10.1080/10643389.2021.1885325

Singh, N., Tang, Y., Zhang, Z., & Zheng, C. (2020). COVID-19 waste management: Effective and successful measures in Wuhan, China. *Resources, Conservation and Recycling*, *163*, 105071. doi:10.1016/j.resconrec.2020.105071 PMID:32834489

SMART Goals: How to make your goals achievable. (2019). *Mind Tools*. https://www.mindtools.com/pages/article/smart-goals.htm

Sokołowski, M. M. (2019). When black meets green: A review of the four pillars of India's energy policy. *Energy Policy*, *130*, 60–68. doi:10.1016/j.enpol.2019.03.051

Sonja, A.-K. (2020). *How can we protect refugees during the COVID-19 crisis? Our world.* United Nations University. https://ourworld.unu.edu/en/how-can-we-protect-refugees-during-the-covid-19-crisis

Source and types of waste. (2022, January 20). *BYJUS*. https://byjus.com/chemistry/waste/

Sow, A., Mehrtash, M., Rousse, D. R., & Haillot, D. (2019). Economic analysis of residential solar photovoltaic electricity production in Canada. *Sustain Energy Techn*, *33*, 83–94. doi:10.1016/j.seta.2019.03.003

Stiglitz, J. (1998). The private uses of public interests: Incentives and institutions. *The Journal of Economic Perspectives*, *12*(2), 3–22. doi:10.1257/jep.12.2.3

Stock, J. H., & Watson, M. W. (1993). A Simple Estimator of Cointegrating Vectors in Higher Order Integrated Systems. *Econometrica*, *61*(4), 783–820. doi:10.2307/2951763

Summers, G. H., Lefebvre, J. F., Black, F. A., Davies, E. S., Gibson, E. A., Pullerits, T., Wood, C. J., & Zidek, K. (2016). Design and characterisation of bodipy sensitizers for dye-sensitized NiO solar cells. *Physical Chemistry Chemical Physics*, *18*(2), 1059–1070. doi:10.1039/C5CP05177K PMID:26660278

Surbhi, S. (2015, June 22). Difference between strategy and policy (with comparison chart). *Key Differences*. https://keydifferences.com/difference-between-strategy-and-policy.html

Szyja, P. (2016). The role of the state in creating green economy. *Oeconomia Copernicana*, *7*(2), 207–222. doi:10.12775/OeC.2016.013

Taylor, W. D. (2018). Interpolating between Hilbert–Samuel and Hilbert–Kunz multiplicity. *Journal of Algebra*, *509*, 212–239. doi:10.1016/j.jalgebra.2018.05.015

The energy mix and energy transition: Items of content in this feature report. (2015). *Planete Energies*. https://www.planete-energies.com/en/medias/close/about-energy-mix

The energy transformation scenarios. (2020). *Shell International B.V.* https://www.shell.com/promos/energy-and-innovation/download-full-report/_jcr_content.stream/1627553067906/fba2959d9759c5ae806a03acfb187f1c33409a91/energy-transformation-scenarios.pdf

The World Bank. (2021). *World Bank Open Data*. Retrieved July 22, 2021, from https://data.worldbank.org/

Three components of policy systems. (n.d.). *AGRIlink2019*. Retrieved November 20, 2019, from https://www.agrilinks.org/post/three-components-policy-systems

Thumuganti, G., Gupta, V., & Singh, S. P. (2019). New dithienosilole- and dithienogermole-based BODIPY for solar cell applications. *New Journal of Chemistry*, *43*(22), 8735–8740. doi:10.1039/C9NJ00550A

Tong, M., Cho, S., Rogers, J. T., Schmidt, K., Hsu, B. B. Y., Moses, D., Coffin, R. C., Kramer, E. J., Bazan, G. C., & Heeger, A. J. (2010). Higher Molecular Weight Leads to Improved Photoresponsivity, Charge Transport and Interfacial Ordering in a Narrow Bandgap Semiconducting Polymer. *Advanced Functional Materials*, *20*(22), 3959–3965. doi:10.1002/adfm.201001271

Truelove, S., Abrahim, O., Altare, C., Lauer, S. A., Grantz, K. H., Azman, A. S., & Spiegel, P. (2020). The potential impact of COVID-19 in refugee camps in Bangladesh and beyond: A modeling study. *PLoS Medicine*, *17*(6), e1003144. doi:10.1371/journal.pmed.1003144 PMID:32544156

Tsong, C.-C., Lee, C.-F., Tsai, L.-J., & Hu, T.-C. (2016). The Fourier approximation and testing for the null of cointegration. *Empirical Economics*, *51*(3), 1085–1113. doi:10.100700181-015-1028-6

TUIK. (2020). *Sera Gazı Emisyon İstatistikleri: 1990-2018*. https://data.tuik.gov.tr/Bulten/Index?p=Sera-Gazi-Emisyon-Istatistikleri-1990-2018-33624

Ulrich, G., Ziessel, R., & Harriman, A. (2008). The chemistry of fluorescent bodipy dyes: Versatility unsurpassed. *Angewandte Chemie International Edition*, *47*(7), 1184–1201. doi:10.1002/anie.200702070 PMID:18092309

UN. (2020). *The Sustainable Development Goals Report 2020*. United Nations. Retrieved from https://unstats.un.org/sdgs/report/2020/The-Sustainable-Development-Goals-Report-2020.pdf

UN. (2021). *SDG: Goal 10*. Retrieved September 4, 2021, from https://sdgs.un.org/goals/goal10

United Nations (UN). (2020, March 24). *Waste management an essential public service in the fight to beat COVID-19*. UN Environment. https://www.unep.org/news-and-stories/press-release/waste-management-essential-public-service-fight-beat-covid-19

United Nations (UN). (2022, January 22). *COP26: Together for our planet*. https://www.un.org/en/climatechange/cop26

United Nations Development Programme (UNDP). (2020). *COVID-19 and the SDGs*. https://feature.undp.org/covid-19-and-the-sdgs/

United Nations Environment Programme (UNEP). (2021, August 9). *Time to get serious about climate change. On a warming planet, no one is safe*. https://www.unep.org/news-and-stories/speech/time-get-serious-about-climate-change-warming-planet-no-one-safe

United States Chamber of Commerce. (2014). *Index of U.S. Energy Security Risk: Assessing America's Vulnerabilities in a Global Energy Market*. Retrieved from https://www.globalenergyinstitute.org/sites/default/files/2020-10/024036 Global Energy Institute US Index_Web.pdf

Usman, M., & Jahanger, A. (2021). Heterogeneous effects of remittances and institutional quality in reducing environmental deficit in the presence of EKC hypothesis: A global study with the application of panel quantile regression. *Environmental Science and Pollution Research International*, *28*(28), 37292–37310. Advance online publication. doi:10.100711356-021-13216-x PMID:33713269

Valverde-Chávez, D.A., Ponseca, C. S., Stoumpos, C., Yartsev, A., Kanatzidis, M. G., Sundström, V., & Cooke, D. G. (2015). *Optical pump - multi-thz probe spectroscopy of a single crystal organic hybrid lead halide perovskite.* Optical Society of America.

Vedung, E. (n.d.). *Public Policy and Program Evaluation* (1st ed.). Tylor & Francis Group. doi:10.4324/9781315127767

Villanthenkodath, M. A., & Mahalik, M. K. (2020). Technological innovation and environmental quality nexus in India: Does inward remittance matter? *Journal of Public Affairs*. Advance online publication. doi:10.1002/pa.2291

Vlachopoulos, N., Liska, P., Augustynski, J., & Graetzel, M. (1988). Very efficient visible light energy harvesting and conversion by spectral sensitization of high surface area polycrystalline titanium dioxide films. *Journal of the American Chemical Society*, *110*(4), 1216–1220. doi:10.1021/ja00212a033

von Hippel, D., Suzuki, T., Williams, J. H., Savage, T., & Hayes, P. (2011). Energy security and sustainability in Northeast Asia. *Energy Policy*, *39*(11), 6719–6730. doi:10.1016/j.enpol.2009.07.001

Vukovic, N., Pobedinsky, V., Mityagin, S., Drozhzhin, A., & Mingaleva, Z. (2019). A study on green economy indicators and modeling: Russian context. *Sustainability*, *11*(17), 4629. doi:10.3390u11174629

Vuola, M., Korkeakoski, M., Vähäkari, N., Dwyer, M. B., Hogarth, N. J., Kaivo-oja, J., Luukkanen, J., Chea, E., Thuon, T., & Phonhalath, K. (2020). What is a green economy? Review of national-level green economy policies in Cambodia and Lao PDR. *Sustainability*, *12*(16), 6664. doi:10.3390u12166664

Wang, Q., & Wu, N. (2012). Long-Run Covariance and its Applications in Cointegration Regression. *The Stata Journal: Promoting Communications on Statistics and Stata*, *12*(3), 515–542. doi:10.1177/1536867X1201200312

Warnan, J., Buchet, F., Pellegrin, Y., Blart, E., & Odobel, F. (2011). Panchromatic trichromophoric sensitizer for dye-sensitized solar cells using antenna effect. *Organic Letters*, *13*(15), 3944–3947. doi:10.1021/ol2014686 PMID:21721515

Water, sanitation, hygiene, and waste management for the COVID-19 virus. (2020). *World Health Organization and the United Nations Children's Fund (UNICEF)*. https://apps.who.int/iris/bitstream/handle/10665/331846/WHO-2019-nCoV-%20IPC_WASH-2020.3-eng.pdf?sequence=1&isAllowed=y

Watson, B. L., Moore, T. A., Moore, A. L., & Gust, D. (2017). Synthesis of a novel building block for the preparation of multi-chromophoric sensitizers for panchromatic dyesensitized solar cells. *Dyes and Pigments*, *136*, 893–897. doi:10.1016/j.dyepig.2016.09.037

Wawrzyniak, D., & Doryń, W. (2020). Does the quality of institutions modify the economic growth-carbon dioxide emissions nexus? Evidence from a group of emerging and developing countries. *Economic Research-Ekonomska Istraživanja, 33*(1), 124–144. doi:10.1080/1331677X.2019.1708770

WDI. (2021). *The World Bank's World Development Indicators*. Retrieved August 29, 2021, from https://databank.worldbank.org/source/world-development-indicators

Weijermars, R., Taylor, P., Bahn, O., Das, S. R., & Wei, Y.-M. (2012). Review of models and actors in energy mix optimization–can leader visions and decisions align with optimum model strategies for our future energy systems? *Energy Strategy Reviews, 1*(1), 5–18. doi:10.1016/j.esr.2011.10.001

Whitford, M. (2009). A framework for the development of event public policy: Facilitating regional development. *Tourism Management, 30*(5), 674–682. doi:10.1016/j.tourman.2008.10.018

Wies, R. (1994). Policy Definition and Classification: Aspects, Criteria and Examples. *Proceedings of the IFIP/IEEE International Workshop on Distributed Systems: Operation and Management*, 10–12.

Woldemariam, K., & Yiheyis, Z. (2017). Remittances and Economic Development in Africa: A Review of Empirical Evidence. In Foreign Capital Flows and Economic Development in Africa (pp. 401–421). Palgrave Macmillan US. doi:10.1057/978-1-137-53496-5_18

Word Vision. (2021, June 21). *How climate change impacts poverty*. https://www.worldvision.ca/stories/climate-change/how-climate-change-impacts-poverty

World Bank. (2020). *World Development Indicators*. Retrieved August 9, 2020, from https://databank.worldbank.org/data/home.aspx

World Health Organization (WHO). (2021, September 13). *Lessons learned from the COVID-19 pandemic – achieving resilience for the next health emergency*. https://www.euro.who.int/en/health-topics/health-emergencies/coronavirus-covid-19/news/news/2021/9/lessons-learned-from-the-covid-19-pandemic-achieving-resilience-for-the-next-health-emergency

World Health Organization (WHO). (2022, January 20). *Climate change and health*. https://www.who.int/news-room/fact-sheets/detail/climate-change-and-health

World Health Organization (WHO). (2022, January 20). *Coronavirus (COVID-19) dashboard*. https://covid19.who.int

Writing a policy: The structure and components of a policy. (2019). *Online Learning for Sports Management*. http://www.leoisaac.com/policy/top125.htm

Xie, G. (2016). Cooperative strategies for sustainability in a decentralized supply chain with competing suppliers. *Journal of Cleaner Production, 113*, 807–821. doi:10.1016/j.jclepro.2015.11.013

Xu, L., Wang, Y., Solangi, Y. A., Zameer, H., & Shah, S. A. A. (2019). Off-grid solar PV power generation system in Sindh, Pakistan: A techno-economic feasibility analysis. *Processes (Basel, Switzerland), 7*(5), 308. doi:10.3390/pr7050308

Xu, X., Yu, J., Zhang, D., & Ji, Q. (2021). Energy insecurity, economic growth, and the role of renewable energy: A cross-country panel analysis. *The Singapore Economic Review, 66*(2), 323–343. doi:10.1142/S021759081943001X

Yadav, D. K., Bhatti, T. S., & Verma, A. (2017). Study of integrated rural electrification system using wind-biogas based hybrid system and limited grid supply system. *International Journal of Renewable Energy Research, 7*(1), 1–11.

Yadav, S. K., & Bajpai, U. (2019). Energy, economic and environmental performance of a solar rooftop photovoltaic system in India. *International Journal of Sustainable Energy*, 1–16.

Yang, B., Jahanger, A., & Ali, M. (2021). Remittance inflows affect the ecological footprint in BICS countries: Do technological innovation and financial development matter? *Environmental Science and Pollution Research International, 28*(18), 23482–23500. doi:10.100711356-021-12400-3 PMID:33449318

Yang, B., Jahanger, A., & Khan, M. A. (2020). Does the inflow of remittances and energy consumption increase CO2 emissions in the era of globalization? A global perspective. *Air Quality, Atmosphere & Health, 13*(11), 1313–1328. doi:10.100711869-020-00885-9

Yenilenebilir Enerji Kaynaklarının Elektrik Enerjisi Üretimi Amaçlı Kullanımına İlişkin Kanun. (2005). *Kanun No. 5346.* Resmi Gazete (Sayı: 25819). Accessed October 2020, https://www.resmigazete.gov.tr/eskiler/2005/05/20050518-1.htm

You, J., Dou, L., Yoshimura, K., Kato, T., Ohya, K., Moriarty, T., Emery, K., Chen, C. C., Gao, J., Li, G., & Yang, Y. (2013). A polymer tandem solar cell with 10.6% power conversion efficiency. *Nature Communications, 4*(1), 1446. doi:10.1038/ncomms2411 PMID:23385590

Zhang, B., Wang, B., & Wang, Z. (2017). Role of renewable energy and non-renewable energy consumption on EKC: Evidence from Pakistan. *Journal of Cleaner Production, 156*, 855–864. doi:10.1016/j.jclepro.2017.03.203

Zhang, F., Shi, W., Luo, J., Pellet, N., Yi, C., Li, X., Zhao, X., Dennis, T. J. S., Li, X., Wang, S., Xiao, Y., Zakeeruddin, S. M., Bi, D., & Gratzel, M. (2017). Isomer-pure bis-pcbm-assisted crystal engineering of perovskite solar cells showing excellent efficiency and stability. *Advanced Materials, 29*(17), 1606806. doi:10.1002/adma.201606806 PMID:28240401

Zhang, J., Lu, F., Qi, S., Zhao, Y., Wang, K., Zhang, B., & Feng, Y. (2016). Influence of various electron-donating triarylamine groups in BODIPY sensitizers on the performance of dye-sensitized solar cells. *Dyes and Pigments, 128*, 296–303. doi:10.1016/j.dyepig.2016.02.008

Zhang, X., Li, M., Ge, Y., & Li, G. (2016). Techno-economic feasibility analysis of solar photovoltaic power generation for buildings. *Applied Thermal Engineering, 108*, 1362–1371. doi:10.1016/j.applthermaleng.2016.07.199

Zhang, X., Zhang, Y., Chen, L., & Xiao, Y. (2015). Star-shaped carbazole-based BODIPY derivatives with improved hole transportation and near-infrared absorption for small-molecule organic solar cells with high open-circuit voltages. *RSC Advances*, *5*(41), 32283–32289. doi:10.1039/C5RA02414E

Zhao, M., Liu, F., Song, Y., & Geng, J. (2020). Impact of air pollution regulation and technological investment on sustainable development of green economy in Eastern China: Empirical analysis with panel data approach. *Sustainability*, *12*(8), 3073. doi:10.3390u12083073

Zhu, H., Wei, L., & Niu, P. (2020). The novel coronavirus outbreak in Wuhan, China. *Global Health Research and Policy*, *5*(1), 6. doi:10.118641256-020-00135-6 PMID:32226823

Zhu, N., & Wong, P. K. (2020). Advances in viral diagnostic technologies for combating COVID-19 and future pandemics. *SLAS Technology*, *25*(6), 513–521. doi:10.1177/2472630320953798 PMID:32833548

Related References

To continue our tradition of advancing academic research, we have compiled a list of recommended IGI Global readings. These references will provide additional information and guidance to further enrich your knowledge and assist you with your own research and future publications.

Ajiboye, O. E., & Yusuff, O. S. (2017). Foreign Land Acquisition: Food Security and Food Chains – The Nigerian Experience. In I. Management Association (Ed.), Natural Resources Management: Concepts, Methodologies, Tools, and Applications (pp. 1524-1545). Hershey, PA: IGI Global. https://doi.org/doi:10.4018/978-1-5225-0803-8.ch072

Alapiki, H. E., & Amadi, L. A. (2018). Sustainable Food Consumption in the Neoliberal Order: Challenges and Policy Implications. In A. Obayelu (Ed.), *Food Systems Sustainability and Environmental Policies in Modern Economies* (pp. 90–123). Hershey, PA: IGI Global. doi:10.4018/978-1-5225-3631-4.ch005

Altaş, A. (2018). Geographical Information System Applications Utilized in Museums in Turkey Within the Scope of the Cultural Heritage Tourism: A Case Study of Mobile Application of Müze Asist. In S. Chaudhuri & N. Ray (Eds.), *GIS Applications in the Tourism and Hospitality Industry* (pp. 42–60). Hershey, PA: IGI Global. doi:10.4018/978-1-5225-5088-4.ch002

Andreea, I. R. (2018). Beyond Macroeconomics of Food and Nutrition Security. *International Journal of Sustainable Economies Management*, *7*(1), 13–22. doi:10.4018/IJSEM.2018010102

Anwar, J. (2017). Reproductive and Mental Health during Natural Disaster: Implications and Issues for Women in Developing Nations – A Case Example. In I. Management Association (Ed.), Gaming and Technology Addiction: Breakthroughs in Research and Practice (pp. 446-472). Hershey, PA: IGI Global. https://doi.org/ doi:10.4018/978-1-5225-0778-9.ch021

Awadh, H., Aksissou, M., Benhardouze, W., Darasi, F., & Snaiki, J. (2018). Socioeconomic Status of Artisanal Fishers in the West Part of Moroccan Mediterranean. *International Journal of Social Ecology and Sustainable Development*, *9*(1), 40–52. doi:10.4018/IJSESD.2018010104

Aye, G. C., & Haruna, R. F. (2018). Effect of Climate Change on Crop Productivity and Prices in Benue State, Nigeria: Implications for Food Security. In V. Erokhin (Ed.), *Establishing Food Security and Alternatives to International Trade in Emerging Economies* (pp. 244–268). Hershey, PA: IGI Global. doi:10.4018/978-1-5225-2733-6.ch012

Azizan, S. A., & Suki, N. M. (2017). Consumers' Intentions to Purchase Organic Food Products. In T. Esakki (Ed.), *Green Marketing and Environmental Responsibility in Modern Corporations* (pp. 68–82). Hershey, PA: IGI Global. doi:10.4018/978-1-5225-2331-4.ch005

Barakabitze, A. A., Fue, K. G., Kitindi, E. J., & Sanga, C. A. (2017). Developing a Framework for Next Generation Integrated Agro Food-Advisory Systems in Developing Countries. In I. Management Association (Ed.), Agri-Food Supply Chain Management: Breakthroughs in Research and Practice (pp. 47-67). Hershey, PA: IGI Global. doi:10.4018/978-1-5225-1629-3.ch004

Beachcroft-Shaw, H., & Ellis, D. (2018). Using Successful Cases to Promote Environmental Sustainability: A Social Marketing Approach. In I. Management Association (Ed.), Sustainable Development: Concepts, Methodologies, Tools, and Applications (pp. 936-953). Hershey, PA: IGI Global. doi:10.4018/978-1-5225-3817-2.ch042

Behnassi, M., Kahime, K., Boussaa, S., Boumezzough, A., & Messouli, M. (2017). Infectious Diseases and Climate Vulnerability in Morocco: Governance and Adaptation Options. In I. Management Association (Ed.), Public Health and Welfare: Concepts, Methodologies, Tools, and Applications (pp. 91-109). Hershey, PA: IGI Global. https://doi.org/ doi:10.4018/978-1-5225-1674-3.ch005

Bekele, F., & Bekele, I. (2017). Social and Environmental Impacts on Agricultural Development. In W. Ganpat, R. Dyer, & W. Isaac (Eds.), *Agricultural Development and Food Security in Developing Nations* (pp. 21–56). Hershey, PA: IGI Global. doi:10.4018/978-1-5225-0942-4.ch002

Benaouda, A., & García-Peñalvo, F. J. (2018). Towards an Intelligent System for the Territorial Planning: Agricultural Case. In F. García-Peñalvo (Ed.), *Global Implications of Emerging Technology Trends* (pp. 158–178). Hershey, PA: IGI Global. doi:10.4018/978-1-5225-4944-4.ch010

Bhaskar, A., Rao, G. B., & Vencatesan, J. (2017). Characterization and Management Concerns of Water Resources around Pallikaranai Marsh, South Chennai. In P. Rao & Y. Patil (Eds.), *Reconsidering the Impact of Climate Change on Global Water Supply, Use, and Management* (pp. 102–121). Hershey, PA: IGI Global. doi:10.4018/978-1-5225-1046-8.ch007

Bhyan, P., Shrivastava, B., & Kumar, N. (2022). Requisite Sustainable Development Contemplating Buildings: Economic and Environmental Sustainability. In A. Hussain, K. Tiwari, & A. Gupta (Eds.), *Addressing Environmental Challenges Through Spatial Planning* (pp. 269–288). IGI Global. https://doi.org/10.4018/978-1-7998-8331-9.ch014

Bogataj, D., & Drobne, D. (2017). Control of Perishable Goods in Cold Logistic Chains by Bionanosensors. In S. Joo (Ed.), *Applying Nanotechnology for Environmental Sustainability* (pp. 376–402). Hershey, PA: IGI Global. doi:10.4018/978-1-5225-0585-3.ch016

Bogataj, D., & Drobne, D. (2017). Control of Perishable Goods in Cold Logistic Chains by Bionanosensors. In I. Management Association (Ed.), Materials Science and Engineering: Concepts, Methodologies, Tools, and Applications (pp. 471-497). Hershey, PA: IGI Global. https://doi.org/ doi:10.4018/978-1-5225-1798-6.ch019

Bogueva, D., & Marinova, D. (2018). What Is More Important: Perception of Masculinity or Personal Health and the Environment? In D. Bogueva, D. Marinova, & T. Raphaely (Eds.), *Handbook of Research on Social Marketing and Its Influence on Animal Origin Food Product Consumption* (pp. 148–162). Hershey, PA: IGI Global. doi:10.4018/978-1-5225-4757-0.ch010

Bouzid, M. (2017). Waterborne Diseases and Climate Change: Impact and Implications. In M. Bouzid (Ed.), *Examining the Role of Environmental Change on Emerging Infectious Diseases and Pandemics* (pp. 89–108). Hershey, PA: IGI Global. doi:10.4018/978-1-5225-0553-2.ch004

Bowles, D. C. (2017). Climate Change-Associated Conflict and Infectious Disease. In M. Bouzid (Ed.), *Examining the Role of Environmental Change on Emerging Infectious Diseases and Pandemics* (pp. 68–88). Hershey, PA: IGI Global. doi:10.4018/978-1-5225-0553-2.ch003

Buck, J. J., & Lowry, R. K. (2017). Oceanographic Data Management: Quills and Free Text to the Digital Age and "Big Data". In P. Diviacco, A. Leadbetter, & H. Glaves (Eds.), *Oceanographic and Marine Cross-Domain Data Management for Sustainable Development* (pp. 1–22). Hershey, PA: IGI Global. doi:10.4018/978-1-5225-0700-0.ch001

Buse, C. G. (2017). Are Climate Change Adaptation Policies a Game Changer?: A Case Study of Perspectives from Public Health Officials in Ontario, Canada. In M. Bouzid (Ed.), *Examining the Role of Environmental Change on Emerging Infectious Diseases and Pandemics* (pp. 230–257). Hershey, PA: IGI Global. doi:10.4018/978-1-5225-0553-2.ch010

Calderon, F. A., Giolo, E. G., Frau, C. D., Rengel, M. G., Rodriguez, H., Tornello, M., ... Gallucci, R. (2018). Seismic Microzonation and Site Effects Detection Through Microtremors Measures: A Review. In N. Ceryan (Ed.), *Handbook of Research on Trends and Digital Advances in Engineering Geology* (pp. 326–349). Hershey, PA: IGI Global. doi:10.4018/978-1-5225-2709-1.ch009

Carfì, D., Donato, A., & Panuccio, D. (2018). A Game Theory Coopetitive Perspective for Sustainability of Global Feeding: Agreements Among Vegan and Non-Vegan Food Firms. In I. Management Association (Ed.), Game Theory: Breakthroughs in Research and Practice (pp. 71-104). Hershey, PA: IGI Global. https://doi.org/doi:10.4018/978-1-5225-2594-3.ch004

Castagnolo, V. (2018). Analyzing, Classifying, Safeguarding: Drawing for the Borgo Murattiano Neighbourhood of Bari. In G. Carlone, N. Martinelli, & F. Rotondo (Eds.), *Designing Grid Cities for Optimized Urban Development and Planning* (pp. 93–108). Hershey, PA: IGI Global. doi:10.4018/978-1-5225-3613-0.ch006

Chekima, B. (2018). The Dilemma of Purchase Intention: A Conceptual Framework for Understanding Actual Consumption of Organic Food. *International Journal of Sustainable Economies Management, 7*(2), 1–13. doi:10.4018/IJSEM.2018040101

Chen, Y. (2017). Sustainable Supply Chains and International Soft Landings: A Case of Wetland Entrepreneurship. In B. Christiansen & F. Kasarcı (Eds.), *Corporate Espionage, Geopolitics, and Diplomacy Issues in International Business* (pp. 232–247). Hershey, PA: IGI Global. doi:10.4018/978-1-5225-1031-4.ch013

Çıtak, L., Akel, V., & Ersoy, E. (2018). Investors' Reactions to the Announcement of New Constituents of BIST Sustainability Index: An Analysis by Event Study and Mean-Median Tests. In M. Risso & S. Testarmata (Eds.), *Value Sharing for Sustainable and Inclusive Development* (pp. 270–289). Hershey, PA: IGI Global. doi:10.4018/978-1-5225-3147-0.ch012

D'Aleo, V., D'Aleo, F., & Bonanno, R. (2018). New Food Industries Toward a New Level of Sustainable Supply: Success Stories, Business Models, and Strategies. In V. Erokhin (Ed.), *Establishing Food Security and Alternatives to International Trade in Emerging Economies* (pp. 74–97). Hershey, PA: IGI Global. doi:10.4018/978-1-5225-2733-6.ch004

Dagevos, H., & Reinders, M. J. (2018). Flexitarianism and Social Marketing: Reflections on Eating Meat in Moderation. In D. Bogueva, D. Marinova, & T. Raphaely (Eds.), *Handbook of Research on Social Marketing and Its Influence on Animal Origin Food Product Consumption* (pp. 105–120). Hershey, PA: IGI Global. doi:10.4018/978-1-5225-4757-0.ch007

Danisman, G. O. (2022). What Drives Eco-Design Innovations in European SMEs? In U. Akkucuk (Ed.), *Disruptive Technologies and Eco-Innovation for Sustainable Development* (pp. 191–206). IGI Global. https://doi.org/10.4018/978-1-7998-8900-7.ch011

Deenapanray, P. N., & Ramma, I. (2017). Adaptations to Climate Change and Climate Variability in the Agriculture Sector in Mauritius: Lessons from a Technical Needs Assessment. In I. Management Association (Ed.), Natural Resources Management: Concepts, Methodologies, Tools, and Applications (pp. 655-680). Hershey, PA: IGI Global. https://doi.org/ doi:10.4018/978-1-5225-0803-8.ch030

Deenapanray, P. N., & Ramma, I. (2017). Adaptations to Climate Change and Climate Variability in the Agriculture Sector in Mauritius: Lessons from a Technical Needs Assessment. In I. Management Association (Ed.), Natural Resources Management: Concepts, Methodologies, Tools, and Applications (pp. 655-680). Hershey, PA: IGI Global. https://doi.org/ doi:10.4018/978-1-5225-0803-8.ch030

Deshpande, S., Basu, S. K., Li, X., & Chen, X. (2017). Smart, Innovative and Intelligent Technologies Used in Drug Designing. In I. Management Association (Ed.), Pharmaceutical Sciences: Breakthroughs in Research and Practice (pp. 1175-1191). Hershey, PA: IGI Global. https://doi.org/ doi:10.4018/978-1-5225-1762-7.ch045

Dlamini, P. N. (2017). Use of Information and Communication Technologies Tools to Capture, Store, and Disseminate Indigenous Knowledge: A Literature Review. In P. Ngulube (Ed.), *Handbook of Research on Theoretical Perspectives on Indigenous Knowledge Systems in Developing Countries* (pp. 225–247). Hershey, PA: IGI Global. doi:10.4018/978-1-5225-0833-5.ch010

Dolejsova, M., & Kera, D. (2017). The Fermentation GutHub Project and the Internet of Microbes. In S. Konomi & G. Roussos (Eds.), *Enriching Urban Spaces with Ambient Computing, the Internet of Things, and Smart City Design* (pp. 25–46). Hershey, PA: IGI Global. doi:10.4018/978-1-5225-0827-4.ch002

Dolunay, O. (2018). A Paradigm Shift: Empowering Farmers to Eliminate the Waste in the Form of Fresh Water and Energy Through the Implementation of 4R+T. In I. Management Association (Ed.), Sustainable Development: Concepts, Methodologies, Tools, and Applications (pp. 882-892). Hershey, PA: IGI Global. https://doi.org/doi:10.4018/978-1-5225-3817-2.ch039

Dube, P., Heijman, W. J., Ihle, R., & Ochieng, J. (2018). The Potential of Traditional Leafy Vegetables for Improving Food Security in Africa. In V. Erokhin (Ed.), *Establishing Food Security and Alternatives to International Trade in Emerging Economies* (pp. 220–243). Hershey, PA: IGI Global. doi:10.4018/978-1-5225-2733-6.ch011

Duruji, M. M., & Urenma, D. F. (2017). The Environmentalism and Politics of Climate Change: A Study of the Process of Global Convergence through UNFCCC Conferences. In I. Management Association (Ed.), *Natural Resources Management: Concepts, Methodologies, Tools, and Applications* (pp. 77-108). Hershey, PA: IGI Global. 10.4018/978-1-5225-0803-8.ch004

Dutta, U. (2017). Agro-Geoinformatics, Potato Cultivation, and Climate Change. In S. Londhe (Ed.), *Sustainable Potato Production and the Impact of Climate Change* (pp. 247–271). Hershey, PA: IGI Global. doi:10.4018/978-1-5225-1715-3.ch012

Edirisinghe, R., Stranieri, A., & Wickramasinghe, N. (2017). A Taxonomy for mHealth. In N. Wickramasinghe (Ed.), *Handbook of Research on Healthcare Administration and Management* (pp. 596–615). Hershey, PA: IGI Global. doi:10.4018/978-1-5225-0920-2.ch036

Ekpeni, N. M., & Ayeni, A. O. (2018). Global Natural Hazard and Disaster Vulnerability Management. In A. Eneanya (Ed.), *Handbook of Research on Environmental Policies for Emergency Management and Public Safety* (pp. 83–104). Hershey, PA: IGI Global. doi:10.4018/978-1-5225-3194-4.ch005

Ene, C., Voica, M. C., & Panait, M. (2017). Green Investments and Food Security: Opportunities and Future Directions in the Context of Sustainable Development. In M. Mieila (Ed.), *Measuring Sustainable Development and Green Investments in Contemporary Economies* (pp. 163–200). Hershey, PA: IGI Global. doi:10.4018/978-1-5225-2081-8.ch007

Escamilla, I., Ruíz, M. T., Ibarra, M. M., Soto, V. L., Quintero, R., & Guzmán, G. (2018). Geocoding Tweets Based on Semantic Web and Ontologies. In M. Lytras, N. Aljohani, E. Damiani, & K. Chui (Eds.), *Innovations, Developments, and Applications of Semantic Web and Information Systems* (pp. 372–392). Hershey, PA: IGI Global. doi:10.4018/978-1-5225-5042-6.ch014

Escribano, A. J. (2018). Marketing Strategies for Trendy Animal Products: Sustainability as a Core. In F. Quoquab, R. Thurasamy, & J. Mohammad (Eds.), *Driving Green Consumerism Through Strategic Sustainability Marketing* (pp. 169–203). Hershey, PA: IGI Global. doi:10.4018/978-1-5225-2912-5.ch010

Eudoxie, G., & Roopnarine, R. (2017). Climate Change Adaptation and Disaster Risk Management in the Caribbean. In W. Ganpat & W. Isaac (Eds.), *Environmental Sustainability and Climate Change Adaptation Strategies* (pp. 97–125). Hershey, PA: IGI Global. doi:10.4018/978-1-5225-1607-1.ch004

Farmer, L. S. (2017). Data Analytics for Strategic Management: Getting the Right Data. In V. Wang (Ed.), *Encyclopedia of Strategic Leadership and Management* (pp. 810–822). Hershey, PA: IGI Global. doi:10.4018/978-1-5225-1049-9.ch056

Fattal, L. R. (2017). Catastrophe: An Uncanny Catalyst for Creativity. In R. Shin (Ed.), *Convergence of Contemporary Art, Visual Culture, and Global Civic Engagement* (pp. 244–262). Hershey, PA: IGI Global. doi:10.4018/978-1-5225-1665-1.ch014

Forti, I. (2017). A Cross Reading of Landscape through Digital Landscape Models: The Case of Southern Garda. In A. Ippolito (Ed.), *Handbook of Research on Emerging Technologies for Architectural and Archaeological Heritage* (pp. 532–561). Hershey, PA: IGI Global. doi:10.4018/978-1-5225-0675-1.ch018

Gharbi, A., De Runz, C., & Akdag, H. (2017). Urban Development Modelling: A Survey. In S. Faiz & K. Mahmoudi (Eds.), *Handbook of Research on Geographic Information Systems Applications and Advancements* (pp. 96–124). Hershey, PA: IGI Global. doi:10.4018/978-1-5225-0937-0.ch004

Ghosh, I., & Ghoshal, I. (2018). Implications of Trade Liberalization for Food Security Under the ASEAN-India Strategic Partnership: A Gravity Model Approach. In V. Erokhin (Ed.), *Establishing Food Security and Alternatives to International Trade in Emerging Economies* (pp. 98–118). Hershey, PA: IGI Global. doi:10.4018/978-1-5225-2733-6.ch005

Glaves, H. M. (2017). Developing a Common Global Framework for Marine Data Management. In P. Diviacco, A. Leadbetter, & H. Glaves (Eds.), *Oceanographic and Marine Cross-Domain Data Management for Sustainable Development* (pp. 47–68). Hershey, PA: IGI Global. doi:10.4018/978-1-5225-0700-0.ch003

Godulla, A., & Wolf, C. (2018). Future of Food: Transmedia Strategies of National Geographic. In R. Gambarato & G. Alzamora (Eds.), *Exploring Transmedia Journalism in the Digital Age* (pp. 162–182). Hershey, PA: IGI Global. doi:10.4018/978-1-5225-3781-6.ch010

Gomes, P. P. (2018). Food and Environment: A Review on the Sustainability of Six Different Dietary Patterns. In A. Obayelu (Ed.), *Food Systems Sustainability and Environmental Policies in Modern Economies* (pp. 15–31). Hershey, PA: IGI Global. doi:10.4018/978-1-5225-3631-4.ch002

Gonzalez-Feliu, J. (2018). Sustainability Evaluation of Green Urban Logistics Systems: Literature Overview and Proposed Framework. In A. Paul, D. Bhattacharyya, & S. Anand (Eds.), *Green Initiatives for Business Sustainability and Value Creation* (pp. 103–134). Hershey, PA: IGI Global. doi:10.4018/978-1-5225-2662-9.ch005

Goodland, R. (2017). A Fresh Look at Livestock Greenhouse Gas Emissions and Mitigation. In I. Management Association (Ed.), Natural Resources Management: Concepts, Methodologies, Tools, and Applications (pp. 124-139). Hershey, PA: IGI Global. doi:10.4018/978-1-5225-0803-8.ch006

Goundar, S., & Appana, S. (2018). Mainstreaming Development Policies for Climate Change in Fiji: A Policy Gap Analysis and the Role of ICTs. In I. Management Association (Ed.), Sustainable Development: Concepts, Methodologies, Tools, and Applications (pp. 402-432). Hershey, PA: IGI Global. doi:10.4018/978-1-5225-3817-2.ch020

Granell-Canut, C., & Aguilar-Moreno, E. (2018). Geospatial Influence in Science Mapping. In M. Khosrow-Pour, D.B.A. (Ed.), Encyclopedia of Information Science and Technology, Fourth Edition (pp. 3473-3483). Hershey, PA: IGI Global. doi:10.4018/978-1-5225-2255-3.ch302

Grigelis, A., Blažauskas, N., Gelumbauskaitė, L. Ž., Gulbinskas, S., Suzdalev, S., & Ferrarin, C. (2017). Marine Environment Data Management Related to the Human Activity in the South-Eastern Baltic Sea (The Lithuanian Sector). In P. Diviacco, A. Leadbetter, & H. Glaves (Eds.), *Oceanographic and Marine Cross-Domain Data Management for Sustainable Development* (pp. 282–302). Hershey, PA: IGI Global. doi:10.4018/978-1-5225-0700-0.ch012

Guma, I. P., Rwashana, A. S., & Oyo, B. (2018). Food Security Policy Analysis Using System Dynamics: The Case of Uganda. *International Journal of Information Technologies and Systems Approach, 11*(1), 72–90. doi:10.4018/IJITSA.2018010104

Guma, I. P., Rwashana, A. S., & Oyo, B. (2018). Food Security Indicators for Subsistence Farmers Sustainability: A System Dynamics Approach. *International Journal of System Dynamics Applications, 7*(1), 45–64. doi:10.4018/IJSDA.2018010103

Gupta, P., & Goyal, S. (2017). Wildlife Habitat Evaluation. In A. Santra & S. Mitra (Eds.), *Remote Sensing Techniques and GIS Applications in Earth and Environmental Studies* (pp. 258–264). Hershey, PA: IGI Global. doi:10.4018/978-1-5225-1814-3.ch013

Hanson, T., & Hildebrand, E. (2018). GPS Travel Diaries in Rural Transportation Research: A Focus on Older Drivers. In I. Management Association (Ed.), Intelligent Transportation and Planning: Breakthroughs in Research and Practice (pp. 609-625). Hershey, PA: IGI Global. doi:10.4018/978-1-5225-5210-9.ch027

Hartman, M. B. (2017). Research-Based Climate Change Public Education Programs. In I. Management Association (Ed.), Natural Resources Management: Concepts, Methodologies, Tools, and Applications (pp. 992-1003). Hershey, PA: IGI Global. doi:10.4018/978-1-5225-0803-8.ch046

Hashim, N. (2017). Zanzibari Seaweed: Global Climate Change and the Promise of Adaptation. In I. Management Association (Ed.), Natural Resources Management: Concepts, Methodologies, Tools, and Applications (pp. 365-391). Hershey, PA: IGI Global. https://doi.org/ doi:10.4018/978-1-5225-0803-8.ch019

Herrera, J. E., Argüello, L. V., Gonzalez-Feliu, J., & Jaimes, W. A. (2017). Decision Support System Design Requirements, Information Management, and Urban Logistics Efficiency: Case Study of Bogotá, Colombia. In G. Jamil, A. Soares, & C. Pessoa (Eds.), *Handbook of Research on Information Management for Effective Logistics and Supply Chains* (pp. 223–238). Hershey, PA: IGI Global. doi:10.4018/978-1-5225-0973-8.ch012

Huizinga, T., Ayanso, A., Smoor, M., & Wronski, T. (2017). Exploring Insurance and Natural Disaster Tweets Using Text Analytics. *International Journal of Business Analytics, 4*(1), 1–17. doi:10.4018/IJBAN.2017010101

Hung, K., Kalantari, M., & Rajabifard, A. (2017). An Integrated Method for Assessing the Text Content Quality of Volunteered Geographic Information in Disaster Management. *International Journal of Information Systems for Crisis Response and Management, 9*(2), 1–17. doi:10.4018/IJISCRAM.2017040101

Husnain, A., & Avdic, A. (2018). Identifying the Contemporary Status of E-Service Sustainability Research. In I. Management Association (Ed.), Sustainable Development: Concepts, Methodologies, Tools, and Applications (pp. 467-485). Hershey, PA: IGI Global. https://doi.org/ doi:10.4018/978-1-5225-3817-2.ch022

Iarossi, M. P., & Ferro, L. (2017). "The Past is Never Dead. It's Not Even Past": Virtual Archaeological Promenade. In A. Ippolito & M. Cigola (Eds.), *Handbook of Research on Emerging Technologies for Digital Preservation and Information Modeling* (pp. 228–255). Hershey, PA: IGI Global. doi:10.4018/978-1-5225-0680-5.ch010

Ignjatijević, S., & Cvijanović, D. (2018). Analysis of Serbian Honey Production and Exports. In *Exploring the Global Competitiveness of Agri-Food Sectors and Serbia's Dominant Presence: Emerging Research and Opportunities* (pp. 109–139). Hershey, PA: IGI Global. doi:10.4018/978-1-5225-2762-6.ch005

Jana, S. K., & Karmakar, A. K. (2017). Globalization, Governance, and Food Security: The Case of BRICS. In I. Management Association (Ed.), Natural Resources Management: Concepts, Methodologies, Tools, and Applications (pp. 692-712). Hershey, PA: IGI Global. https://doi.org/ doi:10.4018/978-1-5225-0803-8.ch032

Jana, S. K., & Karmakar, A. K. (2017). Food Security in Asia: Is There Convergence? In I. Management Association (Ed.), Natural Resources Management: Concepts, Methodologies, Tools, and Applications (pp. 109-123). Hershey, PA: IGI Global. https://doi.org/ doi:10.4018/978-1-5225-0803-8.ch005

John, J., & Kumar, S. (2018). A Locational Decision Making Framework for Shipbreaking Under Multiple Criteria. In I. Management Association (Ed.), Operations and Service Management: Concepts, Methodologies, Tools, and Applications (pp. 504-527). Hershey, PA: IGI Global. doi:10.4018/978-1-5225-3909-4.ch024

John, J., & Srivastava, R. K. (2018). Decision Insights for Shipbreaking using Environmental Impact Assessment: Review and Perspectives. *International Journal of Strategic Decision Sciences, 9*(1), 45–62. doi:10.4018/IJSDS.2018010104

Joshi, Y., & Rahman, Z. (2018). Determinants of Sustainable Consumption Behaviour: Review and Conceptual Framework. In A. Paul, D. Bhattacharyya, & S. Anand (Eds.), *Green Initiatives for Business Sustainability and Value Creation* (pp. 239–262). Hershey, PA: IGI Global. doi:10.4018/978-1-5225-2662-9.ch011

Juma, D. W., Reuben, M., Wang, H., & Li, F. (2018). Adaptive Coevolution: Realigning the Water Governance Regime to the Changing Climate. In I. Management Association (Ed.), Hydrology and Water Resource Management: Breakthroughs in Research and Practice (pp. 346-357). Hershey, PA: IGI Global. https://doi.org/ doi:10.4018/978-1-5225-3427-3.ch014

K., S., & Tripathy, B. K. (2018). Neighborhood Rough-Sets-Based Spatial Data Analytics. In M. Khosrow-Pour, D.B.A. (Ed.), *Encyclopedia of Information Science and Technology, Fourth Edition* (pp. 1835-1844). Hershey, PA: IGI Global. https:// doi.org/ doi:10.4018/978-1-5225-2255-3.ch160

Kabir, F. (2018). Towards a More Gender-Inclusive Climate Change Policy. In N. Mahtab, T. Haque, I. Khan, M. Islam, & I. Wahid (Eds.), *Handbook of Research on Women's Issues and Rights in the Developing World* (pp. 354–369). Hershey, PA: IGI Global. doi:10.4018/978-1-5225-3018-3.ch022

Kabir, F. (2018). Towards a More Gender-Inclusive Climate Change Policy. In I. Management Association (Ed.), Climate Change and Environmental Concerns: Breakthroughs in Research and Practice (pp. 525-540). Hershey, PA: IGI Global. https://doi.org/ doi:10.4018/978-1-5225-5487-5.ch027

Kanyamuka, J. S., Jumbe, C. B., & Ricker-Gilbert, J. (2018). Making Agricultural Input Subsidies More Effective and Profitable in Africa: The Role of Complementary Interventions. In A. Obayelu (Ed.), *Food Systems Sustainability and Environmental Policies in Modern Economies* (pp. 172–187). Hershey, PA: IGI Global. doi:10.4018/978-1-5225-3631-4.ch008

Karimi, H., & Gholamrezafahimi, F. (2017). Study of Integrated Coastal Zone Management and Its Environmental Effects: A Case of Iran. In R. Singh, A. Singh, & V. Srivastava (Eds.), *Environmental Issues Surrounding Human Overpopulation* (pp. 64–88). Hershey, PA: IGI Global. doi:10.4018/978-1-5225-1683-5.ch004

Kaya, I. R., Hutabarat, J., & Bambang, A. N. (2018). "Sasi": A New Path to Sustain Seaweed Farming From Up-Stream to Down-Stream in Kotania Bay, Molucass. *International Journal of Social Ecology and Sustainable Development, 9*(2), 28–36. doi:10.4018/IJSESD.2018040103

Khader, V. (2018). Technologies for Food, Health, Livelihood, and Nutrition Security. In I. Management Association (Ed.), Food Science and Nutrition: Breakthroughs in Research and Practice (pp. 94-112). Hershey, PA: IGI Global. https://doi.org/doi:10.4018/978-1-5225-5207-9.ch005

Kocadağlı, A. Y. (2017). The Temporal and Spatial Development of Organic Agriculture in Turkey. In W. Ganpat, R. Dyer, & W. Isaac (Eds.), *Agricultural Development and Food Security in Developing Nations* (pp. 130–156). Hershey, PA: IGI Global. doi:10.4018/978-1-5225-0942-4.ch006

Koundouri, P., Giannouli, A., & Souliotis, I. (2017). An Integrated Approach for Sustainable Environmental and Socio-Economic Development Using Offshore Infrastructure. In I. Management Association (Ed.), Renewable and Alternative Energy: Concepts, Methodologies, Tools, and Applications (pp. 1581-1601). Hershey, PA: IGI Global. doi:10.4018/978-1-5225-1671-2.ch056

Kumar, A., & Dash, M. K. (2017). Sustainability and Future Generation Infrastructure on Digital Platform: A Study of Generation Y. In N. Ray (Ed.), *Business Infrastructure for Sustainability in Developing Economies* (pp. 124–142). Hershey, PA: IGI Global. doi:10.4018/978-1-5225-2041-2.ch007

Kumar, A., Mukherjee, A. B., & Krishna, A. P. (2017). Application of Conventional Data Mining Techniques and Web Mining to Aid Disaster Management. In A. Kumar (Ed.), *Web Usage Mining Techniques and Applications Across Industries* (pp. 138–167). Hershey, PA: IGI Global. doi:10.4018/978-1-5225-0613-3.ch006

Kumar, C. P. (2017). Impact of Climate Change on Groundwater Resources. In I. Management Association (Ed.), Natural Resources Management: Concepts, Methodologies, Tools, and Applications (pp. 1094-1120). Hershey, PA: IGI Global. doi:10.4018/978-1-5225-0803-8.ch052

Kumari, S., & Patil, Y. (2017). Achieving Climate Smart Agriculture with a Sustainable Use of Water: A Conceptual Framework for Sustaining the Use of Water for Agriculture in the Era of Climate Change. In P. Rao & Y. Patil (Eds.), *Reconsidering the Impact of Climate Change on Global Water Supply, Use, and Management* (pp. 122–143). Hershey, PA: IGI Global. doi:10.4018/978-1-5225-1046-8.ch008

Kumari, S., & Patil, Y. (2018). Achieving Climate Smart Agriculture With a Sustainable Use of Water: A Conceptual Framework for Sustaining the Use of Water for Agriculture in the Era of Climate Change. In I. Management Association (Ed.), Climate Change and Environmental Concerns: Breakthroughs in Research and Practice (pp. 111-133). Hershey, PA: IGI Global. doi:10.4018/978-1-5225-5487-5.ch006

Kursah, M. B. (2017). Least-Cost Pipeline using Geographic Information System: The Limit to Technicalities. *International Journal of Applied Geospatial Research, 8*(3), 1–15. doi:10.4018/ijagr.2017070101

Lahiri, S., Ghosh, D., & Bhakta, J. N. (2017). Role of Microbes in Eco-Remediation of Perturbed Aquatic Ecosystem. In J. Bhakta (Ed.), *Handbook of Research on Inventive Bioremediation Techniques* (pp. 70–107). Hershey, PA: IGI Global. doi:10.4018/978-1-5225-2325-3.ch004

Lallo, C. H., Smalling, S., Facey, A., & Hughes, M. (2018). The Impact of Climate Change on Small Ruminant Performance in Caribbean Communities. In I. Management Association (Ed.), Climate Change and Environmental Concerns: Breakthroughs in Research and Practice (pp. 193-218). Hershey, PA: IGI Global. doi:10.4018/978-1-5225-5487-5.ch010

Laurini, R. (2017). Nature of Geographic Knowledge Bases. In S. Faiz & K. Mahmoudi (Eds.), *Handbook of Research on Geographic Information Systems Applications and Advancements* (pp. 29–60). Hershey, PA: IGI Global. doi:10.4018/978-1-5225-0937-0.ch002

Lawrence, J., Simpson, L., & Piggott, A. (2017). Protected Agriculture: A Climate Change Adaptation for Food and Nutrition Security. In I. Management Association (Ed.), Natural Resources Management: Concepts, Methodologies, Tools, and Applications (pp. 140-158). Hershey, PA: IGI Global. doi:10.4018/978-1-5225-0803-8.ch007

Leadbetter, A., Cheatham, M., Shepherd, A., & Thomas, R. (2017). Linked Ocean Data 2.0. In P. Diviacco, A. Leadbetter, & H. Glaves (Eds.), *Oceanographic and Marine Cross-Domain Data Management for Sustainable Development* (pp. 69–99). Hershey, PA: IGI Global. doi:10.4018/978-1-5225-0700-0.ch004

Lucas, M. R., Rego, C., Vieira, C., & Vieira, I. (2017). Proximity and Cooperation for Innovative Regional Development: The Case of the Science and Technology Park of Alentejo. In L. Carvalho (Ed.), *Handbook of Research on Entrepreneurial Development and Innovation Within Smart Cities* (pp. 199–228). Hershey, PA: IGI Global. doi:10.4018/978-1-5225-1978-2.ch010

Ma, X., Beaulieu, S. E., Fu, L., Fox, P., Di Stefano, M., & West, P. (2017). Documenting Provenance for Reproducible Marine Ecosystem Assessment in Open Science. In P. Diviacco, A. Leadbetter, & H. Glaves (Eds.), *Oceanographic and Marine Cross-Domain Data Management for Sustainable Development* (pp. 100–126). Hershey, PA: IGI Global. doi:10.4018/978-1-5225-0700-0.ch005

Mabe, L. K., & Oladele, O. I. (2017). Application of Information Communication Technologies for Agricultural Development through Extension Services: A Review. In T. Tossy (Ed.), *Information Technology Integration for Socio-Economic Development* (pp. 52–101). Hershey, PA: IGI Global. doi:10.4018/978-1-5225-0539-6.ch003

Malomo, B. I. (2018). A Review of Psychological Resilience as a Response to Natural Hazards in Nigeria. In A. Eneanya (Ed.), *Handbook of Research on Environmental Policies for Emergency Management and Public Safety* (pp. 147–165). Hershey, PA: IGI Global. doi:10.4018/978-1-5225-3194-4.ch008

Manchiraju, S. (2018). Predicting Behavioral Intentions Toward Sustainable Fashion Consumption: A Comparison of Attitude-Behavior and Value-Behavior Consistency Models. In I. Management Association (Ed.), Fashion and Textiles: Breakthroughs in Research and Practice (pp. 1-21). Hershey, PA: IGI Global. doi:10.4018/978-1-5225-3432-7.ch001

Manzella, G. M., Bartolini, R., Bustaffa, F., D'Angelo, P., De Mattei, M., Frontini, F., ... Spada, A. (2017). Semantic Search Engine for Data Management and Sustainable Development: Marine Planning Service Platform. In P. Diviacco, A. Leadbetter, & H. Glaves (Eds.), *Oceanographic and Marine Cross-Domain Data Management for Sustainable Development* (pp. 127–154). Hershey, PA: IGI Global. doi:10.4018/978-1-5225-0700-0.ch006

Mbonigaba, J. (2018). Comparing the Effects of Unsustainable Production and Consumption of Food on Health and Policy Across Developed and Less Developed Countries. In A. Obayelu (Ed.), *Food Systems Sustainability and Environmental Policies in Modern Economies* (pp. 124–158). Hershey, PA: IGI Global. doi:10.4018/978-1-5225-3631-4.ch006

McKeown, A. E. (2017). Nurses, Healthcare, and Environmental Pollution and Solutions: Breaking the Cycle of Harm. In I. Management Association (Ed.), Natural Resources Management: Concepts, Methodologies, Tools, and Applications (pp. 392-415). Hershey, PA: IGI Global. https://doi.org/ doi:10.4018/978-1-5225-0803-8.ch020

Mili, B., Barua, A., & Katyaini, S. (2017). Climate Change and Adaptation through the Lens of Capability Approach: A Case Study from Darjeeling, Eastern Himalaya. In I. Management Association (Ed.), Natural Resources Management: Concepts, Methodologies, Tools, and Applications (pp. 1351-1365). Hershey, PA: IGI Global. https://doi.org/ doi:10.4018/978-1-5225-0803-8.ch064

Mir, S. A., Shah, M. A., Mir, M. M., & Iqbal, U. (2017). New Horizons of Nanotechnology in Agriculture and Food Processing Industry. In B. Nayak, A. Nanda, & M. Bhat (Eds.), *Integrating Biologically-Inspired Nanotechnology into Medical Practice* (pp. 230–258). Hershey, PA: IGI Global. doi:10.4018/978-1-5225-0610-2.ch009

Moallem, M., Sterrett, W. L., Gordon, C. R., Sukhera, S. M., Mahmood, A., & Bashir, A. (2021). An Investigation of the Effects of Integrating Computing and Project- or Problem-Based Learning in the Context of Environmental Sciences: A Case of Pakistani STEM Teachers. In S. Schroth & J. Daniels (Eds.), *Building STEM Skills Through Environmental Education* (pp. 49–89). IGI Global. https://doi.org/10.4018/978-1-7998-2711-5.ch003

Mujere, N., & Moyce, W. (2017). Climate Change Impacts on Surface Water Quality. In W. Ganpat & W. Isaac (Eds.), *Environmental Sustainability and Climate Change Adaptation Strategies* (pp. 322–340). Hershey, PA: IGI Global. doi:10.4018/978-1-5225-1607-1.ch012

Mukherjee, A. B., Krishna, A. P., & Patel, N. (2018). Geospatial Technology for Urban Sciences. In *Geospatial Technologies in Urban System Development: Emerging Research and Opportunities* (pp. 99–120). Hershey, PA: IGI Global. doi:10.4018/978-1-5225-3683-3.ch005

Nagarajan, S. K., & Sangaiah, A. K. (2017). Vegetation Index: Ideas, Methods, Influences, and Trends. In N. Kumar, A. Sangaiah, M. Arun, & S. Anand (Eds.), *Advanced Image Processing Techniques and Applications* (pp. 347–386). Hershey, PA: IGI Global. doi:10.4018/978-1-5225-2053-5.ch016

Naraine, L., & Meehan, K. (2017). Strengthening Food Security with Sustainable Practices by Smallholder Farmers in Lesser Developed Economies. In W. Ganpat, R. Dyer, & W. Isaac (Eds.), *Agricultural Development and Food Security in Developing Nations* (pp. 57–81). Hershey, PA: IGI Global. doi:10.4018/978-1-5225-0942-4.ch003

Naraine, L., & Meehan, K. (2017). Strengthening Food Security with Sustainable Practices by Smallholder Farmers in Lesser Developed Economies. In W. Ganpat, R. Dyer, & W. Isaac (Eds.), *Agricultural Development and Food Security in Developing Nations* (pp. 57–81). Hershey, PA: IGI Global. doi:10.4018/978-1-5225-0942-4.ch003

Naraine, L., & Meehan, K. (2017). Strengthening Food Security with Sustainable Practices by Smallholder Farmers in Lesser Developed Economies. In W. Ganpat, R. Dyer, & W. Isaac (Eds.), *Agricultural Development and Food Security in Developing Nations* (pp. 57–81). Hershey, PA: IGI Global. doi:10.4018/978-1-5225-0942-4.ch003

Nikolaou, K., Tsakiridou, E., Anastasiadis, F., & Mattas, K. (2017). Exploring Alternative Distribution Channels of Agricultural Products. *International Journal of Food and Beverage Manufacturing and Business Models*, *2*(2), 36–66. doi:10.4018/IJFBMBM.2017070103

Nishat, K. J., & Rahman, M. S. (2018). Disaster, Vulnerability, and Violence Against Women: Global Findings and a Research Agenda for Bangladesh. In N. Mahtab, T. Haque, I. Khan, M. Islam, & I. Wahid (Eds.), *Handbook of Research on Women's Issues and Rights in the Developing World* (pp. 235–250). Hershey, PA: IGI Global. doi:10.4018/978-1-5225-3018-3.ch014

O'Hara, S., Jones, D., & Trobman, H. B. (2018). Building an Urban Food System Through UDC Food Hubs. In A. Burtin, J. Fleming, & P. Hampton-Garland (Eds.), *Changing Urban Landscapes Through Public Higher Education* (pp. 116–143). Hershey, PA: IGI Global. doi:10.4018/978-1-5225-3454-9.ch006

Obayelu, A. E. (2018). Integrating Environment, Food Systems, and Sustainability in Feeding the Growing Population in Developing Countries. In A. Obayelu (Ed.), *Food Systems Sustainability and Environmental Policies in Modern Economies* (pp. 1–14). Hershey, PA: IGI Global. doi:10.4018/978-1-5225-3631-4.ch001

Othman, R., Nath, N., & Laswad, F. (2018). Environmental Reporting and Accounting: Sustainability Hybridisation. In G. Azevedo, J. da Silva Oliveira, R. Marques, & A. Ferreira (Eds.), *Handbook of Research on Modernization and Accountability in Public Sector Management* (pp. 130–158). Hershey, PA: IGI Global. doi:10.4018/978-1-5225-3731-1.ch007

Ouadi, A., & Zitouni, A. (2021). Phasor Measurement Improvement Using Digital Filter in a Smart Grid. In A. Recioui & H. Bentarzi (Eds.), *Optimizing and Measuring Smart Grid Operation and Control* (pp. 100–117). IGI Global. https://doi.org/10.4018/978-1-7998-4027-5.ch005

Padigala, B. S. (2018). Traditional Water Management System for Climate Change Adaptation in Mountain Ecosystems. In I. Management Association (Ed.), Climate Change and Environmental Concerns: Breakthroughs in Research and Practice (pp. 630-655). Hershey, PA: IGI Global. doi:10.4018/978-1-5225-5487-5.ch033

Panda, C. K. (2018). Mobile Phone Usage in Agricultural Extension in India: The Current and Future Perspective. In F. Mtenzi, G. Oreku, D. Lupiana, & J. Yonazi (Eds.), *Mobile Technologies and Socio-Economic Development in Emerging Nations* (pp. 1–21). Hershey, PA: IGI Global. doi:10.4018/978-1-5225-4029-8.ch001

Pandian, S. L., Yarrakula, K., & Chaudhury, P. (2018). GIS-Based Decision Support System for Village Level: A Case Study in Andhra Pradesh. In S. Chaudhuri & N. Ray (Eds.), *GIS Applications in the Tourism and Hospitality Industry* (pp. 275–295). Hershey, PA: IGI Global. doi:10.4018/978-1-5225-5088-4.ch012

Quaranta, G., & Salvia, R. (2017). Social-Based Product Innovation and Governance in The Milk Sector: The Case of Carciocacio and Innonatura. In T. Tarnanidis, M. Vlachopoulou, & J. Papathanasiou (Eds.), *Driving Agribusiness With Technology Innovations* (pp. 293–310). Hershey, PA: IGI Global. doi:10.4018/978-1-5225-2107-5.ch015

Rahman, M. K., Schmidlin, T. W., Munro-Stasiuk, M. J., & Curtis, A. (2017). Geospatial Analysis of Land Loss, Land Cover Change, and Landuse Patterns of Kutubdia Island, Bangladesh. *International Journal of Applied Geospatial Research*, *8*(2), 45–60. doi:10.4018/IJAGR.2017040104

Rajack-Talley, T. A. (2017). Agriculture, Trade Liberalization and Poverty in the ACP Countries. In W. Ganpat, R. Dyer, & W. Isaac (Eds.), *Agricultural Development and Food Security in Developing Nations* (pp. 1–20). Hershey, PA: IGI Global. doi:10.4018/978-1-5225-0942-4.ch001

Rajamanickam, S. (2018). Exploring Landscapes in Regional Convergence: Environment and Sustainable Development in South Asia. In I. Management Association (Ed.), Sustainable Development: Concepts, Methodologies, Tools, and Applications (pp. 1051-1087). Hershey, PA: IGI Global. https://doi.org/ doi:10.4018/978-1-5225-3817-2.ch047

Rizvi, S. M., & Dearden, A. (2018). KHETI: ICT Solution for Agriculture Extension and Its Replication in Open and Distance Learning. In U. Pandey & V. Indrakanti (Eds.), *Open and Distance Learning Initiatives for Sustainable Development* (pp. 163–174). Hershey, PA: IGI Global. doi:10.4018/978-1-5225-2621-6.ch008

Roşu, L., & Macarov, L. I. (2017). Management of Drought and Floods in the Dobrogea Region. In I. Management Association (Ed.), Agri-Food Supply Chain Management: Breakthroughs in Research and Practice (pp. 372-403). Hershey, PA: IGI Global. https://doi.org/ doi:10.4018/978-1-5225-1629-3.ch016

Rouzbehani, K., & Rouzbehani, S. (2018). Mapping Women's World: GIS and the Case of Breast Cancer in the US. *International Journal of Public Health Management and Ethics*, *3*(1), 14–25. doi:10.4018/IJPHME.2018010102

Roy, D. (2018). Success Factors of Adoption of Mobile Applications in Rural India: Effect of Service Characteristics on Conceptual Model. In M. Khosrow-Pour, D.B.A. (Ed.), Green Computing Strategies for Competitive Advantage and Business Sustainability (pp. 211-238). Hershey, PA: IGI Global. https://doi.org/ doi:10.4018/978-1-5225-5017-4.ch010

Sajeva, M., Lemon, M., & Sahota, P. S. (2017). Governance for Food Security: A Framework for Social Learning and Scenario Building. *International Journal of Food and Beverage Manufacturing and Business Models*, *2*(2), 67–84. doi:10.4018/IJFBMBM.2017070104

Sambhanthan, A., & Potdar, V. (2017). A Study of the Parameters Impacting Sustainability in Information Technology Organizations. *International Journal of Knowledge-Based Organizations*, *7*(3), 27–39. doi:10.4018/IJKBO.2017070103

Sanga, C., Kalungwizi, V. J., & Msuya, C. P. (2017). Bridging Gender Gaps in Provision of Agricultural Extension Service Using ICT: Experiences from Sokoine University of Agriculture (SUA) Farmer Voice Radio (FVR) Project in Tanzania. In I. Management Association (Ed.), Discrimination and Diversity: Concepts, Methodologies, Tools, and Applications (pp. 682-697). Hershey, PA: IGI Global. https://doi.org/ doi:10.4018/978-1-5225-1933-1.ch031

Santra, A., & Mitra, D. (2017). Role of Remote Sensing in Potential Fishing Zone Forecast. In A. Santra & S. Mitra (Eds.), *Remote Sensing Techniques and GIS Applications in Earth and Environmental Studies* (pp. 243–257). Hershey, PA: IGI Global. doi:10.4018/978-1-5225-1814-3.ch012

Schaap, D. (2017). SeaDataNet: Towards a Pan-European Infrastructure for Marine and Ocean Data Management. In P. Diviacco, A. Leadbetter, & H. Glaves (Eds.), *Oceanographic and Marine Cross-Domain Data Management for Sustainable Development* (pp. 155–177). Hershey, PA: IGI Global. doi:10.4018/978-1-5225-0700-0.ch007

Seckin-Celik, T. (2017). Sustainability Reporting and Sustainability in the Turkish Business Context. In U. Akkucuk (Ed.), *Ethics and Sustainability in Global Supply Chain Management* (pp. 115–132). Hershey, PA: IGI Global. doi:10.4018/978-1-5225-2036-8.ch006

Segbefia, A. Y., Barnes, V. R., Akpalu, L. A., & Mensah, M. (2018). Environmental Location Assessment for Seaweed Cultivation in Ghana: A Spatial Multi-Criteria Approach. *International Journal of Applied Geospatial Research*, *9*(1), 51–64. doi:10.4018/IJAGR.2018010104

Sen, Y. (2018). How to Manage Sustainability: A Framework for Corporate Sustainability Tools. In I. Management Association (Ed.), Sustainable Development: Concepts, Methodologies, Tools, and Applications (pp. 568-589). Hershey, PA: IGI Global. doi:10.4018/978-1-5225-3817-2.ch026

Shamshiry, E., Abdulai, A. M., Mokhtar, M. B., & Komoo, I. (2015). Regional Landfill Site Selection with GIS and Analytical Hierarchy Process Techniques: A Case Study of Langkawi Island, Malaysia. In P. Thomas, M. Srihari, & S. Kaur (Eds.), *Handbook of Research on Cultural and Economic Impacts of the Information Society* (pp. 248–282). Hershey, PA: IGI Global. doi:10.4018/978-1-4666-8598-7.ch011

Sharma, Y. K., Mangla, S. K., Patil, P. P., & Uniyal, S. (2018). Analyzing Sustainable Food Supply Chain Management Challenges in India. In M. Ram & J. Davim (Eds.), *Soft Computing Techniques and Applications in Mechanical Engineering* (pp. 162–180). Hershey, PA: IGI Global. doi:10.4018/978-1-5225-3035-0.ch008

Silvestrelli, P. (2018). The Impact of Events: To Which Extent Are They Sustainable for Tourist Destinations? Some Evidences From Expo Milano 2015. In M. Risso & S. Testarmata (Eds.), *Value Sharing for Sustainable and Inclusive Development* (pp. 185–204). Hershey, PA: IGI Global. doi:10.4018/978-1-5225-3147-0.ch008

Silvius, G. (2018). Sustainability Evaluation of IT/IS Projects. In I. Management Association (Ed.), Sustainable Development: Concepts, Methodologies, Tools, and Applications (pp. 26-40). Hershey, PA: IGI Global. https://doi.org/ doi:10.4018/978-1-5225-3817-2.ch002

Singh, R., Srivastava, P., Singh, P., Upadhyay, S., & Raghubanshi, A. S. (2017). Human Overpopulation and Food Security: Challenges for the Agriculture Sustainability. In R. Singh, A. Singh, & V. Srivastava (Eds.), *Environmental Issues Surrounding Human Overpopulation* (pp. 12–39). Hershey, PA: IGI Global. doi:10.4018/978-1-5225-1683-5.ch002

Singh, R., Srivastava, P., Singh, P., Upadhyay, S., & Raghubanshi, A. S. (2017). Human Overpopulation and Food Security: Challenges for the Agriculture Sustainability. In R. Singh, A. Singh, & V. Srivastava (Eds.), *Environmental Issues Surrounding Human Overpopulation* (pp. 12–39). Hershey, PA: IGI Global. doi:10.4018/978-1-5225-1683-5.ch002

Srivastava, N. (2017). Climate Change Mitigation: Collective Efforts and Responsibly. In I. Management Association (Ed.), Natural Resources Management: Concepts, Methodologies, Tools, and Applications (pp. 64-76). Hershey, PA: IGI Global. https://doi.org/ doi:10.4018/978-1-5225-0803-8.ch003

Stanganelli, M., & Gerundo, C. (2017). Understanding the Role of Urban Morphology and Green Areas Configuration During Heat Waves. *International Journal of Agricultural and Environmental Information Systems*, *8*(2), 50–64. doi:10.4018/IJAEIS.2017040104

Stewart, M. K., Hagood, D., & Ching, C. C. (2017). Virtual Games and Real-World Communities: Environments that Constrain and Enable Physical Activity in Games for Health. *International Journal of Game-Based Learning*, *7*(1), 1–19. doi:10.4018/IJGBL.2017010101

Stone, R. J. (2017). Modelling the Frequency of Tropical Cyclones in the Lower Caribbean Region. In W. Ganpat & W. Isaac (Eds.), *Environmental Sustainability and Climate Change Adaptation Strategies* (pp. 341–349). Hershey, PA: IGI Global. doi:10.4018/978-1-5225-1607-1.ch013

Syed, A., & Jabeen, U. A. (2018). Climate Change Impact on Agriculture and Food Security. In A. Eneanya (Ed.), *Handbook of Research on Environmental Policies for Emergency Management and Public Safety* (pp. 223–237). Hershey, PA: IGI Global. doi:10.4018/978-1-5225-3194-4.ch012

Tam, G. C. (2017). The Global View of Sustainability. In *Managerial Strategies and Green Solutions for Project Sustainability* (pp. 1–24). Hershey, PA: IGI Global. doi:10.4018/978-1-5225-2371-0.ch001

Tam, G. C. (2017). Understanding Project Sustainability. In *Managerial Strategies and Green Solutions for Project Sustainability* (pp. 110–139). Hershey, PA: IGI Global. doi:10.4018/978-1-5225-2371-0.ch005

Tam, G. C. (2017). Perspectives on Sustainability. In *Managerial Strategies and Green Solutions for Project Sustainability* (pp. 53–76). Hershey, PA: IGI Global. doi:10.4018/978-1-5225-2371-0.ch003

Tang, M., & Karunanithi, A. T. (2018). Visual Logic Maps (vLms). In *Advanced Concept Maps in STEM Education: Emerging Research and Opportunities* (pp. 108–149). Hershey, PA: IGI Global. doi:10.4018/978-1-5225-2184-6.ch005

Taşçıoğlu, M., & Yener, D. (2018). The Value and Scope of GIS in Marketing and Tourism Management. In S. Chaudhuri & N. Ray (Eds.), *GIS Applications in the Tourism and Hospitality Industry* (pp. 189–211). Hershey, PA: IGI Global. doi:10.4018/978-1-5225-5088-4.ch009

Thanh Tung, B. (2021). Pharmacology and Therapeutic Applications of Resveratrol. In A. Hussain & S. Behl (Eds.), *Treating Endocrine and Metabolic Disorders With Herbal Medicines* (pp. 321–333). IGI Global. https://doi.org/10.4018/978-1-7998-4808-0.ch014

Tianming, G. (2018). Food Security and Rural Development on Emerging Markets of Northeast Asia: Cases of Chinese North and Russian Far East. In V. Erokhin (Ed.), *Establishing Food Security and Alternatives to International Trade in Emerging Economies* (pp. 155–176). Hershey, PA: IGI Global. doi:10.4018/978-1-5225-2733-6.ch008

Tiftikçigil, B. Y., Yaşgül, Y. S., & Güriş, B. (2017). Sustainability of Foreign Trade Deficit in Energy: The Case of Turkey. In N. Ray (Ed.), *Business Infrastructure for Sustainability in Developing Economies* (pp. 94–109). Hershey, PA: IGI Global. doi:10.4018/978-1-5225-2041-2.ch005

Tiwari, S., Vaish, B., & Singh, P. (2018). Population and Global Food Security: Issues Related to Climate Change. In I. Management Association (Ed.), Climate Change and Environmental Concerns: Breakthroughs in Research and Practice (pp. 41-64). Hershey, PA: IGI Global. https://doi.org/ doi:10.4018/978-1-5225-5487-5.ch003

Toujani, A., & Achour, H. (2018). A Data Mining Framework for Forest Fire Mapping. In I. Management Association (Ed.), Information Retrieval and Management: Concepts, Methodologies, Tools, and Applications (pp. 771-794). Hershey, PA: IGI Global. doi:10.4018/978-1-5225-5191-1.ch033

Tripathy, B., & K., S. B. (2017). Rough Fuzzy Set Theory and Neighbourhood Approximation Based Modelling for Spatial Epidemiology. In I. Management Association (Ed.), *Public Health and Welfare: Concepts, Methodologies, Tools, and Applications* (pp. 1257-1268). Hershey, PA: IGI Global. https://doi.org/ doi:10.4018/978-1-5225-1674-3.ch058

Trukhachev, A. (2017). New Approaches to Regional Branding through Green Production and Utilization of Existing Natural Advantages. In I. Management Association (Ed.), Advertising and Branding: Concepts, Methodologies, Tools, and Applications (pp. 1758-1778). Hershey, PA: IGI Global. doi:10.4018/978-1-5225-1793-1.ch081

Tsobanoglou, G. O., & Vlachopoulou, E. I. (2017). Social-Ecological Systems in Local Fisheries Communities. In G. Korres, E. Kourliouros, & M. Michailidis (Eds.), *Handbook of Research on Policies and Practices for Sustainable Economic Growth and Regional Development* (pp. 306–316). Hershey, PA: IGI Global. doi:10.4018/978-1-5225-2458-8.ch026

Tuydes-Yaman, H., & Karatas, P. (2018). Evaluation of Walkability and Pedestrian Level of Service. In I. Management Association (Ed.), Intelligent Transportation and Planning: Breakthroughs in Research and Practice (pp. 264-291). Hershey, PA: IGI Global. doi:10.4018/978-1-5225-5210-9.ch012

Uddin, S., Chakravorty, S., Ray, A., & Sherpa, K. S. (2018). Optimal Location of Sub-Station Using Q-GIS and Multi-Criteria Decision Making Approach. *International Journal of Decision Support System Technology, 10*(2), 65–79. doi:10.4018/IJDSST.2018040104

Uddin, S., Chakravorty, S., Sherpa, K. S., & Ray, A. (2018). Power Distribution System Planning Using Q-GIS. *International Journal of Energy Optimization and Engineering, 7*(2), 61–75. doi:10.4018/IJEOE.2018040103

Uzun, F. V. (2018). Natural Resources Management. In A. Eneanya (Ed.), *Handbook of Research on Environmental Policies for Emergency Management and Public Safety* (pp. 1–21). Hershey, PA: IGI Global. doi:10.4018/978-1-5225-3194-4.ch001

V, M., Agrawal, R., Sharma, V., & T.N., K. (2018). Supply Chain Social Sustainability and Manufacturing. In I. Management Association (Ed.), *Technology Adoption and Social Issues: Concepts, Methodologies, Tools, and Applications* (pp. 226-252). Hershey, PA: IGI Global. https://doi.org/ doi:10.4018/978-1-5225-5201-7.ch011

van der Vliet-Bakker, J. M. (2017). Environmentally Forced Migration and Human Rights. In C. Akrivopoulou (Ed.), *Defending Human Rights and Democracy in the Era of Globalization* (pp. 146–180). Hershey, PA: IGI Global. doi:10.4018/978-1-5225-0723-9.ch007

Vargas-Hernández, J. G. (2022). A Comprehensive Entrepreneurship Model for the Internationalization of Green Innovation Businesses. In U. Akkucuk (Ed.), *Disruptive Technologies and Eco-Innovation for Sustainable Development* (pp. 131–149). IGI Global. https://doi.org/10.4018/978-1-7998-8900-7.ch008

Vaskov, A. G., Lin, Z. Y., Tyagunov, M. G., Shestopalova, T. A., & Deryugina, G. V. (2018). Design of Renewable Sources GIS for ASEAN Countries. In V. Kharchenko & P. Vasant (Eds.), *Handbook of Research on Renewable Energy and Electric Resources for Sustainable Rural Development* (pp. 1–25). Hershey, PA: IGI Global. doi:10.4018/978-1-5225-3867-7.ch001

Vázquez, D. G., & Gil, M. T. (2017). Sustainability in Smart Cities: The Case of Vitoria-Gasteiz (Spain) – A Commitment to a New Urban Paradigm. In L. Carvalho (Ed.), *Handbook of Research on Entrepreneurial Development and Innovation Within Smart Cities* (pp. 248–268). Hershey, PA: IGI Global. doi:10.4018/978-1-5225-1978-2.ch012

Wahab, I. N., & Soonthodu, S. (2018). Geographical Information System in Eco-Tourism. In S. Chaudhuri & N. Ray (Eds.), *GIS Applications in the Tourism and Hospitality Industry* (pp. 61–75). Hershey, PA: IGI Global. doi:10.4018/978-1-5225-5088-4.ch003

Weiss-Randall, D. (2018). Cultivating Environmental Justice. In *Utilizing Innovative Technologies to Address the Public Health Impact of Climate Change: Emerging Research and Opportunities* (pp. 110–143). Hershey, PA: IGI Global. doi:10.4018/978-1-5225-3414-3.ch004

Weiss-Randall, D. (2018). Cultivating Resilience. In *Utilizing Innovative Technologies to Address the Public Health Impact of Climate Change: Emerging Research and Opportunities* (pp. 204–235). Hershey, PA: IGI Global. doi:10.4018/978-1-5225-3414-3.ch007

Weiss-Randall, D. (2018). Climate Change Solutions: Where Do We Go From Here? In *Utilizing Innovative Technologies to Address the Public Health Impact of Climate Change: Emerging Research and Opportunities* (pp. 236–268). Hershey, PA: IGI Global. doi:10.4018/978-1-5225-3414-3.ch008

Whyte, K. P., List, M., Stone, J. V., Grooms, D., Gasteyer, S., Thompson, P. B., . . . Bouri, H. (2018). Uberveillance, Standards, and Anticipation: A Case Study on Nanobiosensors in U.S. Cattle. In I. Management Association (Ed.), Biomedical Engineering: Concepts, Methodologies, Tools, and Applications (pp. 577-596). Hershey, PA: IGI Global. https://doi.org/ doi:10.4018/978-1-5225-3158-6.ch025

Wulff, E. (2017). Data and Operational Oceanography: A Review in Support of Responsible Fisheries and Aquaculture. In P. Diviacco, A. Leadbetter, & H. Glaves (Eds.), *Oceanographic and Marine Cross-Domain Data Management for Sustainable Development* (pp. 303–324). Hershey, PA: IGI Global. doi:10.4018/978-1-5225-0700-0.ch013

Yener, D. (2017). Geographic Information Systems and Its Applications in Marketing Literature. In S. Faiz & K. Mahmoudi (Eds.), *Handbook of Research on Geographic Information Systems Applications and Advancements* (pp. 158–172). Hershey, PA: IGI Global. doi:10.4018/978-1-5225-0937-0.ch006

Yu, K., Liu, Y., & Sharma, A. (2021). Analyze the Effectiveness of the Algorithm for Agricultural Product Delivery Vehicle Routing Problem Based on Mathematical Model. *International Journal of Agricultural and Environmental Information Systems, 12*(3), 26–38. https://doi.org/10.4018/IJAEIS.2021070103

Zhou, X., Sharma, A., & Mohindru, V. (2021). Research on Linear Programming Algorithm for Mathematical Model of Agricultural Machinery Allocation. *International Journal of Agricultural and Environmental Information Systems*, *12*(3), 1–12. https://doi.org/10.4018/IJAEIS.2021070101

About the Contributors

Mir Sayed Shah Danish (Dr. Eng., MBA, CEng., SMIEEE, MIET) has been an engineering and technology expert and an academician for several years. He demonstrates a simple, in-depth narration style, turning concepts into measurable endeavors and systematically exploring interdisciplinary coverage. Apart from being a scientific scholar, he brings together multidisciplinary skills and expertise (energy, environment, business, and management), providing integrated solutions. He is the author of several academic and technical textbooks, guidebooks, training manuals, and other books in English and Dari (Persian) Languages. These publications have enabled him to link industry with academia, and he has achieved recognition with several awards and expressions of appreciation. Since 2004, He has been involved in multidisciplinary engineering and technology by leading several projects in those fields while continuing as an active scholar and educator. He is an assistant professor at the University of the Ryukyus, Japan; founder and chair of the IEEE-Sustainable Energy and Intelligent Engineering Society (SEIES-PES & FRID joint chapter, Fukuoka Chapter); founder and facilitator of several academic branches of IET, ACM; and member of the board of directors, REPA - Research and Education Promotion Association. In addition, Dr. Danish has worked with national and international organizations and companies as an urban electric power planner, team leader, technical advisor, department head, educational manager, and director. He is a chartered engineer, UK (CEng.), senior member of IEEE, member of IET (MIET), and holds membership in many other academic societies. He received his bachelor's degree in electrical and electronic engineering (B.Sc.) in 2009 from Kabul University, Afghanistan); the first master's degree in energy and electrical engineering (M.Sc.) in 2015 from the University of the Ryukyus, Japan; the other in business administration (MBA) in 2016 from the National Institute of Business Management, India); and a doctorate in sustainable energy (Ph.D.) in 2018 from the University of the Ryukyus. He chaired and has been a committee member of several conferences and symposiums and has more than 100 publications. His main research interest is sustainable energy (policy, economics, market, environment, and management), hydrogen energy, smart cities and housing, storage systems, aquaponics and aquaculture, voltage stability, and related areas.

Tomonobu Senjyu was born in Saga, Japan, in 1963. He received the B.S. and M.S. degrees from the University of the Ryukyus, Okinawa, Japan, in 1986 and 1988, respectively, and the Ph.D. degree from Nagoya University, Aichi, Japan, in 1994, all in electrical engineering. Since 1988, he has been with the Department of Electrical and Electronics Engineering, Faculty of Engineering, University of the Ryukyus, where he is currently working as a Professor. His research interests include stability of ac machines, advanced control of electrical machines, renewable energy, and power electronics. Prof. Senjyu is a member of the Institute of Electrical Engineers of Japan.

* * *

Mikaeel Ahmadi is a Postdoctoral Researcher, Doctor of Engineering at the University of the Ryukyus in the Interdisciplinary Intelligent Systems Engineering department.

Emre Akusta is an research assistant of Economics at Kirklareli University, Turkey. He earned his bachelor's degree in Economics from Istanbul University in 2015 and received a master's degree in Economics from Kirklareli University in 2019. His areas of special interest are energy, energy economics and macro economics.

Raif Cergibozan is an associate professor of Economics at Kirklareli University, Turkey. He earned his bachelor's degree in Economics from Marmara University in 2009 and received a master's degree in Economics from Istanbul Technical University in 2011. Dr. Cergibozan holds two PhD degrees in Economics. He completed his first PhD in Economics at Galatasaray University in 2015 and got the second PhD in Economics at Paris Nanterre University in 2018. His areas of special interest are financial crises, energy economics, labor markets, international trade, time series analysis, and macro econometric models.

Seda Cetindere is a research assistant of Chemistry Department at Gebze Technical University. She received her BSc. degree in Chemistry from Abant Izzet Baysal University in Turkey in 2006. She completed her M.Sc. in Gebze Institute of Technology (former name of GTU) in Chemistry and her Ph.D. in Gebze Technical University in 2017. She completed her post-doctoral research in Ulm University Institute of Inorganic Chemistry I (Materials and Catalysis) in 2019. Due to the successes in her education she was awarded by Izzet Baysal Foundation with a scholarship between 2002-2006 years. She was awarded by Scientific and Technological Research Council of Turkey (TUBITAK) with 2219-Postdoctoral Research Scholarship Programme between 2018-2019 years for her post-doctoral studies. Her research interests are organic-inorganic hybrid systems, phosphazene compounds, boron-dipyrromethene (BODIPY) dyes, polyoxometalates (POMs), photosensitizer (PS)-catalyst systems for light-driven hydrogen evolution. Dr. Cetindere has co-authored in SCI, SCI-expanded journal articles, book chapters and conference papers and has taken part in many national projects.

María Jesús García is a professor at the University of Valencia. After finishing her degree in Law, she was awarded a doctoral research scholarship by the Spanish Department of Education, and she went on to receive her PhD in Law from the University of Valencia. Her area of expertise is Public Law and European Union Law.

Abdul Matin Ibrahimi is a researcher at the University of Ryukyus. He received a M.S. degree in electrical and electronics engineering and a Ph.D. in interdisciplinary intelligent systems engineering from University of the Ryukyus, Okinawa Japan, in 2015 and 2020 respectively. His fields of interest are power system optimization, microgrid, renewable energy, energy storage, and sustainable energy development.

Sebak Kumar Jana is currently Professor of Economics in the Department of Economics, Vidyasagar University, West Bengal, India. Dr. Jana was graduated from Presidency College, Kolkata in 1989 and did his post graduation in Economics from University of Calcutta in 1991. He obtained M.Phil. and Ph.D. degree in Economics from Jadavpur University, Kolkata. His major area of academic interest includes environmental and resource economics, rural development and economics of education. He has about eighty publications in Journals and in edited books. He has published six books as author or editor including Tank Irrigation in the Dry Zones in India, Education in West Bengal – Looking Beyond Schools, Development in Development in Developing Economies (Edited). He, as Principal Investigator, has completed a few research projects funded by ICSSR, UGC and NABARD, IWMI, and IFPRI.

Alper Karasoy is an assistant professor (Ph.D.) in the Department of Economics at Afyon Kocatepe University, Turkey. His research interests include, but are not limited to, environmental economics, energy economics, applied econometrics, the drivers, and the impacts of remittances. His research has been published in Technology in Society, International Migration, Environmental Science and Pollution Research, Management of Environmental Quality: An International Journal, Natural Resources Forum, and OPEC Energy Review.

Alexey Mikhaylov, Ph.D., holds a lecturer position in Financial Markets and Banks Department in Financial University under the Government of the Russian Federation, Russia. He is an author of 30 scientific publications and conference papers indexed in SCOPUS and Web of Science, author of more than 80 scientific publications indexed in relevant scientific databases and author of 8 scientific monographs. He also is Head of laboratory of Financial markets and Banks department in Financial University under the Government of the Russian Federation. He specializes in Energy; Economics, Econometrics and Finance; Business, Management and Accounting; Environmental Science; Materials Science; Mathematics; Computer Science; Psychology; Social Sciences; Neuroscience. He has won numerous awards such as Rector's Award 2020 (Financial University). He is currently the Guest Editor in Metals (Q1 journal) and International Journal of Economics, Finance and Management Sciences. He has filed patent: Modified machine learning algorithm based on linear regression for forecasting oil prices, taking into account the influence of macroeconomic indicators. He is a member of Young Scientist Board of Financial University under the Government of the Russian Federation, ACI Russia – Financial Markets Association, Council on professional qualifications at the financial market, Russian Association of Crypto Industry and Blockchain.

Nima Norouzi does research in techno-economic, techno-environmental, techno-social, and legal aspects of Energy and natural resources economics.

Najib Rahman Sabory graduated from Electrical and Electronics department of engineering school at Kabul University in 2001. Since then he has been teaching in this department. From 2009-2011, he served as the deputy dean for the Engineering school at Kabul University. After a Fulbright scholarship was awarded to him in 2011, he completed a master's degree in Sustainable Energy and a graduate certificate in Project Management from A. James Clark School of Engineering at University of Maryland College Park. He has also earned an MBA in 2014 from World Wide Science Academy in Malaysia. He is also PhD candidate at the university of the Ryukyus, Japan, with research focus on energy policy and sustainable energy. Mr. Sabory was awarded a UNITAR (United Nations Institute for Training and Research) fellowship in 2007 for a period of seven months. He learned many advanced concepts in project management, leadership and change management. He later remained in this program for 2008 and 2009 cycles as coach and resource person. Mr. Sabory has participated in a series of workshops and symposiums about leadership and entrepreneurship back in US. In 2011, Mr. Sabory led an Inter-ministerial Commission for Energy (ICE) committee for preparing capacity building strategy for energy sector of Afghanistan. Currently Mr. Sabory serves as Associate professor and head of the Energy Department of engineering school at Kabul University. He is also the chairman of Afghanistan Renewable Energy Union (AREU) since early 2017.

Shawkatullah Shams is a Teaching Assistant at Kabul University.

Khushboo Singh is a former M.Phil. student, Department of Economics, Vidyasagar University, Midnapore, West Bengal, India.

Index

B

BODIPY 119, 121-142

C

Canonical Cointegrating Regression 154
Circularity 47
Climate Change 1-2, 16, 20, 29-30, 32, 37, 39, 41, 45, 47, 49-51, 56-57, 60-61, 67, 71, 99-100, 102-103, 108, 111-112, 116, 194
COVID-19 45-46, 50-55, 58, 60-63, 65-67, 99-103, 106-109, 111-112, 114-116, 167

D

Dashboard of Sustainability (DS) 56, 67, 103, 116
Direct Material Consumption (DMC) 56, 67, 104, 116
Disaster 55, 57, 60-61, 67-68, 99-100, 108, 111, 116-117
DSSC 122-123, 125, 132
Dynamic Least Squares 161, 170

E

Eco Commerce 48
Ecological Economics 42-43, 48
Ecological Footprint (EF) 56, 68, 103, 117
Eco-Tariffs 48
Egypt 34, 59, 143-146, 149, 152, 154, 161-163, 165
Emission Reduction 71, 73
Emissions Trading 48
Energy 3, 5-7, 9-14, 16-34, 36-39, 41-43, 45-48, 51, 55-58, 63-69, 71-74, 78-79, 81, 84-85, 87, 89, 95-99, 102-103, 106, 109, 112-115, 117-122, 124-130, 132, 134-146, 148-151, 153-156, 160-168, 172-173, 175, 177-179, 181, 183-189, 191, 193-203
Energy Economics 42
Energy Efficiency 38-41, 64, 113, 115, 149, 172-173, 176-179, 181-185, 187-188
Energy Insecurity 143, 145, 148-149, 162-163, 168
Energy Policy Development 1, 3, 5, 7, 11, 14, 17, 20
Energy Policy Framework 12, 25
Energy Strategy 8, 14, 27
Environmental Enterprise 48
Environmental Impact 39, 41, 122, 152, 164, 172-173, 175
Environmental Re-engineering 99, 101
Environmental Sustainability Index (ESI) 56, 68, 103, 117

F

Fully Modified Least Squares 161, 170

G

Global Pandemic 60, 100-101
Green Economy 28-38, 40-46, 48, 55, 109
Green Politics 48
Greenhouse Gases 28, 37, 49, 87, 94, 176

Gross Domestic Product (GDP) 56, 68, 102-103, 117, 145

H

Hazard 68, 117
Human Development Index (HDI) 56, 68, 103, 117

I

India 26, 34, 39, 59, 97, 151-153, 166-168, 189-191, 193-196, 200-202

L

Living Planet Index (LPI) 56, 68, 104, 117
Low-Carbon Economy 49

M

Mitigation 22, 42, 50-51, 61, 64, 68, 102, 108, 113, 117
Morocco 144-146, 149, 154, 162-163

N

Natural Resource Economics 49

O

OSC 129, 132

P

Pandemic 46, 50-55, 57-58, 60-63, 65-66, 68, 99-103, 106-109, 111-112, 114-117, 167
Panel Data Analysis 152, 196, 201
Photovoltaic 40, 70, 95-98, 119-120, 122-124, 129, 132-133, 135, 137, 140, 182, 202
Policy Components 3-4, 16
PSC 130-132

R

Remittances 143, 145-149, 151-154, 160-168
Renewable 2-3, 8-9, 11, 21-26, 29-34, 36-39, 41-42, 45-46, 56, 63-64, 68, 71-74, 78-79, 84, 95-98, 113, 115, 117, 119-120, 126, 140-141, 151, 162-164, 168, 172, 176-179, 181, 183-185, 189-203
Renewable Energies 8, 28-29, 40, 176-179, 183-184
Renewable Energy 2-3, 9, 11, 22-26, 28-34, 36-39, 41-42, 46, 56, 63-64, 68, 70-74, 78-79, 84, 95-96, 98, 113, 115, 117, 119-120, 126, 140, 151, 162-164, 168, 172, 176-177, 181, 183-185, 189-191, 193-203
Renewable Energy Disparity 190
Renewable Energy Market 40, 193
Renewable Energy Potential 3, 190, 196-199, 201
Resources Sustainability 58
RETScreen 70, 73, 79, 84, 87, 95-96
Risk 9, 43, 51, 54, 56-58, 62, 68, 102, 106-112, 117, 143-149, 153-156, 161-163, 165, 167

S

Salience, Credibility, and Legitimacy of GDP and HDI 56, 68, 117
Solar 2, 24-25, 29, 38, 40-41, 48, 70, 72-74, 78-82, 84-87, 89, 91, 93-98, 120-122, 126-142, 163, 182-184, 187, 189, 191, 193-194, 196, 200, 202
Solar Energy 29, 40-41, 70, 72, 74, 81, 84, 87, 89, 97, 119-122, 132, 135, 137, 140, 183-184, 187, 200, 202
Solar Power Plant 70, 73-74, 78-79, 82, 84, 87, 91, 93-94, 97
Sustainability 13, 16, 20, 23, 25, 27, 31-32, 42-44, 46, 49-50, 54-56, 58, 60, 63-65, 67-69, 96, 101-105, 111, 113, 115-118, 132, 136-138, 144, 146-147, 165-166, 168, 173-176, 181, 184, 187-189, 201
Sustainability criteria 50
Sustainability in Terms of Energy and Environment 68, 117
Sustainability Indicators 23, 146

Sustainability Pillars 16, 50, 68, 104, 117
Sustainable Development 1, 14, 30-32, 37, 39, 42-44, 46, 48-49, 53-57, 60-63, 65-66, 100, 109, 111, 114, 143, 147, 166-167, 173-175, 190, 200
Sustainable Energy 1, 13, 23, 25-26, 31, 37, 64, 66, 95-97, 112, 140-141, 189, 191, 196, 201-202

T

Tunisia 144-146, 149, 154, 161-163
Turkey 34, 70-75, 78-84, 86, 91, 94-98, 119, 143

U

Urban Energy 189
Urban Environments 172
Urban Rehabilitation 173, 175-176, 178, 180, 186, 188-189
Urbanism 172, 174-175, 183

W

Waste Management 23, 50-51, 58-60, 65, 67, 105-109, 112, 114-116
Well-Being Index (WI) 56, 69, 103, 118

www.ingramcontent.com/pod-product-compliance
Lightning Source LLC
LaVergne TN
LVHW080856300926
849162LV00003B/431